T0344882

ADVANCED CONTROL OF DOUBLY FED INDUCTION GENERATOR FOR WIND POWER SYSTEMS

ADVANCED CONTROL OF DOUBLY FED INDUCTION GENERATOR FOR WIND POWER SYSTEMS

DEHONG XU
FREDE BLAABJERG
WENJIE CHEN
NAN ZHU

IEEE
PRESS
SERIES
ON POWER
ENGINEERING

IEEE PRESS

WILEY

Published by John Wiley & Sons, Inc., Hoboken, New Jersey.
Published simultaneously in Canada.

No part of this publication may be reproduced, stored in a retrieval system, or transmitted in any form or by any means, electronic, mechanical, photocopying, recording, scanning, or otherwise, except as permitted under Section 107 or 108 of the 1976 United States Copyright Act, without either the prior written permission of the Publisher, or authorization through payment of the appropriate per-copy fee to the Copyright Clearance Center, Inc., 222 Rosewood Drive, Danvers, MA 01923, (978) 750-8400, fax (978) 750-4470, or on the web at www.copyright.com. Requests to the Publisher for permission should be addressed to the Permissions Department, John Wiley & Sons, Inc., 111 River Street, Hoboken, NJ 07030, (201) 748-6011, fax (201) 748-6008, or online at http://www.wiley.com/go/permission.

Limit of Liability/Disclaimer of Warranty: While the publisher and author have used their best efforts in preparing this book, they make no representations or warranties with respect to the accuracy or completeness of the contents of this book and specifically disclaim any implied warranties of merchantability or fitness for a particular purpose. No warranty may be created or extended by sales representatives or written sales materials. The advice and strategies contained herein may not be suitable for your situation. You should consult with a professional where appropriate. Neither the publisher nor author shall be liable for any loss of profit or any other commercial damages, including but not limited to special, incidental, consequential, or other damages.

For general information on our other products and services or for technical support, please contact our Customer Care Department within the United States at (800) 762-2974, outside the United States at (317) 572-3993 or fax (317) 572-4002.

Wiley also publishes its books in a variety of electronic formats. Some content that appears in print may not be available in electronic formats. For more information about Wiley products, visit our web site at www.wiley.com.

Library of Congress Cataloging-in-Publication Data:

ISBN: 978-1-119-17206-2

Printed in the United States of America.

V10002172_070818

CONTENTS

PART III OPERATION OF DFIG UNDER DISTORTED GRID VOLTAGE

CHAPTER 6 ANALYSIS OF DFIG UNDER DISTORTED GRID VOLTAGE 141

CHAPTER 7 MULTIPLE-LOOP CONTROL OF DFIG UNDER DISTORTED GRID VOLTAGE 167

PART V *DFIG TEST BENCH*

PREFACE

FOR THE PAST 10 YEARS, we have seen significant progress in the development of wind power technology and applications. Wind power generation has become one of the most important renewable energy generations. Instead of the fixed-speed wind power systems, variable-speed wind turbine systems prevail due to their significant performance enhancements such as energy-harvesting ability, friendliness to the grid, reliability, etc. Two most popular variable-speed wind turbine configurations are doubly fed induction generator (DFIG) and synchronous generator (SG). Nowadays, DFIG wind system is dominating the market. The knowledge of wind generation is related to electrical engineering, such as electric machine, power electronics, control theory, electric power systems, etc. Now there are thousands of published papers about wind power systems. This book tries to give readers an overview of the progress of doubly fed induction generation systems. It is more focused on modeling and control of DFIG wind power conversion system.

This book primarily emphasizes on the advanced control of the DFIG wind power system, which is realized by the power electronics converters and aims to improve grid integration performance. First, this book will give the readers an introduction to the wind power system. It includes an overview of wind power systems, grid codes for wind power systems, modeling of key components in DFIG wind power systems such as electric machines, converters and inverters, and fundamental controls. It will introduce the control schemes of the DFIG, which include the most widely used control strategies nowadays. Second, the book will introduce the advanced control of DFIG wind power systems. It will cover advanced controls of DFIG under the non-ideal grid with the grid voltage harmonic distortion and the grid voltage unbalanced. The dynamic model of the DFIG and converter under grid voltage harmonic distortion and the grid voltage unbalanced will be introduced. Then the stator harmonic current control is used in order to suppress the effect of the stator lower-order harmonics. Afterward, DC fluctuations of the back-to-back converter for DFIG is investigated under the unbalanced grid. To accommodate the wind turbine to the grid fault, the grid low-voltage fault ride-through (LVRT) for the DFIG wind turbine system is studied. Furthermore, the control strategy for DFIG under recurring grid faults is also investigated. In addition, to improve the reliability of the wind turbine, the smart thermal de-rating control of DFIG system is explained. Finally, a DFIG test bench is introduced. It is helpful for the readers to understand how the real system works and can be a guide to build a small-scale test bench in a laboratory.

This book may be helpful for readers who hope to have knowledge of modeling and controlling DFIG wind power systems and a deep understanding of the interaction of the wind turbine and the grid. It is suitable for both undergraduate and graduate levels and may serve as a useful reference for academic researchers, engineers, managers, and other professionals in the industry. Most of the chapters include descriptions of fundamental and advanced concepts, supported by many illustrations.

The authors would like to acknowledge the contribution and kind support of colleagues and former graduate students of Zhejiang University and Aalborg University—Dr. Jun Xu, Dr. Changjin Liu, Dr. Min Chen, Dr. Ke Ma, Mr. Ye Zhu, and others. We acknowledge the tireless efforts and assistance of Wiley Press editorial staff.

DEHONG XU
FREDE BLAABJERG
WENJIE CHEN
NAN ZHU

NOMENCLATURE

Subscripts

w	wind
T	turbine
r	rotor
s	stator
g	grid
sl	slip
m	magnetic/magnetizing
em	electromagnetic
$mech$	mechanical
sw	switching

Superscripts

$\vec{x}^*$	complex conjugated vector
x^*	complex conjugated value
$\vec{x}^{ref}$	reference vector
x^{ref}	reference value

Variables

v	variable voltage
V	constant voltage
$\vec{v}$	voltage vector
$\vec{V}$	voltage phasor
i	variable current
I	constant current
$\vec{i}$	current vector
$\vec{I}$	current phasor
E	electromotive force
$\vec{E}$	electromotive force phasor
ϕ	magnetic flux
Φ	magnetic flux phasor
ψ	magnetic flux linkage
$\vec{\psi}$	magnetic flux linkage vector
$\vec{\Psi}$	magnetic flux linkage phasor

ω angular speed
T torque
P active power
Q reactive power
n_p number of pole pairs
S_l slip ratio

PART *1*

INTRODUCTION TO WIND POWER GENERATION

CHAPTER 1

INTRODUCTION

In this chapter, an overview of wind power generation and the evolution of wind power systems are briefly introduced, and the challenges and trends in wind power generation are discussed.

1.1 GLOBAL WIND POWER DEVELOPMENT

1.1.1 Global Environment Challenge and Energy Crisis

Nowadays, the human society consumes a huge amount of electricity every year. It is reported by the U.S. Energy Information Administration (EIA) that the global net electricity consumption has grown from 10,395 TWh in 1990 to 20,567 TWh in 2015 [1]. Since most of the electricity is generated from fossil fuels, the increase of the electricity net consumption will lead to large greenhouse gas emissions, and this may cause global warming. The Earth's average surface temperature has risen about 0.74°C for the period 1906–2005, which may cause the sea level rise, widespread melting of snow and ice, or some extreme weather challenges. Furthermore, burning of fossil fuels will produce dust and other chemical materials harmful to humans.

On the other hand, the fossil fuel reserves are limited and unsustainable. Oil will be exhausted in a few decades, followed by natural gas, and coal will also be used up in 200–300 years. The energy crisis brought by the exhaustion of fossil fuels is a long-range challenge for human beings. Many efforts have been made worldwide to try to find an alternative energy.

1.1.2 Renewable Energy Development

Renewable energy is defined as the energy that comes from resources that are naturally replenished on a human timescale such as sunlight, wind, rain, tides, waves and geothermal heat. Typically, the renewable energy includes wind power, photovoltaic (PV) power, hydropower, biomass power, and ocean power. As renewable energy is reproducible and has a low footprint of CO_2, it is regarded as a favorable solution to both the global environment challenge and energy crisis. Rapid deployment of renewable energy has been reported in recent years. Global renewable energy policy

Advanced Control of Doubly Fed Induction Generator for Wind Power Systems, First Edition.
Dehong Xu, Frede Blaabjerg, Wenjie Chen, and Nan Zhu.

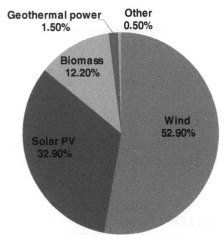

Figure 1.1   Worldwide capacity share of different non-hydro renewable powers by the end of 2016 [2].

multistakeholder network REN21 estimated that by the end of 2016, 30% power generation capacity will come from renewable energy and renewable energy will account for about 24.5% of global electricity generation [2]. Nowadays, the biggest renewable energy generation is from hydropower. However, since the location requirement of the hydropower is limited to lakes or rivers, the worldwide growth of hydropower has become slower in the recent years, which indicates that hydropower is very close to its capacity limit.

The non-hydropower renewable generation, including wind, PV, and biomass, has been growing very fast in the last 10 years. The non-hydropower renewable generation capacity reached 921 GW by the end of 2016, compared to 85 GW in 2004 [2]. The worldwide capacity share of different non-hydro renewable powers by the end of 2016 can be found in Figure 1.1. It is found that wind power has the largest capacity share among the non-hydropower renewable generations. Wind power has reached 56.8% of the non-hydro renewable power capacity.

1.1.3   Wind Energy Development

The wind power generation is regarded as the most widely used non-hydro renewable energy generation. It has a high reserve and is renewable and clean. Besides it produces almost no greenhouse gas emissions. Now at least 83 countries around the world are using wind power to supply their electricity grids [3]. The capacity of wind power installation has grown rapidly for the past 15 years. The statistics show the worldwide total wind power capacity has grown from 24 GW in 2001, to about 487 GW in 2016 [3], as shown in Figure 1.2. China leads the accumulated wind power installation, followed by the United States, Germany, Spain, Indian, etc., as shown in Figure 1.3.

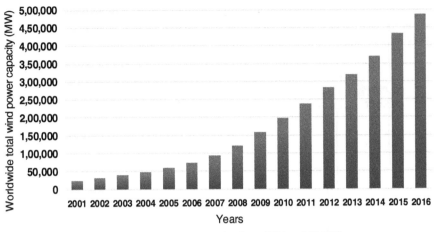

Figure 1.2   Worldwide total wind power capacity from 2001 to 2016 [3].

At the same time, the wind power share in the mix of the power supply also increased in the world, especially in some European countries. In 2014, Denmark set a new world record by reaching a wind power share of 39% in the domestic power supply [4]. Spain has wind power share of more than 15% [5]. Worldwide, the wind energy production has reached around 4% of total worldwide electricity usage in 2014 [6].

1.2   EVOLUTION OF WIND POWER SYSTEM

With the increasing penetration of wind power into the grid, the technology of the wind power generation has undergone a rapid development. One of the typical features is the changing of the wind power system structures. Modern wind power systems are more efficient, more reliable and more intelligent than before.

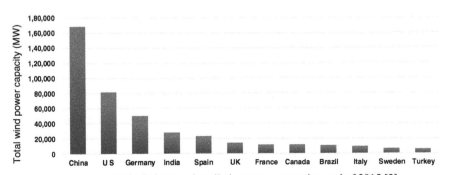

Figure 1.3   Accumulated wind power installation versus countries, end of 2015 [3].

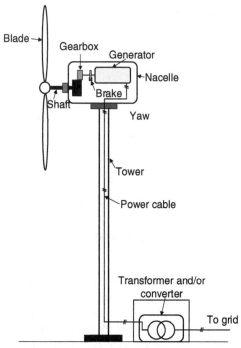

Figure 1.4   Structure of a wind turbine system.

1.2.1   Basic Structure of a Wind Turbine

The mostly used wind turbine is the horizontal wind turbine as shown in Figure 1.4. The blade, the shaft and the nacelle of the wind turbine are installed on a high tower. The blade rotates under wind flow and the wind energy is captured and converted into the mechanical energy in the shaft. The rotating angular speed of the shaft is increased using the gearbox so that it is compatible with the generator. The mechanical energy originated from the wind is converted into electric energy by the generator. Then the electricity is transmitted to the power electronic converter on the ground via the power cable, which is connected to the transformer in the grid. The nacelle provides space for components such as the shaft, the gearbox, and the brake on the tower, and can also target the turbine toward the wind flow direction by the action of the yaw.

1.2.2   Power Flow in the Wind Turbine System

The function of the wind power generation system is to harvest the kinetic energy of the wind flow, convert it into the electrical energy and finally feed into the grid. The configuration of the wind turbine system (WTS), which is composed of the wind turbine, the gearbox, the generator, the power converter, as well as the transformer, can be simplified as shown in Figure 1.5.

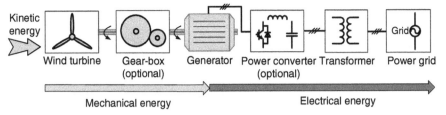

Figure 1.5   Basic configuration of wind power generation system.

Wind Turbine: The kinetic energy in the wind is collected by the wind turbine, and it is converted into mechanical energy on the shaft of the wind turbine. The early wind turbines normally rotate at an almost fixed speed, while the modern wind turbines can adjust the rotation speed with the variations in wind speed in order to increase the wind energy harvesting efficiency [34].

Gearbox: The gearbox is used to adjust the rotating speed of the shaft and make it compatible with the generator. In some cases, for example, in directly driven wind power system with multiple-pole synchronous generators, the gearbox may not be used.

Generator: The generator converts the mechanical energy on the shaft into electrical energy. In different types of WTS, the generator can be caged generator (CG), doubly fed induction generator (DFIG), or permanent magnet synchronous generator (PMSG).

Power Converter: The power converter works as an interface between the generator and the power grid. It converts the original electrical energy from the generator, which may be unstable with respect to amplitude or frequency, into the relatively stable electrical energy, which is more accepted by the power grid. On the other hand, the power converter also controls the generator to cooperate with the wind turbine to achieve better energy harvesting efficiency.

Transformer: The transformer is used to step up the output of the power converter (normally around 690 V) to a higher voltage, and transfers the wind power to the distribution or transmission power lines.

1.2.3   Fixed-Speed Wind Turbine System

The fixed-speed WTSs emerged in the 1970s and were widely used during the 1980s and 1990s. The shaft of the wind turbine is operated at a fixed angular speed, independent of the wind speed. The scheme of the fixed-speed WTS is shown in Figure 1.6. The generator operates with a fixed rotor speed corresponding to the grid frequency. It is directly connected to the grid by a transformer.

The advantage of the fixed-speed WTS is its simplicity of structure. It has a drawback that it cannot realize maximum wind energy tracking according to the variations in the wind speed. Reactive power consumed by the generator needs to be compensated by the capacitor bank. Further, it has no grid fault support capability, which is now needed by the grid operator. It also has higher mechanical stress for the wind turbine.

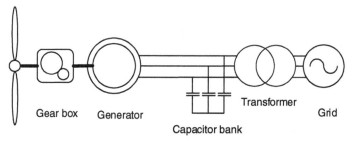

Figure 1.6   Scheme of a fixed-speed wind turbine system.

1.2.4   Variable-Speed Wind Turbine System

The variable-speed WTS is widely used nowadays. Different from the fixed-speed WTS, the variable-speed WTS is able to adjust the rotor speed when the wind speed changes to realize the maximum wind energy harvesting.

The scheme of a variable-speed wind turbine is shown in Figure 1.7. The wind power is captured by a pitch-controlled wind turbine and sent to the generator. The generator is connected to the grid by a power electronic converter. The variable-speed operation of the WTS is achieved by the power electronic converter.

The power electronic converter controls the rotor speed of the generator so that the shaft speed of the blade adjusts when the wind speed changes to realize the highest wind energy harvesting.

When the wind turbine reaches the speed limit or the electric limit, either the mechanical angular speed or electric power can be limited by controlling the power electronic converter. Besides, it can also realize soft start for the wind turbine so that there is less power surge to the grid. When the grid fault happens, the variable-speed WTS can help the grid recover from the fault by feeding reactive power to the grid. The power electronic converters may provide ancillary services to the grid.

Power electronics has been bringing in significant performance improvements for the WTSs. It not only increases the energy yield and reduces the mechanical stress, but also enables the WTS to act like an ideal power source friendlier to the utility [38].

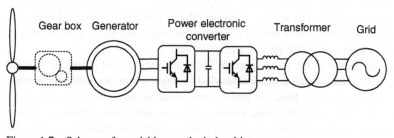

Figure 1.7   Scheme of a variable-speed wind turbine system.

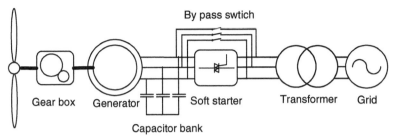

By pass swtich

Gear box   Generator   Soft starter   Transformer   Grid

Capacitor bank

Figure 1.8   Fixed-speed wind turbine with a soft starter.

1.3 POWER ELECTRONICS IN WIND TURBINE SYSTEMS

1.3.1 Power Electronics in Fixed-Speed Wind Turbine System

For the fixed-speed WTS, usually induction machines are used as the generator. Connecting a large induction machine to the power system will cause a large power surge to the utility with a very high inrush current, which results in disturbances to the grid. To limit the starting current of the induction machine, a thyristor soft starter is used in the fixed-speed WTS, as shown in Figure 1.8. The starting current is reduced by gradually increasing the voltage applied to the generator to the grid voltage. The soft starter, based on thyristor technology, typically limits the RMS value of the inrush current to less than two times the rated current of the generator. Once the starting process is over, all thyristors are kept in the on-state. Since the thyristor has a voltage drop when it is conducting and causes power loss, a mechanical switch is used to bypass the thyristor soft starter when the WTS finishes the starting process. Besides reducing the impact on the grid, the soft starter also effectively reduces the torque peak associated with the inrush current during the starting, which is helpful to relieve the mechanical stress on the gearbox.

1.3.2 Power Electronics in Variable-Speed Wind Turbine System

In variable-speed WTSs, the power electronic converter plays an important role as the interface between the WTS and the grid. Two most popular variable-speed wind turbine configurations are DFIG and synchronous generator (SG). The DFIG wind system equipped with partial-scale power converter is dominating the market while the WTS with SG with full-scale power converter has grown in recent years.

1.3.2.1   Doubly Fed Induction Generator

WTSs with DFIG has been used extensively since 2000 and is the most adopted solution nowadays. As shown in Figure 1.9, a back-to-back converter is used in the DFIG system. The stator windings of the DFIG are directly connected to the power grid, while the rotor windings are connected to the back-to-back converter [30]. In this configuration, both the frequency and the current amplitude in the rotor windings

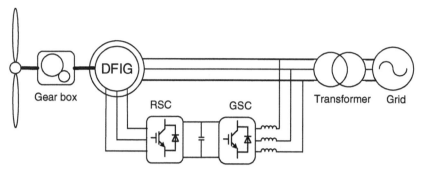

Figure 1.9   Variable-speed wind turbine with a partial-scale power converter and a doubly fed induction generator.

can be freely regulated so that the rotor speed can be changed in a wide range and wind energy harvesting capability is enhanced. Besides, it can realize soft start for the wind turbine and provide the grid fault ride-through ability. It can also reduce the mechanical stress to the wind turbine.

In addition, the DFIG has a special feature that it only needs a back-to-back converter with about 30% capacity of the wind turbine, which is an economical solution at an earlier stage of wind power development when the cost of the power converter was more critical [36–37].

The two-level pulse-width-modulation voltage-source-converter (2L-PWM-VSC) is the mostly used converter topology so far for the DFIG-based wind turbine concept as the power rating requirement for the converter is limited [41]. Normally, two 2L-PWM-VSCs are configured in a back-to-back structure in the WTS, as shown in Figure 1.10, which is called 2L-BTB for convenience. Advantages of the 2L-BTB solution include the full power controllability (four-quadrant operation) with a relatively simple structure and fewer components, which contribute to well-proven robust/reliable performances as well as the advantage of lower cost [29].

1.3.2.2   Asynchronous/Synchronous Generator with Full-Scale Power Converter

The second important configuration that has become popular for the newly developed and installed wind turbines is shown in Figure 1.11. It introduces a full-scale power converter to interconnect the power grid and stator windings of the generator. The reliability enhancement due to the elimination of slip rings and simpler or even

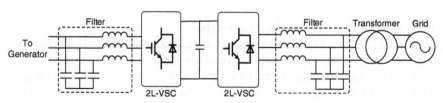

Figure 1.10   Two-level back-to-back (2L-BTB) voltage source converter for a wind turbine.

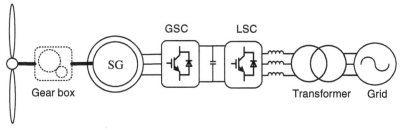

Figure 1.11   Variable-speed wind turbine with a full-scale power converter and synchronous generator.

eliminated gearbox, full power and speed controllability, and better grid support capability are the primary advantages compared to the DFIG-based concept. However, there are some drawbacks such as the high cost of PMSG, the need for a full-power BTB converter as well as the higher power losses in the converter. Instead of PMSG, wound rotor synchronous generator, etc., can be used as the generator.

1.4 CHALLENGES AND TRENDS IN FUTURE WIND POWER TECHNOLOGY

In this section, several emerging technology challenges for the future WTSs are addressed. The discussions will mainly focus on technology issues of power electronic converters in the WTS with respect to cost, reliability, grid integration, new power electronics circuits, etc.

1.4.1 Lower Cost

Cost is one of the most important considerations for the technology which determines the feasibility of certain energy technologies to be widely used in the future. In order to quantify and compare the cost of different energy technologies, levelized cost of energy (LCOE) index is generally used [7]. LCOE represents the price at which the electricity is generated from a specific energy source over the whole lifetime of the generation unit. It is an economic assessment of the cost of the energy-generating system including initial investment, development cost, capital cost, operations and maintenance cost, the cost of fuel, etc. LCOE can be defined in a single formula as [8]:

$$\text{LCOE} = \frac{C_{Dev} + C_{Cap} + C_{O\&M}}{E_{Annual}} \tag{1.1}$$

Here, the initial development cost C_{Dev}, capital cost C_{cap}, and the cost for operation and maintenance $C_{O\&M}$ are first levelized to annual average cost over the lifetime of the generation system, and then divided by the average annual energy production in the whole lifetime E_{Annual}. In order to reduce the cost of energy, one effective way is to reduce the cost of development, capital, operation, and maintenance, and the other

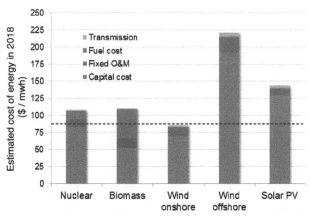

Figure 1.12   Estimated levelized cost of energy for several renewable energy technologies in 2018 [10].

effective way is to increase the lifetime of the generation system. As an example, the LCOE for offshore wind power of Denmark and the United Kingdom is between 140 and 180 EUR/MWh in 2010 according to the studies carried by [9], and this number is expected to be reduced by 50% by 2020 to a range between 67 and 90 EUR/MWh, providing an increase in the lifetime of wind turbines from 20 to 25 years, and other significant cost reductions are achieved.

Figure 1.12 shows another example of US-estimated LCOE for several promising renewable energy technologies in 2018 [10]. It can be seen that the cost distribution of different technologies varies a lot, where the onshore wind power still shows cost advantages compared to other renewable energy sources. It can be also expected that in the United States, the capital cost may still be dominant for most of the renewable energy technologies for the next decade.

As more power electronics are introduced to the energy system to improve the performances of power generation, the cost of the power electronics becomes more important. In the WTS, cost considerations impose challenges for the design and the selection of power electronics.

For instance, the needs for higher power capacity and full-scale power conversion will increase the cost for power semiconductors, passive components, and corresponding thermal management. Due to the limited space in the nacelle, higher power density for the power converters leads to extra cost for the design. Besides, remote locations of the wind turbines increase the cost for installation and maintenance, which demands high reliability, modularity, and redundancy of the system.

1.4.2   Larger Capacity

The size and power generation capacity of the wind turbine has been gradually increasing over the last decades and will be continuously increasing in the future, as shown in Figure 1.13 [11].

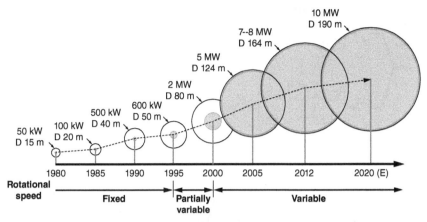

Figure 1.13   Development of wind turbines between 1980 and 2020 (estimated) [11].

Many of the major wind turbine manufacturers have developed high-power and large-scale wind turbine products. Some of the wind turbine product lines of the top wind turbine manufacturers are shown in Table 1.1.

To deal with the growing power capacity, multicell converter topologies have been developed by connecting conventional 2L-BTB converters in parallel or in series. Two of the most adopted multicell solutions are shown in Figure 1.14. One of the advantages of this multicell configuration is that standard and proven converter technologies can be used for higher power capacities. Also, redundancy and modular characteristics can be achieved in this configuration. Such a solution is the state-of-the-art for wind turbines above 3 MW [12, 13] and will likely be utilized in larger-scale wind turbines in the future.

1.4.3   Higher Reliability

The growth of total installation and increasing capacity of the wind turbine make the failures of wind turbines costly. The failures of WTS will not only cause stability problem to the power grid due to the sudden absence of a large amount of power capacity, but also results in high cost for maintenance. In addition, it will cause

TABLE 1.1   Wind turbine product lines of the top wind turbine manufacturers in 2015 [11]

Manufacturer	Rotor diameter (m)	Power range (MW)
Goldwind (China)	70–121	1.5–3
Vestas (Denmark)	90–136/164	1.8–3.45/8
GE Wind (USA)	83–137/150	1.7–3.8/6
Siemens Wind (Denmark/Germany)	101–142/154	2.3–4/6–8
Gamesa (Spain)	80–132/132	2–3.3/5
Enercon (Germany)	44–141	0.8–4.2

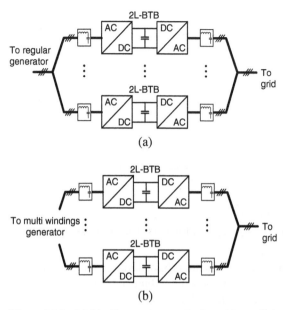

(a)

(b)

Figure 1.14   Multicell converter topologies with parallel connected two-level back-to-back converters: (a) with regular generator; (b) with multiwinding generator.

reduction of annual energy production and thus increase the LCOE. As a result, reliability is a critical design consideration for the next generation wind turbine.

The complex nature of wind speed makes the mission profiles of the wind turbines very complicated. The large wind speed fluctuations may cause thermal cycles which is the main cause of the failures of power electronics components [14–16]. The relationship between the characteristics of thermal cycling and the failures of power semiconductors has been extensively studied, and it is found that the lifetime of power semiconductor will be shorter under thermal cycles with higher fluctuation amplitudes and mean values [11]. With the complicated mission profiles of wind turbines, the power semiconductors in wind power systems may experience many thermal cycles ranging from 15°C to 90°C, and they may cause lifetimes to drop below 20 years according to the life time models for power semiconductor devices [16].

Many efforts have been made to investigate new modeling and testing approaches to evaluate the lifetime consumption of power semiconductors and to apply appropriate control methods to improve the expected lifetime of power semiconductor devices. Reliability improvement of power semiconductors by means of condition monitoring has become a recent focus of research. Many studies have been performed to investigate online condition monitoring methods for wire-bonded power IGBT modules. The on-state voltage drop of the IGBT module has been the most used indicator for condition monitoring [17–22]. However, most of the solutions are dedicated to specific applications or need structure modifications of the power module.

Therefore, making the condition monitoring methods more intelligent and applicable to general applications will be the goal of improvement in the future.

1.4.4   The Application of New Power Semiconductor Devices

Currently, a majority of wind power generators in the market are rated at 690 VAC. Voltage rating for switching devices is 1700 V or 1200 V [23, 24]. SiC MOSFET, JFET, and Schottky barrier diode (SBD) of such voltage ratings are technically mature and commercially available. In [24], SiC devices of 1200 V and 1700 V ratings from the same manufacturer are compared to the state-of-the-art IGBT modules of the same voltage ratings. According to the analysis, a large amount of switching loss reduction can be achieved by applying SiC and hybrid devices.

In [25] and [26], application of SiC MOSFETs and Schottky diodes are investigated in a 1.5 MW full-scale back-to-back converter adopting the two-level topology used for wind power generation based on permanent magnet generator. The 1.5 MW converter is assumed to be composed of ten SiC-based converters that each has twenty 1700 V/10 A SiC MOSFETs in parallel or an Si-based converter with two 1700 V/1200 A IGBT modules in parallel. Since at present, SiC devices with the suitable current rating are not available, a large amount of small-scaled SiC devices are paralleled in this calculation which is unrealistic in practical application. However, as the technology matures, high-current rated SiC MOSFET modules will be developed in the near future. Nevertheless, the calculation results are helpful to give us a glimpse at the advantage of SiC MOSFET in efficiency improvement.

Presently, SiC devices still cost much higher than their Si counterparts. Some studies have used SiC diodes to substitute the free-wheeling diodes in the conventional IGBT modules to form hybrid devices. In [27], an SiC diode module and an IGBT module are used to form a hybrid phase leg. According to the analysis, the inverter adopting the Si IGBT/SiC clamping diode hybrid devices has about 0.6% efficiency increase compared to all-Si IGBT inverter. This is a very cost-effective solution with significant improvement in efficiency and relatively low extra cost. Taking the cost into consideration, hybrid devices may be the compromise between efficiency improvement and cost in the near future.

1.4.5   More Advanced Grid Integration Control

In order to reduce the impact of the wind gust and ensure the security and stability of grid operation, the wind turbines or wind farms are preferred to be configured as distributed generation networks. Its power flows and electrical behaviors are different compared to the traditional centralized generation networks [28, 41]. Therefore, the protection schemes of the future grid utilities with more wind power penetration should be also changed. It results in a more distributed protection structure and may allow the islanding operation of some wind turbine units as microgrids [31].

Moreover, with the growing proportion of wind power in the power grid, more advanced grid requirements are needed. In the case of shutting down of transmission networks, the WTSs may need the abilities to black start [32].

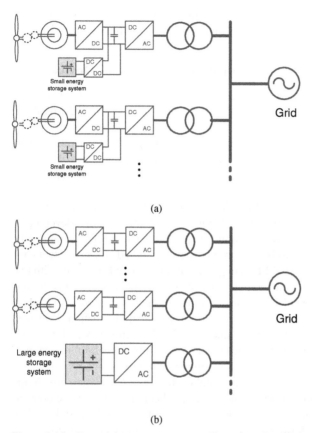

(a)

(b)

Figure 1.15   Potential energy storage configurations for wind power plants to enable virtual power plant operation: (a) distributed energy storage and (b) centralized energy storage.

In order to achieve these more advanced features of grid interconnection, some energy storage systems may be needed for future wind turbines and wind farms. The storage system can be configured locally for each wind turbine unit, as shown in Figure 1.15a, or be configured centrally for several wind turbines/wind farms, as shown in Figure 1.15b. Such WTSs with energy storage will also be ready to operate as a primary controller and may operate as a virtual synchronous machine [39].

1.4.6   Configurations of Wind Power Plants

As the wind power capacity grows, large wind farms which consist of many wind turbines are being developed. These wind farms may have significant impacts to the grids, and therefore they will play an important role in the power quality and the control of the power grid systems. The power electronics technology is again an important part of both the system configurations and the control of the wind farms in order to fulfill the growing grid demands [35]. Some existing and potential configurations of the wind farms are shown in Figure 1.16.

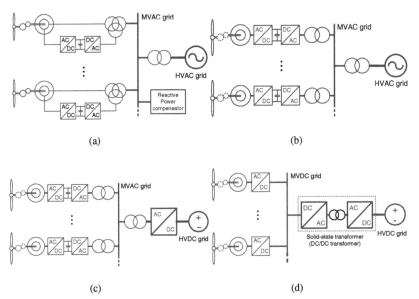

Figure 1.16   Potential wind farm configurations with AC and DC power delivery:
(a) Doubly fed induction generator system with AC grid; (b) full-scale converter system with
AC grid; (c) full-scale converter system with transmission DC grid; (d) full-scale converter
system with both distribution and transmission DC grids.

A wind farm equipped with DFIG-based WTSs is shown in Figure 1.16a. Such
a wind farm system is in operation in Denmark as a 160 MW offshore wind power
station. It is noted that due to the limitation of the reactive power capability, a centralized reactive power compensator like STATCOM may be used in order to fully
satisfy future grid requirements [39].

Figure 1.16b shows another wind farm configuration equipped with a WTS
based on full-scale power converter. Because the reactive power controllability is
significantly extended, the grid-side converter in each of the generation unit can be
used to provide the required reactive power individually, leading to reactive power
compensator-less solutions.

For long-distance power transmission from an offshore wind farm, HVDC may
be an interesting option because the efficiency is improved and no voltage compensators are needed. A typical HVDC transmission solution for wind power is shown in
Figure 1.16c, in which the medium AC voltage of the wind farm output is converted
into a high DC voltage by a boost transformer and high voltage rectifier.

Another possible wind farm configuration with HVDC transmission is shown in
Figure 1.16d, where a solid-state transformer (or DC/DC transformer) is used to convert the low/medium DC voltage from each wind turbine output to the medium/high
DC voltage for transmission, thus a full DC power delivery both in the distribution
and transmission grid can be realized. It is claimed in that the overall efficiency of the
power delivery can be significantly improved compared to the configuration shown

in Figure 1.16c—mainly because of fewer converters and transformers in this DC transmission system. It can be a future solution for large wind farms to increase the overall efficiency of power delivery [33, 40, 42].

1.5 THE TOPICS OF THIS BOOK

This book will focus on the advanced control of the DFIG wind power system, which is realized by the power electronic converters and aims at better grid integration performance. This book is divided into following four parts:

Part I (Chapters 1–3) will provide the basic knowledge of wind power technology. The related grid codes for wind power generator will be introduced. It will make the reader in related areas much easier to understand the following content.

Part II (Chapters 4 and 5) will evaluate the dynamic model of DFIG wind power system and the vector control scheme of the DFIG, which includes the most widely used control strategies nowadays.

Part III (Chapters 6–10) will aim at the advanced control of DFIG under the non-ideal grid, including the grid voltage harmonic distortion and the grid voltage unbalanced. The dynamic model of the DFIG and converter under grid voltage harmonic distortion, and the grid voltage unbalanced will be introduced. The stator harmonic current control is introduced in order to suppress the stator lower-order harmonics. The DC-fluctuations control of the GSC can suppress the DC-bus fluctuations under the unbalanced grid.

Part IV (Chapters 11–13) will introduce the grid fault ride-through of the DFIG WTS such as LVRT of DFIG, the smart thermal derating control of DFIG system, etc. The control strategy for DFIG under recurring faults is also investigated.

In Part V (Chapter 14), a DFIG test bench is introduced. It is helpful for the reader to understand how the real system is built and can be a guide to building a small-scale test bench in a laboratory.

REFERENCES

[1] "International energy statistics," U.S. Energy Information Administration, 2015. [Online]. Available: http://www.eia.gov/
[2] "Renewables 2017 global status report," REN21, 2017. [Online]. Available: http://www.ren21.net/
[3] "Global wind report 2016," Global Wind Energy Council, 2016. [Online]. Available: http://www.gwec.net/
[4] "New record in worldwide wind installations," World Wind Energy Association, 2015, [Online]. Available: http://www.wwindea.org/
[5] "EWEA annual statistics 2015," European Wind Energy Association, 2015. [Online]. Available: http://www.ewea.org/
[6] "Energy development," Wikipedia. [Online]. Available: https://en.wikipedia.org/wiki/ Energy_development
[7] F. Blaabjerg and K. Ma, "Future on power electronics for wind turbine systems," *IEEE J. Emerg. Sel. Topics. Power. Electron*, vol. 1, no. 3, pp. 139–152, Sep. 2013.

[8] "World energy resources: 2013 survey," World Energy Council, 2014. [Online]. Available: http://www.worldenergy.org/

[9] "Denmark-supplier of competitive offshore wind solutions. Megavind's strategy for offshore wind research, development and demonstration," Megavind Secretariat: Danish Wind Industry Association, 2010. [Online]. Available: https://www.windpower.org/download/952/uk_megavind_report_okpdf

[10] "Levelized cost of new generation resources in the annual energy outlook 2013," U.S. Dept. Energy, U.S. Energy Information Administration, 2013. [Online]. Available: http://www.eia.gov/

[11] F. Blaabjerg and K. Ma, "Wind energy systems," in *Proc. IEEE*, vol. PP, no. 99, pp. 1–16.

[12] B. Andresen and J. Birk, "A high power density converter system for the Gamesa G10 × 4,5 MW wind turbine," in *Proc. Eur. Conf. Power Electron. Appl.*, Sep. 2007, pp. 1–7.

[13] R. Jones and P. Waite, "Optimised power converter for multi-MW direct drive permanent magnet wind turbines," in *Proc. Eur. Conf. Power Electron. Appl.*, 2011, pp. 1–10.

[14] C. Busca, R. Teodorescu, F. Blaabjerg, S. Munk-Nielsen, L. Helle, T. Abeyasekerab, and P. Rodriguez, "An overview of the reliability prediction related aspects of high power IGBTs in wind power applications," *Microelectron. Rel.*, vol. 51, nos. 9–11, pp. 1903–1907, 2011.

[15] A. Wintrich, U. Nicolai, and T. Reimann, Semikron Application Manual, p. 128, 2011.

[16] J. Berner, "Load-cycling capability of HiPakIGBT modules," ABB Group, Zürich, Switzerland, Appl. Note 5SYA 2043-02, 2012.

[17] V. Smet, F. Forest, J. J. Huselstein, A. Rashed, and F. Richardeau, "Evaluation of V_{ce} monitoring as a real-time method to estimate aging of bond-wire IGBT modules stressed by power cycling," *IEEE Trans. Ind. Electron.*, vol. 60, no. 7, pp. 2760–2770, Jul. 2013.

[18] P. Ghimire, S. Beczkowski, S. Munk-Nielsen, B. Rannestad, and P. Thogersen, "A review on real time physical measurement techniques and their attempt to predict wear-out status of IGBT," in *Proc. Eur. Conf. Power Electron. Appl.*, Sep. 2013, pp. 1–10.

[19] R. O. Nielsen, J. Due, and S. Munk-Nielsen, "Innovative measuring system for wear-out indication of high power IGBT modules," in *Proc. 2011 IEEE Energy Convers. Congr. Expo.*, Sep. 2011, pp. 1785–1790.

[20] R. I. Davis and D. J. Sprenger, "Methodology and apparatus for rapid power cycle accumulation and in-situ incipient failure monitoring for power electronic modules," in *Proc. 2014 IEEE Electron. Compon. Technol. Conf.*, May 2014, pp. 1996–2002.

[21] S. Beczkowski, P. Ghimre, A. R. de Vega, S. Munk-Nielsen, B. Rannestad, and P. Thogersen, "Online V_{ce} measurement method for wear-out monitoring of high power IGBT modules," in *Proc. 2013 Eur. Conf. Power Electron. Appl.*, Sep. 2013, pp. 1–7.

[22] B. Ji, "In-situ health monitoring of IGBT power modules in EV applications," Ph.D. thesis, School of Electrical and Electronic Engineering, Newcastle University, Newcastle upon Tyne, UK, 2011.

[23] J. He, T. Zhao, X. Jing, and A. O. Demerdash, "Application of wide bandgap devices in renewable energy systems—Benefits and challenges," in *Proc. Int. Conf. Renew. Energy Res. Appl.*, 2014, pp. 749–754.

[24] S. V. Araujo and P. Zacharias, "Perspectives of high-voltage SiC-semiconductors in high power conversion systems for wind and photovoltaic sources," in *Proc. Eur. Conf. Power Electron. Appl.*, 2011, pp. 1–10.

[25] H. Zhang and L. M. Tolbert, "SiC's potential impact on the design of wind generation system," in *Proc. IEEE Annual Conf. Ind. Electron.*, 2008, pp. 2231–2235.

[26] H. Zhang and L. M. Tolbert, "Efficiency impact of silicon carbide power electronics for modern wind turbine full scale frequency converter," *IEEE Trans. Ind. Electron.*, vol. 58, no. 1, pp. 21–28, Jan. 2011.

[27] W. L. Erdman, J. Keller, D. Grider, and E. VanBrunt, "A 2.3-MW medium-voltage, three-level wind energy inverter applying a unique bus structure and 4.5-kV Si/SiC hybrid isolated power modules," in *Proc. IEEE Appl. Power Electron. Conf. Exposition*, 2015, pp. 1282–1289.

[28] Z. Chen, J. M. Guerrero, and F. Blaabjerg, "A review of the state of the art of power electronics for wind turbines," *IEEE Trans. Power. Electron.*, vol. 24, no. 8, pp. 1859–1875, Aug. 2009.

[29] F. Blaabjerg, M. Liserre, and K. Ma, "Power electronics converters for wind turbine systems," *IEEE Trans. Ind. Appl.*, vol. 48, no. 2, pp. 708–719, Mar./Apr. 2012.

[30] R. Datta and V. T. Ranganathan, "Variable-speed wind power generation using doubly fed wound rotor induction machine: A comparison with alternative schemes," *IEEE Trans. Energy Convers.*, vol. 17, no. 3, pp. 414–421, Sep. 2002.

[31] F. Blaabjerg, R. Teodorescu, M. Liserre, and A. V. Timbus, "Overview of control and grid synchronization for distributed power generation systems," *IEEE Trans. Ind. Electron.*, vol. 53, no. 5, pp. 1398–1409, Oct. 2006.

[32] C. Blaabjerg, Z. Chen, and S. B. Kjaer, "Power electronics as efficient interface in dispersed power generation systems," *IEEE Trans. Power Electron.*, vol. 19, no. 5, pp. 1184–1194, Sep. 2004.

[33] J. A. Baroudi, V. Dinavahi, and A. M. Knight, "A review of power converter topologies for wind generators," *Renew. Energy*, vol. 32, no. 14, pp. 2369–2385, 2007.

[34] E. Muljadi and C. P. Butterfield, "Pitch-controlled variable-speed wind turbine generation," *IEEE Trans. Ind. Appl.*, vol. 37, no. 1, pp. 240–246, Jan./Feb. 2001.

[35] R. Teodorescu, M. Liserre, and P. Rodriguez, Grid Converters for Photovoltaic and Wind Power Systems. John Wiley & Sons, 2011.

[36] B. H. Chowdhury and S. Chellapilla, "Double-fed induction generator control for variable speed wind power generation," *Electr. Power Syst. Res.*, vol. 76, no. 9, pp. 786–800, 2006.

[37] S. Muller, M. Deicke, and R. W. De Doncker, "Doubly fed induction generator systems for wind turbines," *IEEE Ind. Appl. Mag.*, vol. 8, no. 3, pp. 26–33, 2002.

[38] H. Li and C. Zhe, "Overview of different wind generator systems and their comparisons," *IET Renew. Power Gener.*, vol. 2, no. 2, pp. 123–138, 2008.

[39] J. M. Carrasco, L. G. Franquelo, J. T. Bialasiewicz, E. Galvan, R. C. PortilloGuisado, M. A. M. Prats, J. I. Leon, and N. Moreno-Alfonso, "Power-electronic systems for the grid integration of renewable energy sources: A survey," *IEEE Trans. Ind. Electron.*, vol. 53, no. 4, pp. 1002–1016, Jun. 2006.

[40] G. Abad, J. Lopez, M. Rodriguez, L. Marroyo, and G. Iwanski, *Doubly Fed Induction Machine: Modeling and Control for Wind Energy Generation*. John Wiley & Sons, 2011.

[41] G. Abad and G. Iwanski, "Properties and control of a doubly fed induction machine," in Power Electronics for Renewable Energy Systems, Transportation and Industrial Application, John Wiley & Sons, 2014.

[42] B. Wu, N. Zargari, and Y. Lang, Power Conversion and Control of Wind Energy Systems. Wiley-IEEE Press, 2011.

CHAPTER 2

BASICS OF WIND POWER GENERATION SYSTEM

In this chapter, the basic knowledge related to modern wind power generation system (WPS) is introduced, especially for the variable-speed WPS. The important parts of the configuration of a WPS are introduced. The steady-state operation conditions of a variable-speed wind turbine are also investigated, and the control of the generator and power converter in different concepts of variable-speed WPSs are introduced. At last, the wind power transmission system is discussed briefly as well as the grid faults and distortions in power systems are analyzed.

2.1 INTRODUCTION

The target of the wind power generation system (WPS) is to collect the kinetic energy of the blowing wind and to convert it into electrical power flow into the power grid. This process is achieved by complicated mechanical and electrical conversion systems in the wind turbines. However, the basic configuration of WPS can be simplified into a few main parts, including the wind turbine, the gearbox, the generator, the power converter, as well as the transformer, as shown in Figure 2.1. The detailed descriptions of the components please refer to Chapter 1.

In this chapter, the basic knowledge related to the modeling of the WPS is introduced, especially for the variable-speed WPS. The important parts in the configuration of the WPS are described and the operation conditions of a variable-speed wind turbine are investigated as well as the control of generator and power converter used in different concepts of variable-speed WPSs are introduced. Finally the wind power transmission system is discussed in brief, where grid faults and distortions in the power systems are also analyzed.

2.2 WIND POWER CONCEPT

As introduced briefly in Chapter 1, an overview of WPSs is shown in Figure 2.2. Basically, there are two kinds of operation methods: fixed-speed wind turbines

Advanced Control of Doubly Fed Induction Generator for Wind Power Systems, First Edition.
Dehong Xu, Frede Blaabjerg, Wenjie Chen, and Nan Zhu.
© 2018 by The Institute of Electrical and Electronics Engineers, Inc. Published 2018 by John Wiley & Sons, Inc.

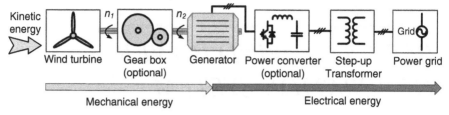

Figure 2.1　Basic configuration of a wind power generation system (WPS).

and variable-speed wind turbines [8–13]. In the earlier years, the wind turbine could only operate at a fixed rotational speed, which is called the fixed-speed concept. It requires a stiff power grid to enable a stable operation and may require a more expensive mechanical construction in order to absorb high mechanical stress, since wind gusts may cause torque pulsations on the drive train. The wind energy utilization efficiency is normally also low in the fixed-speed wind turbines. However, the simplicity of such a concept may also be an advantage in certain applications [8].

The variable-speed wind turbine is able to adjust its rotational speed according to the wind speed and thereby maximize the energy yield. The variable-speed operation of a wind turbine system has many advantages. The wind turbine can increase or decrease its speed according to the variations in the wind speed. This means less wear and tear on the tower, gearbox, and other components in the drive train, and also a higher wind energy utilization efficiency. Also, variable-speed systems can increase the production of energy and reduce the fluctuation of power injected into the grid. In variable-speed systems, the generator is normally connected to the grid through a power electronic system and the power converter will improve the dynamic and steady-state performance, help to control the wind turbine and generator speed, and also decouple the generator from the power grid. However, the disadvantages of variable-speed wind power systems may be higher cost and losses in the power converters [8–9].

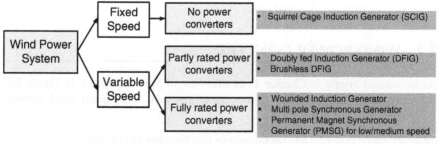

Figure 2.2　Overview of a wind power generation system.

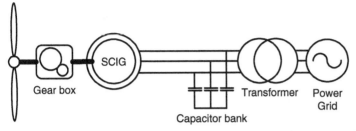

Figure 2.3   Schematic of a fixed-speed wind power generation system using a squirrel cage induction generator (SCIG) [8].

2.2.1   Fixed-Speed Concept

The "Danish concept" of directly connecting a wind turbine to the grid is widely used in the early wind turbine systems [8]. The scheme consists of a squirrel cage induction generator (SCIG), connected via a transformer to the grid and operating at an almost fixed speed. The configuration of a fixed-speed wind turbine system is shown in Figure 2.3. The advantages of wind turbines with induction generators are simple and cheap construction, and no synchronization device is required. These solutions are attractive due to cost and reliability. However, there are some drawbacks like (1) the wind turbine has to operate at a constant speed, (2) it requires a relatively stiff power grid to enable stable operation; and (3) it may require a more expensive mechanical construction in order to absorb high mechanical stress since wind gusts may cause torque pulsations on the drive train. During the 1980s, this concept was extended by adding a capacitor bank for reactive power compensation and also a soft starter for smoother grid connection.

2.2.2   Variable-Speed Concept with Partial Power Converters

The variable-speed wind turbine is able to adjust the rotational speed during wind speed changes and the operation of the variable-speed wind turbine will be introduced in the next part. With the help of a partial-scaled power electronic converter, the wind power system is able to maintain maximum efficiency under different wind speeds. The DFIG wind power concept is the most widely adopted one nowadays, as shown in Figure 2.4. The stator of the DFIG is directly connected to the grid, while the rotor side of the DFIG is controlled by a partial-rated power converters called the rotor-side converter (RSC) and the grid-side converter (GSC). By controlling the rotor current with the RSC, the output power, torque, and rotor speed of the DFIG can be controlled at the synchronous speed. The power rating of the converter is normally only 30% of the rated power of the generator, so this concept is attractive as seen from a cost point of view. However, the main disadvantage of this concept is the sensitivity to grid voltage disturbance, such as grid faults, grid harmonic distortions, and grid unbalance. On the other hand, the ability to operate under grid voltage disturbance is

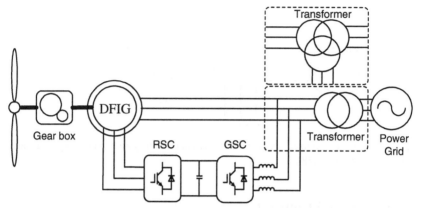

Figure 2.4   Schematic of a doubly fed induction generator (DFIG) generation system. RSC, rotor-side converter; GSC, grid-side converter.

required by the grid codes from many countries and therefore as a result, much effort has been taken on the control strategy of DFIG wind power systems to make them robust against grid voltage disturbance, which is also one of the main topics of this book.

The typical wind power systems in the market using this concept include Vestas V80-2 MW [2], GE-1.5 MW [3], and Gamesa G87-2.0 MW [4], etc.

2.2.3   Variable-Speed Concept with Full-Scale Power Converters

This concept uses a full-scale power electronic converter to isolate the generator from the grid, and this power electronic converter takes full control of the generator so that the generator can operate at variable speeds to maintain maximum efficiency under different wind speeds. The PMSG wind power system is one of the popular full-scale power concepts nqwadays, as shown in Figure 2.5. The rotor of the PMSG is made of permanent magnets, so no magnetizing current is required and the slip rings can be eliminated; thereby reducing maintenance cost. The stator of the PMSG is

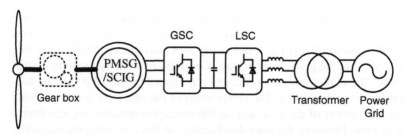

Figure 2.5   Schematic of a permanent magnet synchronous generator (PMSG) or squirrel cage induction generator (SCIG) wind power system using a full-scale power converter. GSC, grid-side converter; LSC, line-side converter.

controlled by the generator-side converter (GSC), and the power generated by the PMSG is transmitted into the grid by the line-side converter (LSC). As the generator is decoupled from the grid, this concept is less sensitive to grid voltage disturbances compared to the DFIG concept. Another advantage is the elimination of slip rings and even the gearbox in some cases. As a result, it is attractive from a reliability point of view. The main drawback of this concept is the higher cost of power converters compared to the partial scale concept. The SCIG or wounded rotor synchronous generator can also be used with a full power converter, so the cost of the generator can be reduced.

Typical wind power systems in the market with this concept include Gold Wind GW-1500 kW [5], Enercon E82-2 MW [6].

2.2.4 Hardware Protection Methods

Recently grid codes in many countries have regulated the fault-ridge-through capabilities of wind turbines. Some hardware protection solutions are used to assist the DFIG ride through serious grid faults. Two of the most-used hardware protection circuits are the rotor-side crowbar and DC chopper [14].

The rotor-side crowbar is one of the commonly used hardware solutions for the DFIG to ride through grid faults. The rotor-side crowbar short-circuits the rotor windings under grid faults, so the transient rotor current can be limited while the RSC can be protected. One of the commonly used structures of the rotor-side crowbar is shown in Figure 2.6.

The DC chopper is a brake resistor in the DC bus, as shown in Figure 2.7. Under voltage dips, the rotor current may flow into the DC bus of the RSC and lead to DC-bus overvoltage. The DC chopper is enabled when the DC-bus voltage is over the safety threshold. Then the energy from the RSC can be dissipated on the resistor of the DC chopper and thereby, the DC-bus voltage can be limited.

The detailed introduction and analysis of the hardware protection solutions will be given in Chapter 12.

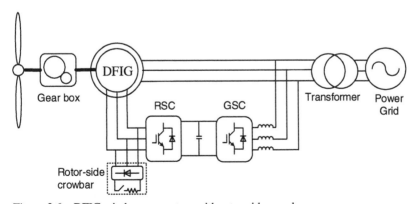

Figure 2.6   DFIG wind power system with rotor-side crowbar.

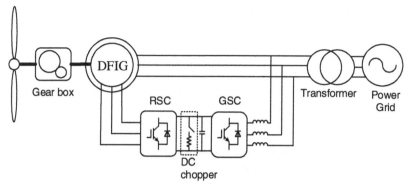

Figure 2.7   DFIG wind power system with DC chopper.

2.3   VARIABLE-SPEED WIND TURBINE

With the variable-speed wind turbine, the wind energy capture ability can be extended with a higher system efficiency. The variable-speed wind turbine is very attractive for a number of reasons, including reduced mechanical stress and increased ability of wind power capture by adjusting the rotor speed along with the wind speed. The operation of the variable-speed wind turbine will be introduced in detail in this section.

2.3.1   Wind Turbine Model

The wind turbine is used for capturing the kinetic power from the wind, and then converting it to mechanical power on the shaft, which then can be directly used by generators. The kinetic power of the wind P_w can be expressed as [15]

$$P_w = \frac{1}{2}\rho\pi r^2 v_w{}^3 \tag{2.1}$$

where ρ is the mass density of air, r is the wind turbine rotor radius, v_w is the wind speed. Unfortunately, the wind turbine cannot extract all the kinetic wind power flowing to the wind turbine. Generally, the mechanical power P_{mech} on the shaft of the wind turbine can be calculated as

$$P_{mech} = C_p(\beta, \lambda)\frac{1}{2}\rho\pi r^2 v_w{}^3 \tag{2.2}$$

where $C_p(\beta, \lambda)$ is the power coefficient decided by the wind turbine's characteristics. It is a function of the turbine's tip speed ratio λ and the pitch angle β. The tip speed ratio λ can be written as

$$\lambda = \frac{2\pi r n_r}{60 v_w} \tag{2.3}$$

where n_r is the rotational speed of the turbine rotor. The relationship among the power coefficient C_p, tip speed ratio λ and the pitch angle β in a 2 MW wind turbine

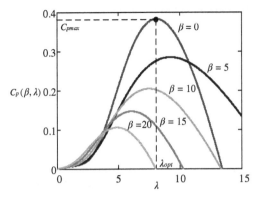

Figure 2.8   Relationship between the power coefficient C_p, tip speed ratio λ, and pitch angle β.

is shown in Figure 2.8. It can be found that the power coefficient C_p is influenced by both the tip speed ratio λ and the pitch angle β. With (2.3), it can be found that at a certain wind speed v_w, the power coefficient C_p can be controlled to be its maximum value by adjusting both the pitch angle β and the rotation speed n_r of the wind turbine. As shown in Figure 2.8, when $\beta = 0$ and $\lambda = \lambda_{opt}$, the power coefficient C_p will reach its maximum value C_{pmax}. The maximum theoretically achievable power coefficient is $C_{pmax} = 59\%$.

The pitch angle β can be adjusted by the pitch control in order to control the harvested power and will be discussed in the next section. The rotational speed n_r of the wind turbine is decided by the mechanical layout and model of the drive train. A simplified two-mass model is usually used for representing the basic relationship between the rotational speed and the torque in the system, as shown in Figure 2.9 [16].

In Figure 2.9, T_{T_ar} and T_{em} are the mechanical torque of the wind turbine in the fast speed shaft (after the gearbox) and the electromagnetic torque of the generator.

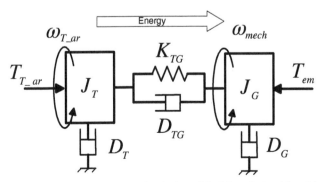

Figure 2.9   A two-mass mechanical model of the wind turbine drive train.

ω_{T_ar} and ω_{mech} are the angular speeds of the wind turbine in the fast speed shaft and the mechanical rotating speed of the generator. If the turbine torque and angular speed are expressed as T_T and ω_T, respectively, their relationship with T_{T_ar} and ω_{T_ar} can be represented as

$$T_{T\_ar} \approx \frac{T_T}{N} \tag{2.4}$$

$$\omega_{T\_ar} \approx N\omega_T \tag{2.5}$$

where N is the gear speed ratio. J_T and J_G are the inertia of the wind turbine and generator. K_{TG} and D_{TG} are the stiffness and damping coefficient, respectively, between the turbine and the generator, D_T and D_G are the friction coefficients of the turbine and the generator, respectively. This model can also be expressed as

$$J_T \frac{d\omega_{T\_ar}}{dt} \approx T_{T\_ar} - D_T\omega_{T\_ar} - T_{em} \tag{2.6}$$

$$J_G \frac{d\omega_{mech}}{dt} \approx T_{em} - D_G\omega_{mech} + T_{T\_ar} \tag{2.7}$$

$$\frac{dT_{em}}{dt} = K_{TG}(\omega_{T\_ar} - \omega_{mech}) + D_{TG}\left(\frac{d\omega_{T\_ar}}{dt} - \frac{d\omega_{mech}}{dt}\right) \tag{2.8}$$

The electromagnetic torque T_{em} can be controlled by the power converter and the mechanical torque T_{T_ar} is influenced by the wind speed and can also be controlled by the wind turbine, so that the ω_{T_ar} and ω_{mech} can be controlled.

2.3.2 Pitch Control

The pitch angle β is the angle between the chord line of the blade and the plane of rotation. The pitch control is able to rotate the turbine blade and change the pitch angle. The pitch angle of each blade can be controlled together or independently. When studying the dynamic control system, a simplified first-order model as shown in Figure 2.10 can be used, where $\tau_{d\beta}$ is the time constant of the pitch control system. Figure 2.10 shows also the pitch control loop. A PI controller is used to generate a reference rate of change of pitch. This changing rate is normally limited to about 10°/s during normal operation and 20°/s for emergencies [16].

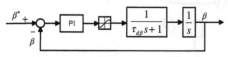

Figure 2.10   Simplified pitch control loop.

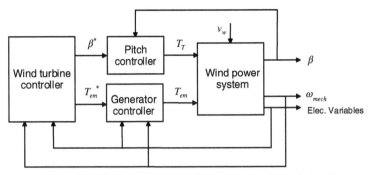

Figure 2.11   Overall scheme of a variable-speed control for wind turbine system with pitch control.

2.3.3   Overall Control Scheme

The overall scheme of a variable-speed wind turbine system with pitch control is shown in Figure 2.11. There are two degrees of control freedom in this system, the mechanical torque T_T, which can be controlled by the pitch controller, and the electromagnetic torque T_{em}, which can be adjusted by the generator controller (inverter control). The overall wind turbine controller makes these two controllers coordinate, so the rotor speed and output power of the wind power system can be controlled at different wind speeds [13]. The wind turbine control strategy will be introduced in the following section, while the generator and inverter control in different variable-speed wind turbine systems will be discussed in a later section.

2.3.4   Operational Range of Wind Turbine Systems

Considering the rotor speed and the output power limitations of a wind power system, the wind turbine controller normally controls the wind turbine to operate like the curves shown in Figure 2.12. The parameters of a commercialized 2 MW wind turbine are shown in Table 2.1.

The relationship of mechanical power P_{mech}, rotor rotational speed n_r, and wind speed v_w are expressed as

$$P_{mech}(v_w) = \begin{cases} 0 & \text{if } v_w < 4 \\[2mm] \frac{1}{2}\rho\pi r^2 C_p(\beta_{opt}, \lambda)v_w^3 & \text{if } 4 \leq v_w \leq v_B \\[2mm] \frac{1}{2}\rho\pi r^2 C_{pmax}v_w^3 & \text{if } v_B \leq v_w \leq v_C \\[2mm] \frac{1}{2}\rho\pi r^2 C_p(\beta_{opt}, \lambda)v_w^3 & \text{if } v_C \leq v_w \leq 12 \\[2mm] 2 \times 10^6 & \text{if } 12 \leq v_w \leq 20 \end{cases} \qquad (2.9)$$

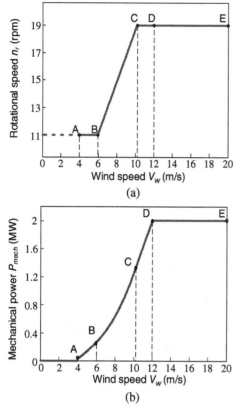

(a)

(b)

Figure 2.12   Operational range of a 2 MW DFIG wind turbine.

$$
n_r(v_w) = \begin{cases}
1050 & \text{if } v_w < v_B \\
\dfrac{60 N \lambda_{opt} v_w}{2\pi r} & \text{if } v_B \le v_w \le v_C \\
1800 & \text{if } v_C \le v_w \le 20
\end{cases} \tag{2.10}
$$

As shown in Figure 2.12, the operation range of a wind turbine can be divided into five stages which will be introduced in detail in the following subsections.

TABLE 2.1   Parameters of a 2 MW wind turbine

Rated power	2 MW	Cut-in wind speed	4 m/s
Wind rotor diameter	82.6 m	Cut-out wind speed	20 m/s
Rated rotational speed	19 rpm	Rated wind speed	12 m/s
Optimal tip speed ratio λ	8.1	Maximum power coefficient C_{pmax}	0.387
Optimal pitch angle	0°		

2.3.5 Wind Turbine Operation Around Cut-In Speed

Stage 1 (Before A): When the wind speed is smaller than the cut-in speed, which is 4 m/s in Figure 2.12, and illustrated as point A, the wind power system is not connected to the power grid, but keeps at minimum rotor rotational speed by adjusting pitch angle before the cut-in wind speed (point A is reached). In this stage, the WPS is working in stand-by mode, waiting for an increase in the wind speed.

Stage 2 (A–B): When the wind speed is larger than the cut-in speed, the wind power system is connected to the power grid, and the mechanical power increases with the wind speed. However, the rotational speed still keeps at minimum rotor rotational speed, as the wind speed is still low. So the power coefficient is not at the maximum value. In other words, the wind turbine cannot capture the wind power as much as possible. The pitch angle in this stage is typically controlled at zero.

2.3.6 MPPT Operation of Wind Turbine

Stage 3 (B–C): Maximum power point tracking (MPPT) is achieved at this stage, by adjusting the rotor rotational speed with the change in wind speed, in order to keep the optimal tip speed ratio λ_{opt}, and thus get the maximum power coefficient. The rotor rotational speed varies proportionally with the wind speed.

The wind turbine is controlled at the optimum turbine rotational speed (ω_{mech} in Figure 2.11) generated by the controller according to the present wind speed. Also, there are some methods using the relation between the electromagnetic torque and the maximum power to indirectly adjust the rotor speed to the optimal value under different wind speeds.

2.3.7 Wind Turbine Operation Around Cut-off Speed

Stage 4 (C–D): At point C, the rotor rotational speed reaches the rated value. Then, the rotational speed is controlled at the rated speed and kept constant to avoid large mechanical stress as well as noise in the wind turbine. So the tip speed ratio is not at the optimal value and the power coefficient is lower than the MPPT stage 3 (B–C). The wind turbine keeps at the optimal pitch angle until the mechanical power reaches the rated value.

Stage 5 (D–E): In this stage, the wind speed is above 12 m/s, wind turbine output mechanical power is controlled at rated value in order to avoid overcurrent in the power converters as well as overloading the whole drive train by adjusting the pitch angle.

2.4 CONTROL OF POWER CONVERTER

The generator control in Figure 2.11 is normally achieved by the power converter in the wind power system and the electromagnetic torque can be controlled by adjusting the rotor speed of the wind turbine at different wind speeds. Also, as an interface between the wind power system and the power grid, the power converter needs to support the power grid as well. For example, the reactive power support is required by the power grid to enhance the voltage stability and control, as well as to operate properly under grid fault [14, 17, 18].

2.4.1 Control of DFIG Power Converter

The DFIG power converter controls the generator by adjusting its rotor current and voltage, and a conventional control scheme of DFIG power converters is shown in Figure 2.13. The control scheme is applied in a *dq* reference frame. The RSC controls the torque (or the active power) generated by the generator and the reactive power generated by the stator side of the generator. The RSC control consists of the power/torque outer loop and the rotor current inner loop. The power/torque outer loop generates the rotor current reference $\vec{i}_{rdq}^{\,ref}$, by open-loop calculations or by closed-loop control, and the rotor current inner loop controls the rotor current $\vec{i}_{rdq}^{\,ref}$ to track its reference.

The GSC provides the DC-bus voltage for the RSC, and the GSC can also provide some reactive power support to the grid. The GSC control consists of the DC voltage and reactive power outer loop and the grid current inner loop. The grid current reference $\vec{i}_{gdq}^{\,ref}$ is generated by the outer loop, and the rotor current inner loop controls the rotor current $\vec{i}_{gdq}^{\,ref}$ to track its reference.

A phase lock loop (PLL) is used to keep the *dq* reference frame synchronized with the stator/grid voltage.

2.4.2 Control of PMSG Power Converter

In the PMSG wind power system using a full-scale power converter, the power converter decouples the generator from the power grid as shown Figure 2.14. The RSC

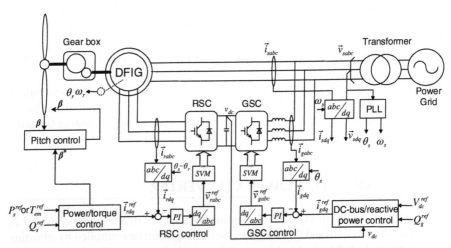

Figure 2.13 Conventional control scheme of power converters for DFIG wind turbine system. SVM, space vector modulation; PLL, phase lock loop [19].

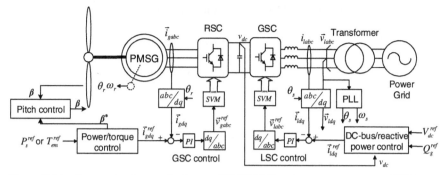

Figure 2.14   Conventional control scheme of PMSG full-scale power converters. SVM, space vector modulation; PLL, phase lock loop [19].

controls the stator current and voltage of the PMSG, so the electromagnetic torque and rotor speed of the PMSG can be adjusted. The outer loop of the RSC control generates the stator current reference $\vec{i}_{gdq}^{ref}$ from the electromagnetic torque or active power reference from wind turbine controller. The inner loop controls the stator current $\vec{i}_{gdq}$ to track its reference. As for PMSG, the magnetizing is provided by the permanent magnets in the rotor, and the RSC normally does not need to provide reactive current to the generator.

The GSC converts the power generated by the PMSG to the grid. The DC voltage of the back-to-back converter and the reactive power are controlled by the outer loop, the line current reference $\vec{i}_{ldq}^{ref}$ is generated, and the inner loop controls the line current $\vec{i}_{ldq}$ to track the reference.

2.4.3   Control of SCIG Power Converter

As PMSGs are relatively expensive, the concept of using an SCIG connected with a full-scale power converter has become popular in recent years. The control scheme of an SCIG full-scale power converter is shown in Figure 2.15. It is similar to the

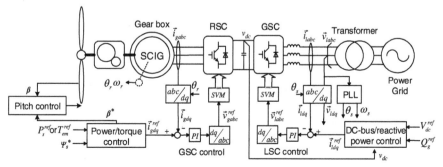

Figure 2.15   Conventional control scheme of SCIG-based full-scale power converters. SVM, space vector modulation; PLL, phase lock loop.

control scheme used for the PMSG power converter, which is shown in Figure 2.14. The GSC controls the output power and torque of the SCIG. It should be noticed that in the SCIG wind power system, the magnetizing current should be provided by the GSC as well, while in the PMSG wind power system, the magnetizing is provided by the permanent magnet in the rotor. So the outer control loop of the SCIG RSC should include the reactive power or magnetizing current loop. The GSC control scheme is basically the same as that in the PMSG wind power converter. The DC voltage of the back-to-back converter and the reactive power are controlled by the outer loop, the line current reference $\vec{i}_{ldq}^{ref}$ is generated, and the inner loop controls the line current $\vec{i}_{ldq}$ to track the reference.

2.5 WIND POWER TRANSMISSION

The electrical power generated by each wind turbine is normally centralized into wind farms, and transmitted to the large scale power grid. The integration of wind power will influence the power system as well as the disturbances of the power system, such as grid faults, voltage unbalance, and distortions will have an impact on the wind power system, especially for the DFIG wind power system [20, 21].

2.5.1 Wind Farm

In the early years, the wind power systems were directly connected to the low voltage distribution grid. As the rated power of the wind power system becomes larger and larger, the wind power systems need to be connected to the medium voltage distribution grid, normally 10–33 kV and a transformer was necessary. In the early stage wind turbines, the transformer, breaker, and protection system were located in shelters near the wind turbine and were shared by several wind turbines. Once the size of the towers became adequate, the transformers were located in the tower. Nowadays, most of the wind turbines are installed in groups called wind farms. They share a common medium or high voltage transformer and a PCC to the grid. Normally, they are directly connected to the transmission or subtransmission grids by means of an electrical substation specially designed for wind farms. Each of the wind turbines in the wind farm can be controlled by a centralized controller. The capacity of modern wind farms are sometimes larger than 400 MW and contains many dozens of MW-rated wind turbines [21]. The basic configuration of a typical wind farm is shown in Figure 2.16.

The substation is an essential component in wind farms. The coupling transformer to the transmission grid is installed in the substation. The primary voltage is normally from 66 to 220 kV which is different from country to country, while the secondary voltage will vary from 10 to 66 kV. The input and output breakers can also be found in the substation. A static VAR compensation such as capacitor banks and inductors, and a dynamic VAR compensation based on SVC or STATCOM, as well as a current limiter, are normally installed in such a substation.

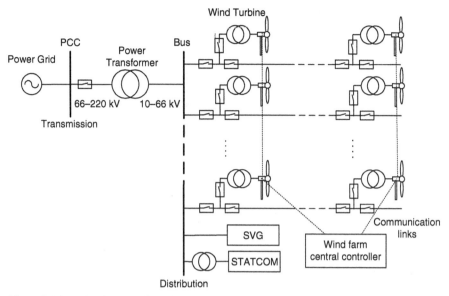

Figure 2.16   Wind farm configuration using a STACOM and an SVG.

The wind turbines are connected to the substation by means of transformers and breakers. The electrical layout of the wind farm is important because it affects the terminal voltages of the wind turbines and has an influence on the turbine behavior.

A centralized wind farm controller controls the active power and reactive power injected into the grid (at the PCC), according to the grid code of the power system, and they will be introduced in the Chapter 3. In order to implement a centralized control, it is necessary to have an effective communication between the wind farm centralized controller and each of the wind turbines. Thus, while each of the wind turbines report to the wind farm centralized controller the active power and reactive power that they can deliver at any moment, the wind farm centralized controller should provide each of the wind turbines with references of active and reactive powers. Also, the VAR compensation in the substation can regulate the reactive power together with the wind turbines [21].

2.5.2   Power System

The power system is the set of infrastructures responsible for the generation, transmission, and distribution of the electrical energy. The power system may be different in different countries, but generally, they share the following characteristics:

- They consist of a three-phase system with constant AC voltage. The power generation and transmission systems are three-phase systems as well as industrial loads. The civilian and commercial loads are configured to achieve equal distribution between each phase so that the three-phase system can be balanced.

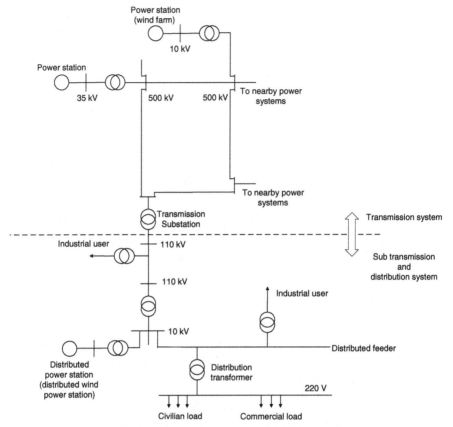

Figure 2.17   An example of a basic configuration of a power system from the power station to load [8, 22].

- Most of the main power sources in the power system are synchronous generators, which are driven by the heat engine, hydro turbine, etc., and convert the mechanical energy from these sources into electrical energy.
- The power system transmits the electrical energy to the loads in different areas, through the power transmission and distribution lines at different voltage levels.

The basic configuration of a power system is shown in Figure 2.17. It is usually divided into the following subsystems [22]:

(a) **Transmission System:** The transmission system connects the major power stations and load centers. It is the backbone of the power system and operates on the highest voltage level of the whole system (usually larger than 230 kV). The large wind farms are normally connected to the transmission system.

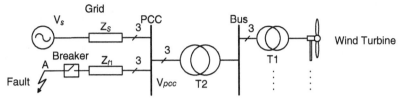

Figure 2.18   Simplified model of a grid fault with wind turbines connected to the grid.

(b) **Subtransmission System:** The subtransmission system transfers the electrical power from the transmission substation to the distribution substation. Large industrial loads are usually directly supplied by the subtransmission system.

(c) **Distribution System:** The distribution system is the last stage to the user of electricity. The primary distribution voltages are normally between 400 V and 34.5 kV and the industrial load is normally supplied with this voltage level. The secondary distribution voltage is typically 110 or 220 V and provides electrical power to the civilian and commercial loads.

2.5.3   Grid Faults

The grid faults are usually caused by short circuits in the power system. They can be introduced by lightning, aging of insulation materials, accidents, etc. The grid faults will normally introduce a complicated dynamic process in the power system. A simplified model of a grid fault in the power system is shown in Figure 2.18, which can be used to derive the terminal voltage of the wind farm and wind power system under grid faults.

In Figure 2.18, the power grid is regarded as an ideal voltage source V_s in series connection with grid impendence Z_s. A grid fault happens at point A, where Z_{f1} is the line impendence between the wind farm terminal and the fault location.

In this case, the grid voltage at PCC ($\vec{v}_{pcc}$) can be regarded as

$$\vec{v}_{pcc} = \frac{Z_{f1}}{Z_{f1} + Z_s} \vec{V}_s \tag{2.11}$$

If the fault is asymmetrical, the voltage on the faulted phase can also be evaluated this way. However, the transformer inside the wind farm (T1, T2 in Figure 2.18) may change the fault type. The grid voltages at the PCC under different fault types are shown in Table 2.2. Here V_s is the normal grid voltage amplitude, p is the voltage dip level, which is mainly determined by the relationship between the impedance Z_s and Z_{f1}.

In less than 1 s after the grid fault, the circuit breaker is normally triggered by the large fault current, and the fault is isolated from the grid, so that the grid voltage can recover. As most breakers in the power system are AC-breakers, which can only be opened when the fault current crosses zero. So, the voltage recovery time in each phase will also be determined by the time when the fault current crosses zero, which

TABLE 2.2 Wind farm terminal grid voltage under different fault types

Fault type	Three-phase short-circuit /to ground fault	Single-phase to ground fault	Two-phase to ground fault	Two-phase short-circuit fault
Voltage vector	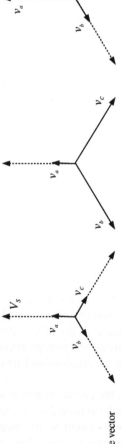		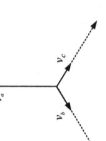	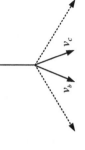
Three-phase voltage	$v_a = (1-p)V_s \cos \omega t$ $v_b = (1-p)V_s \cos\left(\omega t - \frac{2}{3}\pi\right)$ $v_c = (1-p)V_s \cos\left(\omega t - \frac{4}{3}\pi\right)$	$v_a = (1-p)V_s \cos \omega t$ $v_b = V_s \cos\left(\omega t - \frac{2}{3}\pi\right)$ $v_c = V_s \cos\left(\omega t - \frac{4}{3}\pi\right)$	$v_a = V_s \cos \omega t$ $v_b = (1-p)V_s \cos\left(\omega t - \frac{2}{3}\pi\right)$ $v_c = (1-p)V_s \cos\left(\omega t - \frac{4}{3}\pi\right)$	$v_a = V_s \cos \omega t$ $v_b = p'V_s \cos\left(\omega t - \frac{2}{3}\pi - \phi\right)$* $v_c = p'V_s \cos\left(\omega t - \frac{4}{3}\pi + \phi\right)$
Symmetrical component	Positive sequence $|\vec{v}_p| = (1-p)V_s$ Negative sequence $|\vec{v}_{ne}| = 0$	Positive sequence $|\vec{v}_p| = \left(1 - \frac{p}{3}\right)V_s$ Negative sequence $|\vec{v}_{ne}| = \frac{p}{3}V_s$	Positive sequence $|\vec{v}_p| = \left(1 - \frac{2p}{3}\right)V_s$ Negative sequence $|\vec{v}_{ne}| = \frac{2p}{3}V_s$	Positive sequence $|\vec{v}_p| = \left(1 - \frac{p}{2}\right)V_s$ Negative sequence $|\vec{v}_{ne}| = \frac{p}{2}V_s$

*$p' = \sqrt{1 + \frac{3}{4}p^2 - \frac{3}{2}p}$ $\phi = \arcsin\left(\frac{\sqrt{3}}{2}p\right)$

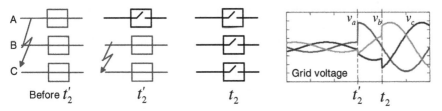

Figure 2.19   Voltage recovery after a three-phase short-circuit fault. t_2', breaker of phase A opens; t_2, breakers of phase B and C open and grid voltage recovers [23].

can be described with the angle between the fault current and voltage, also named as the grid fault angle θ in [7]. When the grid fault happens in the transmission system, the grid fault angle θ is normally between 75° and 85°, while for the grid fault that happens in the distribution system, θ is normally between 45° and 60°. Taking the three-phase short-circuit fault as an example; the voltage recovery after the grid fault is shown in Figure 2.19. The breaker in phase A is open first at t_2', the grid voltage in phase A recovers first. After t_2', the three-phase fault becomes a two-phase fault. In about 5 ms, the fault current between phases B and C crosses zero at t_2 and the breakers in phase B or C are opened, and the three-phase grid voltage recovers to normal.

The voltage recovery after different types of grid faults may come through a few steps like the ones listed in Table 2.3.

TABLE 2.3   Voltage recovery after grid faults with different fault types [23]

Fault type	Stage 1	Stage 2	Stage 3
Three-phase short-circuit fault	Two-phase short-circuit fault	Normal grid	
Three-phase to ground fault	Two-phase to ground fault	Single-phase to ground fault	Normal grid
Two-phase to ground fault	Single-phase to ground fault	Normal grid	
Two-phase short-circuit fault	Normal grid		
Single-phase to ground fault	Normal grid		

It should be noticed that the grid voltages after grid faults or recoveries are influenced by other dynamic effects in the power system, such as the saturation of transformers, the induction machines connected to the fault line, etc.

As most grid faults are temporary, after the faults are isolated, the circuit breaker in the transmission system may be reclosed automatically to improve the service continuity. However, if the fault is permanent, the reclosing of the circuit breaker may introduce a second fault right after the first one. This is one of the main reasons of recurring grid faults, which a wind turbine should also be able to handle.

2.5.4 Unbalanced Grid Voltage

The unbalanced grid voltage in a power system can be generally classified under two conditions. One is the short-time unbalance introduced by asymmetrical grid faults, as introduced in Section 2.5.3. The other is the long-time unbalanced grid voltage introduced by asymmetrical line impendence, or the asymmetrical load, for example, the electrified railway supplied by single-phase power line.

The degree of asymmetry τ_u in a grid is usually used to describe the unbalance of the grid. It is defined as the ratio of the negative-sequence to the positive-sequence grid voltage amplitudes, as shown in (2.12).

$$\tau_u = \frac{\left|\vec{v}_{ne}\right|}{\left|\vec{v}_p\right|} \times 100\% \tag{2.12}$$

where $\left|\vec{v}_{ne}\right|$ and $\left|\vec{v}_p\right|$ are the amplitudes of the negative- and positive-sequence grid voltages, respectively.

For the unbalanced grid voltage introduced by the asymmetrical grid faults, the degree of asymmetry τ_u is relatively large but it will only last for no more than a second. For the unbalanced grid voltage introduced by the asymmetrical line impendence or asymmetrical load, although the degree of asymmetry τ_u may only be 2–4%, it is a steady-state unbalance and it usually lasts longer. As the wind farms are usually located in remote areas, the weak grid may also increase the degree of asymmetry.

2.5.5 Grid Harmonic Voltage

The harmonic voltages in power systems are generated by the voltage source or the loads. For example, the magnetic saturation in generators and transformers will introduce 3rd-order harmonic voltage, and the nonlinear loads will bring 5th, 7th, 11th, and 13th order harmonic voltage into the grid, and the interactions between the converters can also bring harmonics into the system. The harmonic voltage will reduce the power quality of the power system as well as the efficiency of the devices connected to the grid, causing overheating of the transformer or power lines, and sometimes even resonances.

The total harmonic distortion (THD) is used to characterize the power quality of the power systems, and it is defined as the ratio of the RMS voltage of all harmonic components to the RMS voltage of the fundamental component, as shown in (2.13).

$$\mathrm{THD}_n = \frac{\sqrt{V_{2rms}^2 + V_{3rms}^2 + \cdots + V_{nrms}^2}}{V_{1rms}} \times 100\% \qquad (2.13)$$

where V_{1rms} is the RMS grid voltage of the fundamental component, and V_{2rms} to V_{nrms} are the second to nth harmonic voltage components.

As the wind farms are normally located in remote areas, the grid harmonic voltage may also be enlarged in the weak grid, which brings more challenge to the wind power system.

2.6 SUMMARY

In this chapter, the basic knowledge related to modern WPS is introduced, especially for the variable-speed WPS. Three basic wind power concepts are introduced. The modeling of the wind turbine is discussed, and the control of the variable-speed wind turbine is explained. Then, the operation range of a wind turbine is analyzed in detail. The static operating conditions of a variable-speed wind turbine are also investigated.

The basic control of generators and power converters in different concepts of the variable-speed WPS are also introduced. At last, the wind power transmission, the grid faults and distortions in power systems are discussed.

From this chapter, a brief picture of how a wind power system is working, and how the wind energy is transferred to the power grid is established. Since the grid faults and distortions are inevitable in today's power system, grid codes are made to regulate the operation of grid-connected wind power systems in order to improve their grid connection. Different grid codes for the wind power systems in the world will be introduced briefly in Chapter 3.

REFERENCES

[1] F. Blaabjerg, M. Liserre, and K. Ma, "Power electronics converters for wind turbine systems," *IEEE Trans. Ind. Appl.*, vol. 48, no. 2, pp. 708–719, 2012.
[2] Available: http://www.vestas.com/
[3] Available: https://www.gerenewableenergy.com/wind-energy/turbines.html
[4] Available: http://www.gamesacorp.com/recursos/doc/productos-servicios/aerogeneradores/catalogo-g9x-20-mw-eng.pdf
[5] Available: http://www.goldwindamericas.com/sites/default/files/Goldwind-Brochure-1.5-Web.pdf
[6] Available: http://www.enercon.de/en/products/ep-2/e-82/
[7] Z. Chen and F. Blaabjerg, "Wind energy—The world's fastest growing energy source," *IEEE Power Electron. Soc. Newsl.*, vol. 18, no. 3, pp. 15–19, 2006.
[8] Z. Chen, J. M. Guerrero, and F. Blaabjerg, "A review of the state of the art of power electronics for wind turbines," *IEEE Trans. Power Electron.*, vol. 24, no. 8, pp. 1859–1875, 2009.
[9] F. Blaabjerg, F. Iov, T. Kerekes, and R. Teodorescu, "Trends in power electronics and control of renewable energy systems," in *Proc. 14th Int. Power Electron. Motion Control Conf. EPE-PEMC*, Ohrid, 2010, pp. K-1–K-19.

[10] L. H. Hansen, P. H. Madsen, F. Blaabjerg, H. C. Christensen, U. Lindhard, and K. Eskildsen, "Generators and power electronics technology for wind turbines," in *Proc. 27th Annu. Conf. IEEE Ind. Electron. Soc.*, vol. 3, 2001, pp. 2000–2005.

[11] H. Li and Z. Chen, "Overview of different wind generator systems and their comparisons," *IET Renew. Power Gen.*, vol. 2, no. 2, pp. 123–138, 2008.

[12] J. M. Carrasco, L. G. Franquelo, J. T. Bialasiewicz, E. Galvan, R. C. PortilloGuisado, M. A. M. Prats, J. I. Leon, and N. Moreno-Alfonso, "Power-electronic systems for the grid integration of renewable energy sources: A survey," *IEEE Trans. Ind. Electron.*, vol. 53, no. 4, pp. 1002–1016, Jun. 2006.

[13] B. Wu, N. Zargari, and Y. Lang, *Power Conversion and Control of Wind Energy Systems*. Wiley-IEEE Press, 2011.

[14] G. Abad, J. Lopez, M. Rodriguez, L. Marroyo, and G. Iwanski, *Doubly Fed Induction Machine: Modeling and Control for Wind Energy Generation*. John Wiley & Sons, Sep. 2011.

[15] H. Polinder, F. F. A. van der Pijl, G. J. de Vilder, and P. J. Tavner, "Comparison of direct-drive and geared generator concepts for wind turbines," *IEEE Trans. Energy Convers.*, vol. 21, no. 3, pp. 725–733, 2006.

[16] E. Muljadi and C. P. Butterfield, "Pitch-controlled variable-speed wind turbine generation," *IEEE Trans. Ind. Appl.*, vol. 37, no. 1, pp. 240–246, Jan./Feb. 2001.

[17] R. Teodorescu, M. Liserre, and P. Rodriguez, *Grid Converters for Photovoltaic and Wind Power Systems*. John Wiley & Sons, 2011.

[18] G. Abad and G. Iwanski, "Properties and control of a doubly fed induction machine," in *Power Electronics for Renewable Energy Systems, Transportation and Industrial Application*. John Wiley & Sons, Sep. 2014.

[19] F. Blaabjerg and K. Ma, "Future on power electronics for wind turbine systems," *IEEE J. Emerg. Sel. Top. Power Electron.*, vol. 1, no. 3, pp. 139–152, Sep. 2013.

[20] M. Liserre, R. Cardenas, M. Molinas, and J. Rodriguez, "Overview of multi-MW wind turbines and wind parks," *IEEE Trans. Ind. Electron.*, vol. 58, no. 4, pp. 1081–1095, 2011.

[21] A. D. Hansen, P. Sørensen, F. Iov, and F. Blaabjerg, "Centralised power control of wind farm with doubly fed induction generators," *Renew. Energy*, vol. 31, no. 7, pp. 935–951, 2006.

[22] Z. Chen and F. Blaabjerg, "Wind farm—A power source in future power systems," *Renew. Sustainable Energy Rev.*, vol. 13, nos. 6–7, pp. 1288–1300, Aug./Sept. 2009.

[23] M. H. J. Bollen, "Voltage recovery after unbalanced and balanced voltage dips in three-phase systems," *IEEE Trans. Power Deliv.*, vol. 18, no. 4, pp. 1376–1381, Oct. 2003.

CHAPTER 3

GRID CODES FOR WIND POWER GENERATION SYSTEMS

The grid codes for the wind power generation system are the "laws" for wind turbines to be connected to the power grid. They are developed by the power system operators in order to smoothen the effects of high wind power penetration on the power system stability and power quality. One of the main targets of the wind turbine manufacturers is to develop the corresponding control strategies and protection schemes to satisfy the related grid codes and be able to sell the product to the market. In this chapter, the grid codes for wind power generation system connection in several countries are introduced, including the grid codes for steady-state operation, as well as the grid codes under abnormal operations, such as grid faults, unbalanced grid voltage, and harmonic distortions.

3.1 INTRODUCTION

The grid codes (GC) for the wind power generation system are the "laws" for wind power systems (WPS) to be connected to the grid. The GCs are normally developed by the transmission system operators (TSO) or distribution system operators (DSO) according to their experience acquired through the operation of power systems, in order to smoothen and adapt the effects of wind power penetration on the power system stability and power quality. Some of the regulations have also been incorporated into technical standards, for example, IEEE standard 1547 [1].

Before a large amount of the renewable energy was connected into the grid, the power sources in the power system were mainly large synchronous generators in thermal power stations or hydro stations. The technology of large synchronous generators has been well established over many decades. The synchronous generators are able to support the grid by offering inertia, oscillation damping, reactive power generation, short-circuit capability, and fault ride through (FRT). The stability theory of modern power system is also established based on the characteristic of the synchronous generators.

With respect to the wind power integration, unfortunately, as the wind power is not constant and less predictable, the wind generator does not act as synchronous

Advanced Control of Doubly Fed Induction Generator for Wind Power Systems, First Edition.
Dehong Xu, Frede Blaabjerg, Wenjie Chen, and Nan Zhu.

generators in many ways. The general demand to the grid codes for the WPS is that the WPS should behave as synchronous generator-based conventional power plants as much as possible so that the stability of the power system with wind power integration can be ensured. The grid codes are also developed continuously with the development of new wind generation technologies. The first generation of WPS employed was a fixed-speed wind turbine with a caged induction generator directly connected to the grid. The output active and reactive powers cannot be regulated and the FRT was difficult to achieve. Most of the earlier GCs usually did not fulfill the demands. With the development of variable-speed WPS such as doubly fed induction generator (DFIG) and permanent magnet synchronous generator WPS, more degrees of freedom have been achieved with the help of variable-speed wind turbine, especially the power electronics converters. For example, the output active and reactive powers can be regulated in a wide range according to the demands of the TSO, to provide support to the power system, and the FRT can also be achieved. As a result, nowadays, the GCs are specified with more demands for the WPS.

Another important fact is that the wind power penetration has increased rapidly during the last two decades. Now in some countries, for example, in Denmark, it is larger than 40%. Also, the capacity of the wind farms continues to increase. Today, the large wind farms usually reach capacities of hundreds of MW and even GW. So the influence of the WPS on the power system's stability becomes more and more significant. The GC requires the WPS to take the responsibility to ensure the stability of the power system.

Although the GCs in different countries are developed by different TSOs or DSOs, they usually include the following demands for the WPS.

During normal operation, the WPS is required to generate active and reactive powers within a range around the rated voltage and frequency. The active power response speed and the reactive power capacity are restricted as well. In some countries, the WPS should also provide reactive power support according to the variation of the grid voltage and take part in the voltage control.

Under grid faults, the WTS should remain connected to the grid and provide reactive power to support the grid, and the active power generation should be restored in a required time which is also referred to as FRT requirements. In some countries, the behavior of the WPS under voltage swells or recurring grid faults are also restricted by the GCs. Furthermore, the power quality must be ensured within a certain range of grid voltage unbalance and distortions.

This chapter gives an overview of the grid codes for normal and abnormal operations of WPS.

3.2 GRID CODE REQUIREMENTS UNDER NORMAL OPERATION

3.2.1 Frequency and Voltage Deviation

The wind turbine should be a part of the static operation and the requirements demand so that the WTS will operate within a range around the rated voltage and frequency.

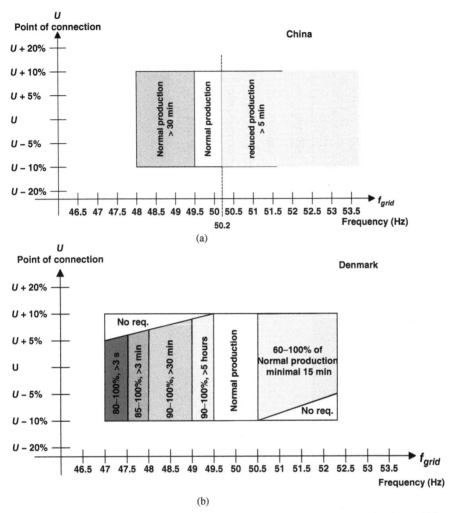

Figure 3.1   Frequency and voltage deviation requirements for wind power plant in (a) China [2]; (b) Denmark [3]; (c) Germany [4]; and (d) Spain [5].

Usually, this requirement can be described using three zones: (a) continuous operation zones, which means the WPS needs to operate normally in these zones, (b) time-limited operation zones, within which the WPS needs to operate for a certain period, and (c) immediate disconnection zones, which means that the WPS should be disconnected immediately in order to protect the turbine and the power system. The frequency and voltage deviation requirements in China, Denmark, Germany, and Spain are shown in Figures 3.1a–3.1d, respectively.

The Chinese GC [2] demands that the WPP should keep normal production within a voltage range of 90–110% of the normal voltage, and within a frequency range of 49.5–50.2 Hz (at the point of common coupling (PCC) for wind farm). When the frequency is between 48 and 49.5 Hz, the WPP should keep working for at least

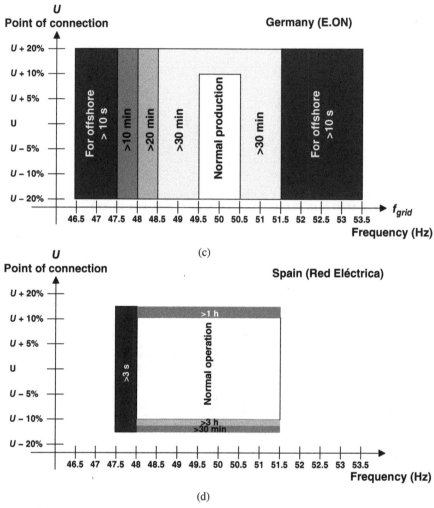

Figure 3.1 (*Cont*).

30 min. When the frequency is higher than 50.2 Hz, the WPP should keep working for at least 5 min, and reduce the active power generation or disconnect according to TSO's grid demands, which is shown in Figure 3.1a.

In Denmark [3], the WPP requires to keep normal production within a voltage range of 90–110% of the normal voltage, and within a frequency range of 49.5–50.5 Hz. When the frequency is between 49.5 and 49 Hz, 48 and 49 Hz, the WPP should keep 90–100% of the normal production for at least 5 hours and 30 min, respectively. When the frequency is between 47.5 and 48 Hz, 47 and 47.5 Hz, the WPP should keep 85–100% normal production for at least 3 min and keep 85–100% normal production for at least 3 s, respectively. When the frequency is between 50.5 and 52 Hz, the WPP

TABLE 3.1   The active power changing rate limitation regulated by the Chinese GC [2]

Wind farm capacity (MW)	Maximum active power change rate in 10 min (MW)	Maximum active power change rate in 1 min (MW)
< 30	10	3
30–150	Capacity/3	Capacity/10
> 150	50	15

is required to keep 60–100% normal production for at least 15 min, which is shown in Figure 3.1b.

The German GC (E.ON) covers a larger frequency and voltage deviation area [4], as shown in Figure 3.1c. It requires the WPP to keep normal production within a voltage range of 80–110% of the normal voltage, and within a frequency range of 49.5–50.5 Hz. When the voltage is between 80% and 120% of the normal voltage, and when the frequency is between 48.5 and 51.5 Hz, the WPP should keep working for at least 30 min. The WPP is also required to keep working for at least 20 and 10 min, when the frequency is between 48 and 48.5 Hz, and between 47.5 and 48 Hz, respectively. For offshore wind farms, the WPP should keep working for at least 10 s when the frequency is between 46.5 and 47.5 Hz, as well as between 51.5 and 53.5 Hz.

In Spain [5], the GC demands the WPP to keep working within a voltage range of 90–110% of the normal voltage, and within a frequency range of 48–51.5 Hz. When the voltage ranges from 110% to 115% of the normal voltage, the WPP should keep working for at least 1 hour. When the voltage is between 85% and 90% of the normal voltage, the WPP should keep working for at least 3 hours to 3 min, and when the frequency is between 47.5 and 48 Hz, the WPP needs to work for at least 3 s.

3.2.2   Active Power Control

The GC requirements on the active power control normally include two aspects. One is that the active power from the wind farms should be limited according to the TSO's demands, and the change rate of the active power should also be regulated to ensure the stability of the power system. The other is that the active power output should be changed during the frequency variation, so the WPP can be involved in the primary control of the power system when the wind is blowing.

The limitation of the active power change rate in the Chinese GC [2] is shown in Table 3.1. It demands the wind farm to limit the change rate in 1 and 10 min, according to the wind farm capacity. In the German grid code, the active power must be reduced at least 10% per min after received the active power command. In the Danish grid code [3], it is required that the wind power plant must be equipped with active power constraint functions. Three constraint functions are demanded, including the absolute power constraint, the delta power constraint, and ramp rate constraint functions. The absolute power constraint is used to limit the active power from a wind

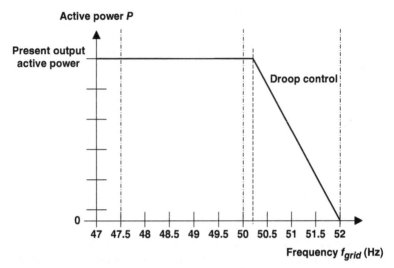

Figure 3.2 Frequency response demands in the Danish grid code [3].

power plant to a set point defining the maximum power limit at the point of connection. The delta power constraint is used to constrain the active power from a wind power plant to a required constant value in proportion to the possible active power. If the set point for the absolute power constraint or delta power constraint needs to be changed, it must be commenced within 2 s and completed no later than 10 s after the receipt of an order to change the set point. The ramp rate constraint is used to limit the maximum speed by which the active power can be changed in the event of changes in wind speed or the active power set points. A ramp rate constraint is typically used for the reasons of system operation in order to prevent the changes in the active power from adversely impacting the stability of the grid. Control using a new ramp rate for the ramp rate production constraint must be commenced within 2 s and completed no later than 10 s after the receipt of an order to change the ramp rate.

In some counties, the output active power of the WPS, or wind farms, should be regulated with a variation of the grid frequency. In Denmark, in the event of frequency deviations in the grid, the WPS must contribute to grid stability by automatically reducing the active power at grid frequencies above 50.20 Hz. The Control must start no later than 2 s after a frequency change is detected and must be completed within 15 s, which is shown in Figure 3.2.

The German grid code (E.ON) [4] demands that when the grid voltage frequency is higher than 50.2 Hz, the WPP should reduce the active power according to the grid frequency. The power reduction ΔP can be expressed as (3.1)

$$\Delta P = 20 P_M \frac{50.2 \text{ Hz} - f_{grid}}{50 \text{ Hz}}, \quad \text{when } 50.2 \text{ Hz} \leq f_{grid} \leq 51.5 \text{ Hz} \qquad (3.1)$$

where P_M is the available power and f_{grid} is the grid frequency. When the grid frequency is between 47.5 and 50.2 Hz, no active power limitation is carried out, and when the frequency is higher than 51.5 Hz or lower than 47.5 Hz, the WPP should be disconnected from the grid, as shown in Figure 3.1.

3.2.3   Reactive Power Control

The reactive power control requirements demand that the WPP should regulate the output reactive power Q independently, or in response to the grid voltage variation to provide the necessary reactive power support to the power system. In general, the reactive power requirement is usually given in three different ways:

- **Q Control:** The reactive power should be controlled independently of the active power at the point of connection, according to the demands from the TSOs.
- **Power Factor Control:** The reactive power is controlled proportionally to the active power at the point of connection.
- **Voltage Control:** It is a function which controls the voltage in the voltage reference point by changing the reactive power generation.

It should be noticed that the Q control and voltage control functions are mutually exclusive, which means that they normally cannot be activated at the same time.

In China, the grid code demands that the power factor of the wind farms should be regulated within a range of 0.95 (lead) to 0.95 (lag). The wind farms should also have the ability of Q control and voltage control. When the power system is under normal operation, the terminal voltage of the wind farms should be controlled within a range of 97–107% of the normal voltage.

In Spain, the WPS should provide the reactive power with a minimum range of 0.15 (lagging) to 0.15 (leading) under normal voltage and for all technical P ranges. Also, the WPS should provide the reactive power with a minimum range of 0.3 (lagging) to 0.3 (leading) as a function of the grid voltage [5], like shown in Figure 3.3a.

In the Danish grid code, the Q control, power factor control, and voltage control are all required for WPS for the installed capacity smaller than 25 MW. The detailed demands are different according to the capacity of the WPS. For the WPS in the capacity range of 1.5–25 MW, the WPS should provide a minimum reactive power capacity of 0.228 pu (leading or lagging), and should be able to control the power factor with a minimum range of 0.975 (leading) to 0.975 (lagging) under normal grid voltage with different active power outputs. The minimum range of the reactive power is also related to the grid voltage variation, as shown in Figure 3.3b. The WPS should further have the ability to stabilize the voltage at the voltage reference point given by the TSOs. Voltage control must have a setting range from minimum to maximum voltage with an accuracy of 0.5%. If the voltage set point needs to be changed, such a change must be commenced within 2 s and completed no later than 10 s after the receipt of an order to change the set point. The individual wind power plant must be capable of performing the control within its dynamic range and voltage limits with the droop characteristic configured as shown in Figure 3.4.

The voltage control reference point is the voltage reference point. When the voltage control has reached the wind power plant's dynamic design limits, the control function must wait for the possible overall control from the tap changer or

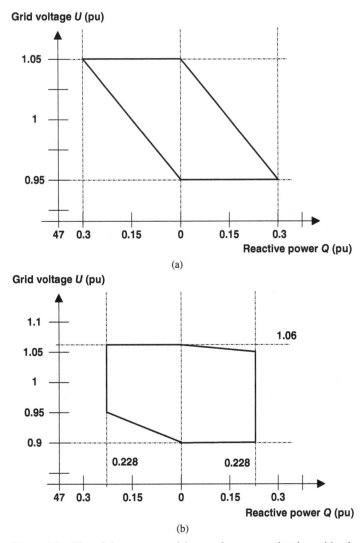

Figure 3.3 The minimum range of the reactive power related to grid voltage in (a) Spanish [5] and (b) Danish grid codes [3].

other voltage control functions. The overall voltage coordination is handled by the electricity supply undertaken in collaboration with the transmission system operator.

3.2.4 Inertial Control and Power System Stabilizer Function

Conventional generators in a power system are fixed-speed, synchronous generators rotating at synchronous speed, and providing a substantial synchronous inertia to keep the frequency during faults. This inertia provides important short-term energy storage so that small deviations in the system frequency result in the spinning inertias only accelerating or decelerating slightly. The large inertia absorbs excess energy from the

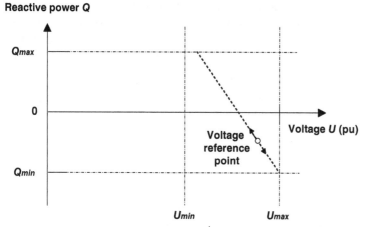

Figure 3.4   Voltage control for a wind power plant in the Danish grid [3].

system or provides additional energy as required. Also, the large inertia can provide emergency energy in the event of a sudden loss of generation.

Power system stabilizer (PSS) functions can be provided by conventional generators as well. The output power of the generator is controlled in response to the frequency deviations, in order to damp out resonances between generators. These resonances are most likely to occur between two groups of large generators separated by a relatively weak interconnection.

The modern WPS are mostly variable-speed generators controlled by power electronics converters. They have a faster dynamic response compared to the conventional generators. However, they have less synchronously connected inertia. As a result, with the increase of the wind power installation into the grid, the inertia of the power system will be smaller and therefore the power system will be harder to control under frequency disturbances and sudden loss of generation.

So, it has been discussed that in future GCs, the WPS would be required to provide a similar inertia with conventional synchronous generators, which can be achieved by the control of power electronics converters. Also, the PSS function can be provided by the variable-speed WPS and it might be required. However, as the variable-speed wind turbines have very little synchronously connected inertia, the risk of such resonances actually reduces as the wind penetration increases. Thus, there is an argument that PSS functions should be provided only by the conventional generation, but it is a topic for continuous discussion.

3.3   GRID CODE REQUIREMENTS UNDER NON-IDEAL GRID

3.3.1   Low Voltage Ride-Through Requirement

It has previously been introduced that the grid faults will introduce voltage dips in power systems. Besides, a sudden connection of large loads and the start of induction machines may also result in voltage dips in the power system.

The earlier GCs usually required that the WPS should be disconnected from the grid experience voltage dips, in order to protect the WPS from overvoltage or overcurrent introduced by the voltage dips. With the increase of the wind power installation into the grid, disconnection of the WPS under voltage dips may lead to the following problems in the power system:

- When the WPS is disconnected from the grid, the protections in the power system (mainly the breakers) were unable to detect faults in the lines near the wind farms, due to the loss of fault current from the wind farms.
- The loss of wind power generation may cause instabilities of the power system and may worsen the grid faults.

As a result, the recent GCs in many countries have required the WPS to stay connected to the grid and provide reactive power support to the grid, also mentioned as low voltage ride-through (LVRT) or FRT requirements. Three requirements are normally requested to the WPS to fulfill the following three aspects of demands [6, 7]:

- Keep connected to the grid under certain levels of voltage dips.
- Provide reactive power support within a required dynamic response dependent on voltage level.
- Restore the active power generation after grid voltage recovery.

3.3.1.1 Voltage Dip Tolerance

This requirement indicates a maximum voltage dip level and duration under which the WPS must stay connected to the grid, for both asymmetrical and symmetrical fault. They are normally described by the grid voltage versus duration curves, as shown in Figure 3.5. The WPS must stay connected to the grid in the grid voltage and the duration areas above these curves, as it is shown by the shadowed areas in Figure 3.5. For the areas below these curves, the WPS can be disconnected from the grid.

The GCs in China, Spain, and Denmark require the WPS to stay connected to the grid when the grid voltage falls to 20% of its normal value for 500–625 ms at most. In Germany, United States, and Australia, the GCs are stricter on the voltage dip tolerance. The WPS should stay connected to the grid even when the grid voltage is down to zero, which means the fault location is just at the terminals of the wind farm. The fault duration is 150 ms at most in these cases.

3.3.1.2 Reactive Current Support

The reactive current support requirements indicated the smallest reactive current which the WPS must provide under voltage dips with different voltage dips level, and in some countries, under different types of voltage dips. The requirements in different countries are shown in Figure 3.6. The German GC requires the WPS to provide 1 pu reactive current when the grid voltage is smaller than 0.8 pu, while in China, the WPS is required to provide 1.4 pu reactive current when the grid voltage is down to 0.2 pu under symmetrical voltage dips. In Denmark and Spain, the WPS

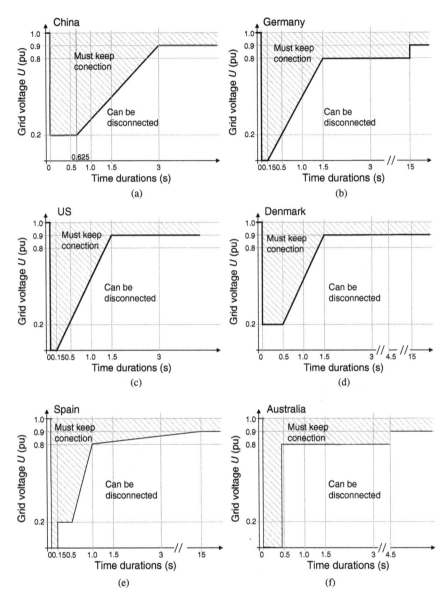

Figure 3.5   Voltage dips tolerance requirement in different countries: (a) China [2]; (b) Germany (E.ON) [4]; (c) United States (WECC); (d) Denmark [3]; (e) Spain (RED) [5]; and (f) Australia (WP).

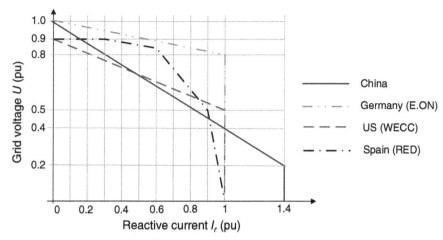

Figure 3.6   Reactive current support requirements in different countries for different voltage dip levels.

is required to provide at least 1 pu reactive current, when the grid is down to 0.5 and 0 pu, respectively.

Another aspect of the reactive current support requirement demands is that the reactive current must be provided within a required response time after the voltage dips. The reactive current response time in different countries are listed in Table 3.2. The German and Spanish grid codes demand the WPS provide the reactive current support in only 30 and 20 ms after the voltage dips, respectively. In China and Denmark, the reactive current response speed is required to be 80 and 100 ms after the voltage dips. This is also a large challenge for the WPS, especially for the DFIG WPS, as the stator time constant of the large scale DFIG is normally larger than 1 s.

Besides for reactive current support, the active current is also required to be maintained after voltage dips. For example, in Denmark, the WPS is demanded to provide active power as much as possible if the reactive current support requirements have been fulfilled. In Australia, the active current should only be reduced by 30% of the previous value under voltage dips.

3.3.1.3   Active Power Restore after Voltage Recovery

The WPS will return to normal operation conditions after the voltage recovers to the normal value. The GC has also related demands that the WPS should restore

TABLE 3.2   Demands to the reactive current response time in different countries

Countries	China [2]	Germany [4]	Denmark [3]	Spain [5]
Reactive response time (ms)	80	30	100	20–150 Decided by the voltage dip level and type of dip

TABLE 3.3   The active power restore speed in different countries

Countries	Requirements
China [2]	The WPS should restore the active power generation within the minimum speed of 0.1 pu per s after the grid voltage recovers.
Germany [4]	The WPS should restore the active power generation within the minimum speed of 0.2 pu per s after the grid voltage recovers.
Denmark [3]	The active power generation should be restored to 100% of the value before the fault in 5s after the grid voltage recovers.
Australia	The active power generation should be restored as soon as possible after the grid voltage recovers. The restoration rate is determined by the TSO and it cannot be smaller than 3% of the value before the fault.

the active power generation and back into normal operation within a required time. The demanded active power restoration speeds in different countries are listed in Table 3.3.

The active power restoration speed demanded by different countries are much slower compared to the reactive power response speed and the WPS are required to restore the active power in more than 1 s after the voltage recovery in many countries.

3.3.2   High Voltage Ride-Through Requirement

A high voltage in the power system can be introduced by large loads switching on and off, single-phase short-term interruption (STI) or overcompensation of the reactive power supplied by the capacitor banks. Many of the voltage swells come along with the grid voltage dips. In recent GCs, the WPS is also required to stay connected to the grid during the voltage swells, and absorb reactive power to help the power system with voltage control. It is also called as high voltage ride-through (HVRT) requirements.

The HVRT requirements for WTSs in Germany and Spain are shown in Figure 3.7. It is required for WTS to stay connected for at least 0.1 s if the voltage at the PCC reaches to 120% of its normal value in Germany; while in Spain, the WTS should stay connected to the grid for 0.15 s under a 130% voltage swell situation.

Similar to the LVRT operation, the WTS have to provide reactive power support to the grid during HVRT. However, instead of delivering reactive power to the grid, the WTS should absorb reactive power during HVRT operation in order to alleviate the voltage rise at the PCC, as it is shown in Figure 3.7b. The German grid code demands the WTS to absorb as least 1.0 pu reactive current during under a 120% voltage swell situation, while in Spain, the requirement is to absorb about 0.73 pu reactive current under a 130% voltage swell condition.

3.3.3   Recurring Fault Ride-Through Requirement

Recurring fault can be introduced by the auto-reclosing of the breaker in the power system after a permanent fault, which is also a challenge for the WTSs. Thus, the

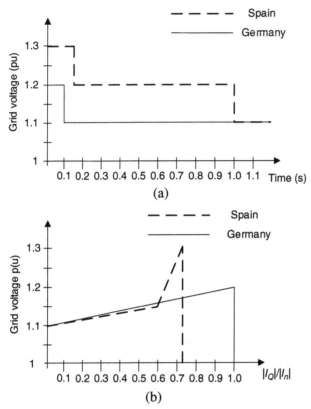

Figure 3.7 HVRT requirements in Spain [5] and Germany [4], (a) voltage swell tolerance and (b) reactive current support. I_Q, reactive current; I_n, nominal current.

Danish GC [3] has defined that the WTSs should withstand recurring faults, besides for the single fault discussed in 3.3.1. The wind power plant and any compensation equipment must stay connected after faults have occurred in the power grid as specified in Table 3.4. The requirements apply at the PCC, but the fault sequence is at a random point in the grid.

TABLE 3.4 Fault types and duration of the recurring fault ride-through requirement in the Danish grid code [3]

Type	Fault duration
Three-phase short circuit	Short circuit for a period of 150 ms.
Two-phase short circuit with/without earth contact	Short circuit for a period of 150 ms followed by a new short-circuit fault of 150 ms after 0.5–3 s.
Single-phase short circuit to earth	Single-phase-to-ground fault of 150 ms followed by a new single-phase-ground fault with the same duration after 0.5–3 s.

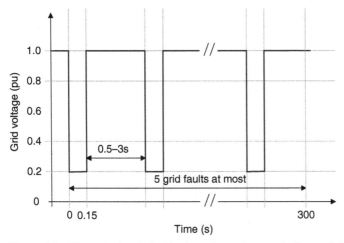

Figure 3.8   The recurring fault ride-through requirement in Denmark [3].

The energy reserves provided by the auxiliary equipment such as emergency supply equipment, the hydraulic system and also the pneumatic system should be sufficient for the wind power plant to meet the requirements in Table 3.4 in the event of at least two independent faults of the specified types occurring within 2 min. The energy reserves provided by the auxiliary equipment such as emergency supply equipment, the hydraulic system, and the pneumatic system should be sufficient in order for the wind power plant to meet the specified requirements in the event of at least six independent faults of the types specified in Table 3.4 occurring at 5-min intervals. This is also illustrated in Figure 3.8.

3.3.4   Unbalanced Grid Operation

The operation of the WPS under short-term unbalanced grid voltage dips introduced by the grid faults has been introduced in the LVRT requirements part. In this part, the unbalanced grid voltage indicates long-term steady unbalanced grid voltage. In the Chinese GC, the WPS is required to keep normal operation when the asymmetrical rate is smaller than 2%, or smaller than 4% for a short time. In Denmark, the GC also limits the asymmetry between each phase to a specific value. Other international standards such as EN50160, IEC1000-3 have also the related requirements to the unbalanced grid operation of the WPS [10, 11].

3.3.5   Harmonic Distortion Requirements

This kind of specification has focused on the output power quality of the WPS. It normally limits the THD of the output current of the WPS or the wind farms to a specific value. Also, it is required that the WPS should keep operation with a high quality under harmonic distorted grid voltage. In China, the GC demands that the

TABLE 3.5 Largest harmonic current injection into the grid in the Chinese GC: Part 1

Rated voltage (kV)	Short-circuit capacity (MVA)	Harmonic order and the largest harmonic current injection into the PCC (A)											
		2	3	4	5	6	7	8	9	10	11	12	13
0.38	10	78	62	39	62	26	44	19	21	16	28	13	24
6	100	43	34	21	34	14	24	11	11	8.5	16	7.1	13
10	100	26	20	13	20	8.5	16	7.1	13	6.1	6.8	5.3	1
35	250	15	12	7.7	12	5.1	8.8	3.8	4.1	3.1	5.6	2.6	4.7
66	500	16	13	8.1	13	5.4	9.3	4.1	4.3	3.3	5.9	2.7	5.0
110	750	12	9.6	6.0	9.6	4.0	6.8	3.0	3.2	2.4	4.3	2.0	3.7

largest harmonic current injection into the grid (PCC) is as shown in Tables 3.5 and 3.6.

Also, the maximum voltage harmonic distortion in the Chinese public grid is shown in Table 3.7. So, the WPS must provide a high quality output power even under harmonic distorted grid voltage, as listed in Table 3.7 for China.

Similar requirements can also be found in Danish GC. For the WPS with the capacity smaller than 1.5 MW, the harmonic current injected into the grid at the PCC should be smaller than the limitation as listed in Table 3.8. For the WPS with a capacity larger than 1.5 MW, the wind farm should make sure that when the WPSs are connected to the grid, the THD of the grid voltage at the PCC is under the limitation regulated by IEC-61000-3-6, as shown in Table 3.9.

In Table 3.9, MV represents medium Voltage (1–35 kV), HV is high voltage (35–230 kV), and EHV represents extremely high voltage (>230 kV) [12].

3.4 GRID CODES FOR DISTRIBUTED WIND POWER GENERATION

Most of the GCs introduced previously are valid for centralized WPS, normally with ten and above MW rated capacity and connected directly to the transmission lines.

TABLE 3.6 Largest harmonic current injection into the grid in the Chinese GC: Part 2

Rated voltage (kV)	Short-circuit capacity (MVA)	Harmonic order and the largest harmonic current injection into the PCC (A)											
		14	15	16	17	18	19	20	21	22	23	24	25
0.38	10	11	12	9.7	18	8.6	16	7.8	8.9	7.1	14	6.5	12
6	100	6.1	6.8	5.3	1	4.7	9.0	4.3	2.9	3.9	7.4	2.6	6.8
10	100	4.7	4.1	32	6.0	2.8	5.4	2.6	2.9	2.3	4.5	2.1	4.1
35	250	2.2	2.5	1.9	3.6	1.7	3.2	1.5	1.9	1.4	2.7	1.3	2.5
66	500	2.3	2.6	2.0	3.8	1.8	3.4	1.6	1.9	1.5	2.8	1.4	2.6
110	750	1.7	1.5	1.5	2.8	1.3	2.5	1.2	1.4	1.1	2.1	1.0	1.9

TABLE 3.7   Maximum voltage harmonic distortion in the Chinese public grid

		Harmonic voltage (%)	
Rated voltage (kV)	THD (%)	Odd order	Even order
0.38	5.0	4.0	2.0
6	4.0	3.2	1.6
10			
35	3.0	2.4	1.2
66			
110	2.0	1.6	0.8

Distributed wind power generation systems are directly connected to the distributed power system (normally with a voltage rating lower than 35 kV). It is more flexible and the transmission cost can be saved. It may be another trend of the wind power development besides the offshore wind power generation [13].

The GCs for distributed power generation are normally made separately from the GC for centralized WPS. The IEEE 1547 [1] is the earliest international standard for distributed power generation systems, including the distributed wind power. The State Grid of China has published the related GC Q/GDW 480-2010 for any distributed generation (including synchronous machine, induction machine, converters, etc.) connected to the distributed grid with the voltage level lower than 10 kV. In Germany, the published GC standards for power plants connected into the low-voltage distributed grid in 2011 (VDE-AR-N 4105:2011-08). A similar grid code can also be found in other countries, such as Great Britain and Canada. They will be introduced in detail as follows [14, 15].

3.4.1   Grid Limitation

As the capacity of the distributed power system is normally much smaller than the transmission power system, the GCs in some countries limit the installed capacity of all kinds of distributed generation system. In China, the GC demands that the total installed capacity of the distributed generation system will not exceed 25% of the front transformer capacity. The ratio between the short current capacity of the PCC and the rated current of the distributed generation system will not be lower than 10:3. A distributed generation system with a capacity smaller than 200 kW is suggested

TABLE 3.8   Largest harmonic current injection into the grid specified in the Danish GC

Voltage level (AC)	Odd-order harmonics h (no multiple of 3)					Even-order harmonics h (no multiple of 3)		
	5	7	11	13	17–49	2	4	8–50
≤ 10 kV	3.6	2.5	1.0	0.7	—	—	—	—
≥ 10 kV	4.0	4.0	2.0	2.0	$400/h^2$	0.8	0.2	0.1

TABLE 3.9   Grid harmonic voltage limitation regulated by IEC-61000-3-6

Odd-order harmonics (no multiple of 3)			Odd-order harmonics (multiples of 3)			Even-order harmonics		
	Harmonic voltage (%)			Harmonic voltage (%)			Harmonic voltage (%)	
Harmonic order h	MV	HV–EHV	Harmonic order h	MV	HV–EHV	Harmonic order h	MV	HV–EHV
5	5	2	3	4	2	2	1.8	1.4
7	4	2	9	1.2	1	4	1	0.8
11	3	1.5	15	0.3	0.3	6	0.5	0.4
13	2.5	1.5	21	0.2	0.2	8	0.5	0.4
11–19	$19 \times 17/ h-1.2$	$12 \times 17/h$	21–45	0.2	0.2	10–50	$0.25 \times 10/ h+0.22$	$0.19 \times 10/ h+0.16$

to be connected to the 380 V grid, while the distributed generation system with the capacity larger than 200 kW is suggested to be connected to 10 kV or higher voltage level grid [16].

The German GC and IEEE 1547 did not limit the integration capacity of the distributed generation system directly. But it is required in IEEE 1547 that the voltage variation introduced by the distributed generation system at the PCC will not exceed 5% of the rated voltage, while in Germany, the voltage variation introduced by the distributed generation system at the PCC is limited to 3%.

3.4.2   Active and Reactive Power Control

In many countries, the GC for distributed generation and for centralized wind power generation may be very different. For Centralized WPS or wind farms, the GC demands that they should join the voltage and frequency control of the power system by generating reactive or active power according to the TSO's command or the current grid voltage and frequency [17]. For distributed generation systems, as the voltage and frequency control is normally achieved in the transmission power system, the voltage and frequency control function is not required in the IEEE 1547 standard, the Canadian grid code or the Chinese grid code. If the grid frequency is out of the normal range, the distributed generation system should be disconnected from the grid, as shown in Table 3.10.

TABLE 3.10   Frequency response in the Chinese GC for distributed generation system

Frequency range (Hz)	Connected into grid at and above 10 kV	Connected to 380 V grid
< 48	Not required	Not required
48–49.5	Keep operating for 10 min	Disconnected in 0.2 s
49.5–50.2	Keep normal operating	Keep normal operating
50.2–50.5	Keep normal operating	Disconnected in 0.2 s
> 50.5	Disconnected immediately	Disconnected in 0.2 s

TABLE 3.11   Frequency response in the German GC for distributed generation system

Frequency range (Hz)	Action
< 47.5	Disconnect immediately
47.5–50.2	Keep normal operating
50.2–51.5	The output active power should be reduced by 40% per Hz with the frequency increase. The active power can only be increased after the frequency is lower than 50.05 Hz.
> 51.5	Disconnect immediately (in 0.2 s)

It should be mentioned that in the German GC for distributed power system, active power and reactive power control under grid frequency and voltage variation are required. The distributed power system should adjust the active power according to the demands from the DSO and to the grid frequency, as listed in Table 3.11. For the reactive power control, the Q control, power factor control, and voltage control are all required for the distributed power system, according to the DSO's commands [8, 9].

3.4.3   Operation under Grid Faults

Although the FRT requirements are very important parts in the GC for centralized WPS, it is not required in the GC for distributed generation system in China, Canada, and in the IEEE 1547 standards. The distributed generation system should disconnect from the grid under grid fault immediately. The German GC for distributed generation system requires that the distributed generation system should have the FRT ability, as shown in Figure 3.9. The distributed generation system is required to stay connected to the grid even if the grid voltage falls to zero, for at least 150 ms, and the reactive current support is also required [18].

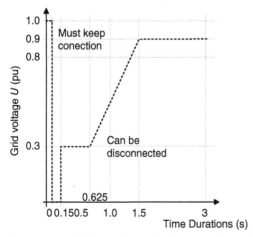

Figure 3.9   LVRT requirement in Germany for the distributed generation system.

Apart from the FRT requirements, an islanding detection is required for the distributed generation system in some countries. The IEEE 1547 indicated that the distributed generation converter should be cut-off from the grid after 10 cycles after the islanding happens, and the distributed generation system should detect the islanding and stop the power converter within 2 s. The Chinese GC for distributed generation system also requires the distributed generation converter to detect the islanding operation and stop the power converter after the islanding happens [19].

3.5 SUMMARY

The grid codes are the "laws" for wind turbines to be connected to the grid. It is also the demand to be fulfilled when designing a control strategy for a WPS. In this chapter, the grid codes for wind power generation system connection in several countries are introduced, including the grid codes related to the voltage and frequency deviation, active power/frequency control, and reactive power/voltage control. The grid codes under abnormal operation, such as grid faults, unbalanced grid voltage, and harmonic distortions are introduced as well. The grid codes for the distributed generation system may be quite different from the centralized WPS and the related grid codes for distributed generation system are also discussed in this chapter.

REFERENCES

[1] IEEE Std 1547-2003, "IEEE standard for interconnecting distributed resources with electric power systems," 2003.

[2] GB/T-19963-2011, "Technical rule for connecting wind farm to power system," The National Standard of China, 2012. [Online]. Available: http://www.sac.gov.cn/

[3] Technical Regulation 3.2.5, "Technical regulation 3.2.5 for wind power plants with a power output greater than 11 kW," Energinet, 2011. [Online]. Available: http://www.energinet.dk

[4] "Grid Code—High and extra high voltage," E.ON Netz GmbH, Bayreuth, Germany, Apr. 2006.

[5] Resolution P.O. 12.3, "Response requirements against voltage dips in wind installations," Red Eléctrica, Mar. 2006. [Online]. Available: http://www.ree.es

[6] Transmission Code 2007, "Networks and system rules of the German transmission system operators," VDN-e.v. beim VDEW, Aug. 2007. [Online]. Available: http://www.vdn-berlin.de

[7] "Requirements for offshore grid connections in the E.ON nets network," E.ON, (2008). [Online]. Available: http://www.eon-netz.com

[8] Q/GDW 480-2010, "Technical rule for connecting distributed generation systems to power system," The Standard of Chinese State Grid, 2010.

[9] VDE-AR-N 4105:2011-08. "Power generation systems connected to the low voltage distribution network," 2011.

[10] M. Altin, Ö. Göksu, R. Teodorescu, P. Rodriguez, B. B. Jensen, and L. Helle, "Overview of recent grid codes for wind power integration," in *Int. Conf. Optim. Electr.Electron. Equip.*, 2010, pp. 1152–1160.

[11] M. Tsili and S. Papathanassiou, "A review of grid code technical requirements for wind farms," *IET Renew. Power Gener.*, vol. 3, no. 3, pp. 308–332, 2009.

[12] F. Iov, R. Teodorescu, F. Blaabjerg, B. Andresen, J. Birk, and J. Miranda, "Grid code compliance of grid-side converter in wind turbine systems," in *Proc. IEEE Power Electron. Specialists Conf.*, 2006, pp. 1–7.

[13] I. Erlich and U. Bachmann, "Grid code requirements concerning connection and operation of wind turbines in Germany," *IEEE Power Eng. Soc. Gen. Meeting*, vol. 2, pp. 1253–1257, 2005.

[14] M. Tsili and S. Papathanassiou, "A review of grid code technical requirements for wind farms," *IET Renew. Power Gener.*, vol. 3, no. 3, pp. 308–332, 2009.

[15] C. Jauch, J. Matevosyan, T. Ackermann, and S. Bolik, "International comparison of requirements for connection of wind turbines to power systems," *Wind Energ.*, vol. 8, pp. 295–306, 2005.

[16] I. Erlich, W. Winter, and A. Dittrich, "Advanced grid requirements for the integration of wind turbines into the German transmission system," in *Proc. IEEE Power Eng. Soc. Gen. Meeting*, 2006.

[17] X. Zhao, S. Zhang, R. Yang, and M. Wang, "Constraints on the effective utilization of wind power in China: An illustration from the northeast China grid," *Renew. Sust. Energ. Rev.*, vol. 16, no. 7, pp. 4508–4514, Aug. 2016.

[18] S. M. Muyeen, R. Takahashi, T. Murata, and J. Tamura, "A variable speed wind turbine control strategy to meet wind farm grid code requirements," *IEEE Trans. Power. Syst.*, vol. 25, no. 1, pp. 331–340, Feb. 2010.

[19] N. R. Ullah, T. Thiringer, and D. Karlsson, "Voltage and transient stability support by wind farms complying with the E.ON netz grid code," *IEEE Trans. Power. Syst.*, vol. 22, no. 4, pp. 1647–1656, Nov. 2007.

PART II

MODELING AND CONTROL OF DFIG

MODELING OF DFIG WIND POWER SYSTEMS

In this chapter, the modeling of the DFIG wind power system is introduced, including the steady-state model of the DFIG, the dynamic model of the DFIG, and the power electronic converter. It may be helpful for the readers to understand the operation of the DFIG wind power system, which are the fundamentals of designing the control systems given in the following chapter.

4.1 INTRODUCTION

The basics of the wind power system and its related grid codes have been introduced in Chapters 2 and 3. The reader should have a brief concept of the wind power system and the requirements of the grid. From this chapter onward, the book will focus on the modeling and control of DFIG wind power system. To design the control of wind power systems, it is necessary to know models for DFIG wind power systems, including the steady-state and dynamic model of both DFIG and power converters. Since these models describe the operation relationship of the variables in the DFIG power system, they are the fundamentals for designing controls for the system in the following chapters.

In this chapter, first, the steady-state model of the DFIG is introduced. The typical operation modes with different rotor speed and active/reactive power output conditions are discussed with corresponding phasor diagrams. Second, the dynamic model of the DFIG and power electronic converter in different reference frames (abc, $\alpha\beta$, and dq) are introduced. The relationship of dynamic models under these different reference frames is explained.

4.2 STEADY-STATE EQUIVALENT CIRCUIT OF A DFIG

As introduced in Chapter 2, a DFIG has two separate three-phase windings: the stator winding and the rotor winding. Apart from the common-used caged-rotor induction

Advanced Control of Doubly Fed Induction Generator for Wind Power Systems, First Edition.
Dehong Xu, Frede Blaabjerg, Wenjie Chen, and Nan Zhu.

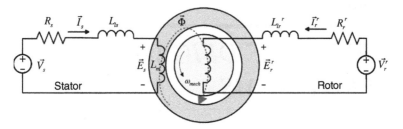

Figure 4.1   One-phase steady-state equivalent circuit of a DFIG.

machine, both the stator and rotor windings are supplied by power sources. The stator winding is supplied by the grid while the rotor winding is usually supplied by a three-phase inverter. To derive the steady-state model of the DFIG, it is assumed that both the stator and rotor windings are connected with the star configuration. The grid is ideal and its outputs are balanced three-phase voltages with constant amplitude and frequency. The rotor voltages supplied by the inverter are also balanced three-phase voltages with constant amplitude and frequency. In this case, DFIG's three phases are symmetric and it can be simplified to analyze characteristics of one phase of the machine [1–7].

4.2.1   Steady-State Equivalent Circuit of a DFIG

One-phase concept figure of a DFIG is shown in Figure 4.1. In Figure 4.1, $\vec{V}_s$ and $\vec{V}_r^r$ are the stator and rotor voltage phasors supplied by the grid and the inverter, respectively. $\vec{I}_s$ and $\vec{I}_r^r$ are the stator and rotor current phasors, respectively. $\vec{E}_s$ and $\vec{E}_r^r$ are the stator and rotor induced electromotive forces (EMFs), respectively. $\vec{\Phi}$ is the mutual flux. L_m is the mutual inductance. L_{ls} and L_{lr}^r are the leakage inductance of the stator winding and rotor winding, R_s and R_r^r are the stator and rotor resistances. Variables with superscript r represent the rotor-side variables. The rotor rotates with a mechanical angular speed ω_{mech}. If the pole pairs of the DFIG is n_p, the electrical angular frequency of the rotor windings ω_r is expressed by

$$\omega_r = n_p \omega_{mech} \tag{4.1}$$

For simplicity, ω_r is simply called the angular frequency instead of electrical angular frequency in this book.

As the stator side of the DFIG is connected to the grid, the angular frequency of the voltages and currents in the stator windings is $\omega_s = 2\pi f_s$ where f_s is the grid frequency and is also called the synchronous frequency. The angular frequency of the induced EMF voltage in the rotor windings depends on both the stator flux rotational speed and rotor rotational speed. It is determined by the following equation:

$$\omega_{sl} = \omega_s - \omega_r \tag{4.2}$$

It is observed that ω_{sl} is equal to the difference between the stator angular frequency and the rotor angular frequency. It is called slip angular frequency of the DFIG and $f_{sl} = \omega_{sl}/2\pi$ is the slip frequency.

According to Figure 4.1, it is easy to write phasor equations for the stator side and rotor side of the DFIG, respectively:

$$\vec{V}_s = \vec{E}_s + (R_s + j\omega_s L_{ls})\vec{I}_s \tag{4.3}$$

$$\vec{V}_r^r = \vec{E}_r^r + (R_r^r + j\omega_{sl} L_{lr}^r)\vec{I}_r^r \tag{4.4}$$

It should be noticed that equation (4.3) is derived based on the synchronous frequency f_s while equation (4.4) is expressed based on the slip frequency f_{sl}. It is common to transform the electrical variables in the rotor side into the stator side so that it is easier to understand the relationship between the stator and rotor variables.

The stator windings and rotor windings of the DFIG are coupled with mutual flux $\vec{\Phi}$. The induced EMF on the stator winding is expressed as

$$\vec{E}_s = j\omega_s k_s N_s \vec{\Phi} = j\omega_s \vec{\Psi} \tag{4.5}$$

where N_s is the number of turns of the one-phase stator winding, k_s is the stator winding factor, and $\vec{\Psi}$ is the flux linkage.

Similarly, the induced EMF on the rotor winding is represented as

$$\vec{E}_r^r = j\omega_{sl} k_r N_r \vec{\Phi} \tag{4.6}$$

where N_r is the number of turns of one-phase rotor windings, and k_r is the rotor winding factor.

With equations (4.5) and (4.6), the relationship between $\vec{E}_s$ and $\vec{E}_r^r$ is derived:

$$\frac{\vec{E}_r^r}{\vec{E}_s} = \frac{\omega_{sl}}{\omega_s} \frac{k_r N_r}{k_s N_s} \approx S_l \frac{1}{n_{sr}} \tag{4.7}$$

where $S_l = f_{sl}/f_s = \omega_{sl}/\omega_s$ is defined to be the slip ratio, $n_{sr} = N_s/N_r$ is the stator to rotor turns ratio. Besides, the approximation $k_r/k_s \approx 1$ is used in equation (4.7).

The relationship between stator-side variables and rotor-side variables is similar to that of a transformer. It is natural to convert the variables in rotor side to stator side by referring the analysis of the transformer.

Equation (4.4) can be rewritten by using equation (4.7).

$$\vec{V}_r^r = S_l \frac{1}{n_{sr}} \vec{E}_s + (R_r^r + j\omega_{sl} L_{lr}^r)\vec{I}_r^r \tag{4.8}$$

After both sides of the equation (4.8) are multiplied by $\frac{1}{S_l} n_{sr}$, the following equation is obtained:

$$\frac{1}{S_l} n_{sr} \vec{V}_r^r = \vec{E}_s + \left(\frac{1}{S_l} n_{sr}^2 R_r^r + j\omega_s n_{sr}^2 L_{lr}^r \right) \vec{I}_r^r \frac{1}{n_{sr}} \tag{4.9}$$

To simplify the expression of the above equation, the following variables or parameters are introduced:

$$\vec{V}_r = n_{sr}\vec{V}_r^r \tag{4.10}$$

$$\vec{I}_r = \frac{1}{n_{sr}}\vec{I}_r^r \tag{4.11}$$

$$R_r = n_{sr}^2 R_r^r \tag{4.12}$$

$$L_{lr} = n_{sr}^2 L_{lr}^r \tag{4.13}$$

where $\vec{V}_r$ represents the rotor voltage $\vec{V}_r^r$ referred to the stator side, $\vec{I}_r$ represents the rotor current $\vec{I}_r^r$ on the stator side, and R_r and L_{lr} represent the rotor resistance R_r^r and leakage inductance L_{lr}^r, respectively, on the stator side.

By using equations (4.10)–(4.13), rotor-side equation (4.9) is changed to

$$\frac{1}{S_l}\vec{V}_r = \vec{E}_s + \left(\frac{1}{S_l}R_r + j\omega_s L_{lr}\right)\vec{I}_r \tag{4.14}$$

In the above equation, rotor voltage and rotor current have been converted to the stator side according to turns ratio of the stator winding to the rotor winding. Rotor winding resistance and leakage inductance have been converted to the stator side. It more clearly describes the relationship between the rotor-induced EMF and stator-induced EMF.

The stator-induced EMF $\vec{E}_s$ will produce magnetic current $\vec{I}_m$ in the motor:

$$\vec{E}_s = j\omega_s L_m \vec{I}_m \tag{4.15}$$

where L_m is magnetizing inductance.

By combining equations (4.5) and (4.15), following equation is derived:

$$\vec{\Psi} = L_m \vec{I}_m \tag{4.16}$$

Magnetic current $\vec{I}_m$ can be expressed by

$$\vec{I}_m = \frac{\vec{\Psi}}{L_m} \tag{4.17}$$

According to equations (4.3), (4.14), and (4.15), the one-phase steady-state equivalent circuit of the DFIG is derived as shown in Figure 4.2. All variables in the rotor side have been converted to the stator side. However, the frequency of the variables in rotor side is different from that of the stator side. The frequency of the variables in stator side is the synchronous frequency f_s while the frequency of the variables in rotor side is the slip frequency f_{sl}.

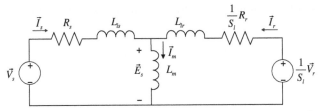

Figure 4.2   One-phase steady-state equivalent circuit of the DFIG referred to the stator side.

4.2.2 Power in the DFIG

Referring to Figure 4.2, the following equation can be obtained according to KCL:

$$\vec{I}_m = \vec{I}_s + \vec{I}_r \tag{4.18}$$

By applying KVL to the left loop in Figure 4.2, the following equation is obtained:

$$\vec{V}_s = R_s\vec{I}_s + jX_{ls}\vec{I}_s + jX_m\vec{I}_m \tag{4.19}$$

where $X_{ls} = \omega_s L_{ls}$, $X_m = \omega_s L_m$.

Similarly, by using KVL to the right loop in Figure 4.2, the following equation is obtained:

$$\frac{\vec{V}_r}{S_l} = \frac{R_r}{S_l}\vec{I}_r + jX_{lr}\vec{I}_r + jX_m\vec{I}_m \tag{4.20}$$

where $X_{lr} = \omega_s L_{lr}$.

Rotor power is described by

$$P_r = \text{Re}(\vec{V}_r\vec{I}_r^*) \tag{4.21}$$

where the superscript "*" stands for the complex conjugate of the vector.

By substituting $\vec{V}_r$ with equation (4.20), the rotor power is expressed as

$$P_r = \text{Re}\left[R_r\vec{I}_r\vec{I}_r^* + jS_lX_{lr}\vec{I}_r\vec{I}_r^* + jS_lX_m\vec{I}_m\vec{I}_r^*\right] \tag{4.22}$$

Since $\vec{I}_r\vec{I}_r^* = I_r^2$,

$$P_r = R_rI_r^2 + \text{Re}(jS_lX_m\vec{I}_m\vec{I}_r^*) \tag{4.23}$$

By substituting equation (4.18), the rotor power is expressed as

$$P_r = R_rI_r^2 + \text{Re}\left[jS_lX_m\vec{I}_m(\vec{I}_m^* - \vec{I}_s^*)\right] \tag{4.24}$$

Since $\text{Re}(jS_lX_m\vec{I}_m\vec{I}_m^*) = 0$,

$$P_r = R_rI_r^2 + \text{Re}\left[jS_lX_m\vec{I}_m(-\vec{I}_s^*)\right] \tag{4.25}$$

By using equation (4.19) to eliminate the term $X_m \vec{I}_m$ in equation (4.25), following equation is obtained:

$$P_r = R_r I_r^2 + S_l \text{Re}((-\vec{V}_s \vec{I}_s^* + R_s \vec{I}_s \vec{I}_s^* + jX_{ls} \vec{I}_s \vec{I}_s^*) \tag{4.26}$$

The above equation is simplified as

$$P_r = R_r I_r^2 + S_l R_s I_s^2 - S_l \text{Re}(\vec{V}_s \vec{I}_s^*) \tag{4.27}$$

As we know, the stator power can be expressed as

$$\vec{P}_s = \text{Re}(\vec{V}_s \vec{I}_s^*) \tag{4.28}$$

Then the rotor power can be described as

$$P_r = R_r I_r^2 + S_l R_s I_s^2 - S_l P_s \tag{4.29}$$

Above equation can be expressed as

$$P_r = P_{rcu} + S_l(P_{scu} - P_s) \tag{4.30}$$

where $P_{scu} = R_s I_s^2$ is the copper loss of the stator winding, $P_{rcu} = R_r I_r^2$ is the copper loss of the rotor winding.

Generally speaking, the copper losses P_{rcu} and P_{scu} are much smaller than both the stator power P_s and the rotor power P_r. Then equation (4.30) can be approximated as

$$P_r \approx -S_l P_s \tag{4.31}$$

It is observed that the rotor power P_r is a fraction of the stator power P_s according to slip ratio S_l. In the wind power system, the slip ratio S_l ranges from -0.3 to 0.3. In this case, the rotor power P_r is less than $1/3$ of the stator power P_s. It means that the stator power can be controlled by the rotor-side power converter with $1/3$ power rating of the rated power of the machine.

Similarly, rotor reactive power is described by

$$Q_r = \text{Im}(\vec{V}_r \vec{I}_r^*) \tag{4.32}$$

By substituting $\vec{V}_r$ according to equation (4.20), the rotor reactive power is expressed as

$$Q_r = \text{Im}\left[R_r \vec{I}_r \vec{I}_r^* + jS_l X_{lr} \vec{I}_r \vec{I}_r^* + jS_l X_m \vec{I}_m \vec{I}_r^*\right] \tag{4.33}$$

Since $\vec{I}_r \vec{I}_r^* = I_r^2$,

$$Q_r = S_l X_{lr} I_r^2 + \text{Im}(jS_l X_m \vec{I}_m \vec{I}_r^*) \tag{4.34}$$

By using equation (4.18) to replace variable $\vec{I}_r$, the above equation is changed to

$$Q_r = S_l X_{lr} I_r^2 + \text{Im}\left[jS_l X_m \vec{I}_m (\vec{I}_m^* - \vec{I}_s^*)\right] \tag{4.35}$$

Since $\vec{I}_m\vec{I}_m^* = I_m^2$,

$$Q_r = S_l X_{lr} I_r^2 + S_l X_m I_m^2 - \text{Im}\left[jS_l X_m \vec{I}_m \vec{I}_s^*\right]$$ (4.36)

By using equation (4.19) to eliminate the term $X_m \vec{I}_m$ in equation (4.36), the following equation is obtained:

$$Q_r = S_l X_{lr} I_r^2 + S_l X_m I_m^2 - \text{Im}\left[S_l(\vec{V}_s - R_s \vec{I}_s - jX_{ls}\vec{I}_s)\vec{I}_s^*\right]$$ (4.37)

Since $\vec{I}_s\vec{I}_s^* = I_s^2$,

$$Q_r = S_l X_{lr} I_r^2 + S_l X_m I_m^2 + S_l X_{ls} I_s^2 - S_l \text{Im}(\vec{V}_s \vec{I}_s^*)$$ (4.38)

As we know, the stator reactive power can be expressed as

$$Q_s = \text{Im}(\vec{V}_s \vec{I}_s^*)$$ (4.39)

Then the rotor reactive power can be described as

$$Q_r = S_l X_{ls} I_s^2 + S_l X_{lr} I_r^2 + S_l X_m I_m^2 - S_l Q_s$$ (4.40)

Above equation can be reorganized as

$$Q_r = S_l(Q_{ls} + Q_{lr} + Q_m - Q_s)$$ (4.41)

where $Q_{ls} = X_{ls} I_s^2$ is the reactive power of stator leakage inductance, $Q_{lr} = X_{lr} I_r^2$ is the reactive power of rotor leakage inductance, and $Q_m = X_m I_m^2$ is the reactive power of magnetizing inductance.

If it is assumed that $|Q_s| \gg Q_{ls} + Q_{lr} + Q_m$, then the equation (4.41) can be approximated as

$$Q_r = -S_l Q_s$$ (4.42)

The above equation means that the reactive power of the stator may be controlled by the rotor side supplied by a power converter with a smaller power rating.

In Figure 4.3, the input power to the rotor of the generator is composed of mechanical power P_{mech} and the power from the rotor converter P_r. To simplify the power flow relationship, we ignore both electric losses and mechanical losses. The following power balance condition is obtained:

$$P_{mech} + P_r = P_{em} = -P_s$$ (4.43)

where P_{em} is transferred electromagnetic power between the stator and the rotor.

By substituting (4.31) in (4.43), the mechanical power P_{mech} can be expressed by

$$P_{mech} = (S_l - 1)P_s$$ (4.44)

For DFIG wind power system, the rotor side is connected to the grid indirectly through back-to-back converter, where P_g is the power of the grid-side converter

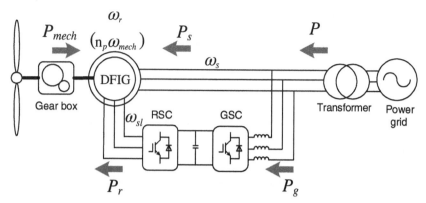

Figure 4.3   Power flow in a DFIG wind system.

(GSC). Since the converter losses are neglected, $P_g = P_r$. Therefore, the total power to grid can be expressed as

$$P = P_s + P_r \tag{4.45}$$

By using the relationship $P_r \approx -S_l P_s$, the total power to grid can be expressed by

$$P = P_s + P_r = (1 - S_l)P_s \tag{4.46}$$

The DFIG has three operating modes: super-synchronous mode, synchronous mode, and sub-synchronous mode.

In the super-synchronous mode, the rotor angular frequency is higher than the synchronous angular frequency, $\omega_r > \omega_s$, and slip $S_l < 0$. Then according to equation (4.46), the total power to grid is $P = (1 + |S_l|)P_s$. Both the stator and the rotor deliver power to the grid.

In the synchronous mode, the rotor angular frequency is equal to the synchronous angular frequency, $\omega_r = \omega_s$, and slip $S_l = 0$. DC current flows into the rotor winding. According to equation (4.46), the total power to grid $P = P_s$. Only the stator delivers power to the grid.

In the sub-synchronous mode, the rotor angular frequency is lower than the synchronous angular frequency, $\omega_r < \omega_s$, and slip $S_l > 0$. According to equation (4.46), the total power to grid $P = (1 - |S_l|)P_s$. The stator delivers power to the grid while the rotor receives power from the grid.

4.3   DYNAMIC MODEL OF A DFIG

The steady-state equivalent circuit of a DFIG is derived with the assumption that the DFIG works at steady state under the ideal grid condition. It is simplified as a one-phase system. It only describes the steady-state performance of the DFIG.

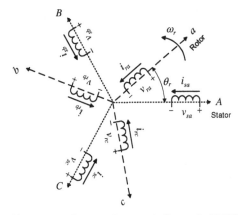

Figure 4.4   Stator and rotor windings of a DFIG.

For dynamic analysis and control system design, the dynamic model of the DFIG is needed [8–11].

DFIG is a high order, nonlinear, and time-varying system. To simplify the dynamic analysis, the following assumption is introduced:

- Three-phase windings of the generator are balanced with Y connection, with 120° phase shift between each other in space. The induced magneto-motive force is distributed in a sinusoidal form along the air gap.
- The magnetic saturation of the stator and the rotor core is neglected.
- The iron loss of both the stator and the rotor core is neglected.
- The stator or rotor winding parasitic resistances do not change with the temperature and the frequency.

To simplify the analysis, all electrical variables on the rotor side are converted to the stator side. Positive reference direction for both the power and the current are selected to be the direction in which the power or the current flows into the stator or the rotor of the DFIG.

4.3.1   ABC (abc) Model

With the assumptions mentioned above, the DFIG is modeled as three static windings sitting on the stator (*ABC*), and three windings sitting on the rotor (*abc*) rotating with angular frequency ω_r, as shown in Figure 4.4. If the initial angle between phase-a rotor winding and phase-A stator winding are zero, the angle θ_r between phase-a rotor winding and phase-A stator winding is described as

$$\theta_r = \omega_r t \qquad (4.47)$$

In Figure 4.4, v_{sa}, v_{sb}, v_{sc} and v_{ra}, v_{rb}, v_{rc} are three-phase stator voltages and three-phase rotor voltages, respectively. i_{sa}, i_{sb}, i_{sc} and i_{ra}, i_{rb}, i_{rc} are three-phase stator currents and three-phase rotor currents, respectively. Stator equations are formulated as follows:

$$v_{sa} = R_s i_{sa} + \frac{d}{dt}\psi_{sa}$$

$$v_{sb} = R_s i_{sb} + \frac{d}{dt}\psi_{sb} \tag{4.48}$$

$$v_{sc} = R_s i_{sc} + \frac{d}{dt}\psi_{sc}$$

where R_s is the resistance of each stator winding, and $\psi_{sa}, \psi_{sb}, \psi_{sc}$ are three-phase stator winding fluxes. Three equations in (4.48) can been expressed with matrix form:

$$\boldsymbol{v}_s = R_s \boldsymbol{i}_s + p\boldsymbol{\psi}_s \tag{4.49}$$

where stator voltage vector $\boldsymbol{v}_s = \begin{bmatrix} v_{sa} & v_{sb} & v_{sc} \end{bmatrix}^{\mathrm{T}}$, stator current vector $\boldsymbol{i}_s = \begin{bmatrix} i_{sa} & i_{sb} & i_{sc} \end{bmatrix}^{\mathrm{T}}$, stator flux vector $\boldsymbol{\psi}_s = \begin{bmatrix} \psi_{sa} & \psi_{sb} & \psi_{sc} \end{bmatrix}^{\mathrm{T}}$, and $p = d/dt$.
Similarly, rotor equations are formulated as follows:

$$v_{ra}^r = R_r i_{ra}^r + \frac{d}{dt}\psi_{ra}^r$$

$$v_{rb}^r = R_r i_{rb}^r + \frac{d}{dt}\psi_{rb}^r \tag{4.50}$$

$$v_{rc}^r = R_r i_{rc}^r + \frac{d}{dt}\psi_{rc}^r$$

where the superscript r indicates that all variables in equation (4.50) are described under abc reference frame sitting on the rotor. Actually, the abc reference frame rotates with the rotor of the machine. R_r is the resistance of each rotor winding. Here, $\psi_{ra}^r, \psi_{rb}^r, \psi_{rc}^r$ are three-phase rotor winding fluxes. Three equations in (4.50) are also expressed with the matrix form:

$$\boldsymbol{v}_r^r = R_r \boldsymbol{i}_r^r + p\boldsymbol{\psi}_r^r \tag{4.51}$$

where rotor voltage vector $\boldsymbol{v}_r^r = \begin{bmatrix} v_{ra}^r & v_{rb}^r & v_{rc}^r \end{bmatrix}^{\mathrm{T}}$, rotor current vector $\boldsymbol{i}_r^r = \begin{bmatrix} i_{ra}^r & i_{rb}^r & i_{rc}^r \end{bmatrix}^{\mathrm{T}}$, rotor flux vector $\boldsymbol{\psi}_r^r = \begin{bmatrix} \psi_{ra}^r & \psi_{rb}^r & \psi_{rc}^r \end{bmatrix}^{\mathrm{T}}$.
The stator flux vector and the rotor flux vector can be described by

$$\boldsymbol{\psi}_s = \boldsymbol{L}_{ss}\boldsymbol{i}_s + \boldsymbol{L}_{sr}\boldsymbol{i}_r^r \tag{4.52}$$

$$\boldsymbol{\psi}_r^r = \boldsymbol{L}_{rs}\boldsymbol{i}_s + \boldsymbol{L}_{rr}\boldsymbol{i}_r^r \tag{4.53}$$

where stator winding mutual inductance matrix is expressed by

$$\boldsymbol{L}_{ss} = \begin{bmatrix} L_{ms} + L_{ls} & -\frac{1}{2}L_{ms} & -\frac{1}{2}L_{ms} \\ -\frac{1}{2}L_{ms} & L_{ms} + L_{ls} & -\frac{1}{2}L_{ms} \\ -\frac{1}{2}L_{ms} & -\frac{1}{2}L_{ms} & L_{ms} + L_{ls} \end{bmatrix} \tag{4.54}$$

Rotor winding mutual inductance matrix is expressed by

$$
\boldsymbol{L}_{rr} = \begin{bmatrix} L_{mr} + L_{lr} & -\dfrac{1}{2}L_{mr} & -\dfrac{1}{2}L_{mr} \\[2mm] -\dfrac{1}{2}L_{mr} & L_{mr} + L_{lr} & -\dfrac{1}{2}L_{mr} \\[2mm] -\dfrac{1}{2}L_{mr} & -\dfrac{1}{2}L_{mr} & L_{mr} + L_{lr} \end{bmatrix} \tag{4.55}
$$

Stator and rotor winding mutual inductance matrix is described by

$$
\boldsymbol{L}_{sr} = \boldsymbol{L}_{rs}^{T} = L_{ms} \begin{bmatrix} \cos\theta_r & \cos\left(\theta_r + 120°\right) & \cos\left(\theta_r - 120°\right) \\[1mm] \cos\left(\theta_r - 120°\right) & \cos\theta_r & \cos\left(\theta_r + 120°\right) \\[1mm] \cos\left(\theta_r + 120°\right) & \cos\left(\theta_r - 120°\right) & \cos\theta_r \end{bmatrix} \tag{4.56}
$$

where L_{ms} and L_{mr} are maximum mutual inductances between a stator winding and a rotor winding. Since the rotor-side variables have been referred to the stator side and both stator windings and rotor windings have the same turns, $L_{ms} = L_{mr}$. L_{ls} and L_{lr} are leakage inductances of the stator and rotor windings, respectively. The stator winding mutual inductance matrix $\boldsymbol{L}_{ss}$ is a constant matrix since the geometry of phase windings A, B, and C of the stator are fixed. It is also true for the rotor winding mutual inductance matrix $\boldsymbol{L}_{rr}$. However, the stator and rotor winding mutual inductance matrix $\boldsymbol{L}_{sr}$ depends on the rotor angle θ_r because the flux coupling between the stator winding and the rotor winding changes with the angle θ_r between phase-A rotor winding and phase-A stator winding.

It is observed that both the stator flux and the rotor flux change with rotor angle θ_r. Therefore, DFIG model in *ABC* (*abc*) frame is a time-variant system. Generally speaking, it is more difficult to analyze its dynamics in *ABC* (*abc*) reference frame.

4.3.2 $\alpha\beta$ Model

For a three-phase power system, usually the three-phase variables $x_a, x_b,$ and x_c satisfy the relationship $x_a + x_b + x_c = 0$. In this case, all the three-dimensional space vectors satisfying this condition are located on the subset in the three-dimensional space, which is actually a plane. Therefore, three-phase variables can be simply expressed as a complex vector $\vec{x}$ in the complex plane, which is called $\alpha\beta$ plane, with the following transformation.

$$
\vec{x} = \frac{2}{3}\left(x_a + \alpha x_b + \alpha^2 x_c\right) = \frac{2}{3}\begin{bmatrix} 1 & \alpha & \alpha^2 \end{bmatrix}\begin{bmatrix} x_a \\ x_b \\ x_c \end{bmatrix} \tag{4.57}
$$

where $\alpha = e^{j2\pi/3}$ and $\alpha^2 = e^{j4\pi/3}$. The complex plane is shown in Figure 4.5. Constant 2/3 of expression (4.57) is chosen to scale the complex vector to have the same amplitude as that of the three-phase variables.

Complex vector $\vec{x}$ is composed of the real part x_α and the imaginary part x_β:

$$
\vec{x} = x_\alpha + jx_\beta \tag{4.58}
$$

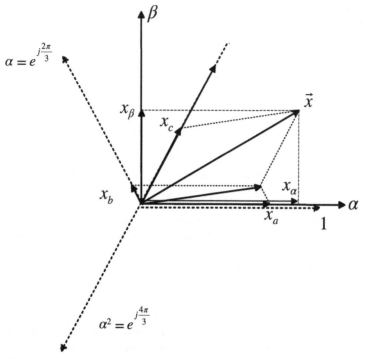

Figure 4.5   Vector representation in complex plane $\alpha\beta$.

According to equations (4.57) and (4.58), the relationship between x_a, x_b, x_c and x_α, x_β are derived:

$$x_\alpha = \frac{2}{3}\left(x_a + x_b \cos\frac{2}{3}\pi + x_c \cos\frac{4}{3}\pi\right)$$

$$x_\beta = \frac{2}{3}\left(x_b \sin\frac{2}{3}\pi + x_c \sin\frac{4}{3}\pi\right) \tag{4.59}$$

Equation (4.59) can be written in the matrix form:

$$\begin{bmatrix} x_\alpha \\ x_\beta \end{bmatrix} = \frac{2}{3}\begin{bmatrix} 1 & -\frac{1}{2} & -\frac{1}{2} \\ 0 & \frac{\sqrt{3}}{2} & -\frac{\sqrt{3}}{2} \end{bmatrix} \cdot \begin{bmatrix} x_a \\ x_b \\ x_c \end{bmatrix} \tag{4.60}$$

The above transformation is called the Clarke transformation. The following matrix is called the Clarke transformation matrix:

$$T_C = \frac{2}{3}\begin{bmatrix} 1 & -\frac{1}{2} & -\frac{1}{2} \\ 0 & \frac{\sqrt{3}}{2} & -\frac{\sqrt{3}}{2} \end{bmatrix} \tag{4.61}$$

The inverse Clarke transformation can be derived as follows:

$$
\begin{bmatrix} x_a \\ x_b \\ x_c \end{bmatrix} = \frac{3}{2} \begin{bmatrix} \frac{2}{3} & 0 \\ -\frac{1}{3} & \frac{\sqrt{3}}{3} \\ -\frac{1}{3} & \frac{\sqrt{3}}{3} \end{bmatrix} \cdot \begin{bmatrix} x_\alpha \\ x_\beta \end{bmatrix}
\tag{4.62}
$$

With regard to the DFIG, the stator voltage equations (4.48) can be changed into the complex vector equation form by using transform (4.57). By multiplying $\frac{2}{3}$ to the first equation, multiplying $\frac{2}{3}\alpha$ to the second equation, and multiplying $\frac{2}{3}\alpha^2$ to the third equation, then by adding these three equations together, stator equation in the complex plane is obtained:

$$
\vec{v}_s = R_s \vec{i}_s + p\vec{\psi}_s
\tag{4.63}
$$

Similarly, rotor voltage equations (4.50) are changed into the complex vector equation form as follows:

$$
\vec{v}_r^r = R_s \vec{i}_r^r + p\vec{\psi}_r^r
\tag{4.64}
$$

where superscript r indicates that the complex vector in equation (4.64) is referred to the complex $\alpha\beta^r$ plane sitting on the rotor, which is rotating with the rotor.

Similarly, by applying the same transformation to stator or rotor fluxes, complex vector equations for stator flux and rotor flux are obtained (See Appendix for the derivation).

$$
\vec{\psi}_s = L_s \vec{i}_s + L_m e^{j\theta_r}\vec{i}_r^r
\tag{4.65}
$$

$$
\vec{\psi}_r^r = L_m e^{-j\theta_r}\vec{i}_s + L_r \vec{i}_r^r
\tag{4.66}
$$

where $L_m = \frac{3}{2}L_{ms}$, $L_s = L_m + L_{ls}$, $L_r = L_m + L_{lr}$. It is observed that flux expressions in the complex plane are simpler than those in the ABC (abc) reference frame. Besides the term $e^{j\theta_r}$, all other coefficients in flux equations (4.65) and (4.66) are constant.

To refer the complex vectors on the rotor side to the same $\alpha\beta$ plane as the stator, the rotor equation (4.64) is multiplied by $e^{j\theta_r}$

$$
\vec{v}_r^r e^{j\theta_r} = R_s \vec{i}_r^r e^{j\theta_r} + e^{j\theta_r} p\vec{\psi}_r^r
\tag{4.67}
$$

According to the relationships between the variable in stator $\alpha\beta$ plane and the variables in rotor $\alpha\beta^r$ plane: $\vec{v}_r^r = \vec{v}_r e^{-j\theta_r}$, $\vec{i}_r^r = \vec{i}_r e^{-j\theta_r}$, and $\vec{\psi}_r^r = \vec{\psi}_r e^{-j\theta_r}$, equation (4.67) can be changed as follows:

$$
\vec{v}_r = R_s \vec{i}_r + e^{j\theta_r} p(\vec{\psi}_r e^{-j\theta_r})
\tag{4.68}
$$

Finally, the rotor equation in stator $\alpha\beta$ plane is obtained.

$$
\vec{v}_r = R_s \vec{i}_r + p\vec{\psi}_r - j\omega_r\vec{\psi}_r
\tag{4.69}
$$

By using the relationship $\vec{i}_r^r = \vec{i}_r e^{-j\theta_r}$, the stator flux equation (4.65) is changed to

$$\vec{\psi}_s = L_s \vec{i}_s + L_m \vec{i}_r \tag{4.70}$$

By multiplying $e^{j\theta_r}$ to the rotor flux equation (4.66), the following equation is obtained:

$$\vec{\psi}_r^r e^{j\theta_r} = L_m e^{-j\theta_r} \vec{i}_s e^{j\theta_r} + L_r \vec{i}_r^r e^{j\theta_r} \tag{4.71}$$

After the simplification of the above equation, rotor flux is expressed in the · stator $\alpha\beta$ reference frame:

$$\vec{\psi}_r = L_m \vec{i}_s + L_r \vec{i}_r \tag{4.72}$$

The dynamic model of the DFIG in stator $\alpha\beta$ reference frame is rewritten together as

$$\vec{v}_s = R_s \vec{i}_s + p\vec{\psi}_s \tag{4.73}$$

$$\vec{v}_r = R_s \vec{i}_r + p\vec{\psi}_r - j\omega_r \vec{\psi}_r \tag{4.74}$$

$$\vec{\psi}_s = L_s \vec{i}_s + L_m \vec{i}_r \tag{4.75}$$

$$\vec{\psi}_r = L_m \vec{i}_s + L_r \vec{i}_r \tag{4.76}$$

where $\vec{v}_s = v_{s\alpha} + jv_{s\beta}$, $\vec{v}_r = v_{r\alpha} + jv_{r\beta}$ are the stator voltage and rotor voltage, respectively, in the $\alpha\beta$ reference frame. $\vec{i}_s = i_{s\alpha} + ji_{s\beta}$, $\vec{i}_r = i_{r\alpha} + ji_{r\beta}$, and $\vec{\psi}_s = \psi_{s\alpha} + j\psi_{s\beta}$, $\vec{\psi}_r = \psi_{r\alpha} + j\psi_{r\beta}$ are the stator current, rotor current, stator flux and rotor flux, respectively, in the $\alpha\beta$ reference frame. Flux expressions are linear with all coefficients constant. Compared with the *ABC* (*abc*) model, the dynamic model in $\alpha\beta$ reference frames is simplified.

The equivalent circuit of the dynamic model in $\alpha\beta$ reference frame can be derived from (4.73)–(4.76), as shown in Figure 4.6.

According to the dynamic model in $\alpha\beta$ reference frame, the active and reactive powers of the stator side and the rotor side are calculated as follows:

$$P_s = \frac{3}{2}\text{Re}\left\{\vec{v}_s \vec{i}_s^*\right\} = \frac{3}{2}(v_{s\alpha}i_{s\alpha} + v_{s\beta}i_{s\beta})$$

$$Q_s = \frac{3}{2}\text{Im}\left\{\vec{v}_s \vec{i}_s^*\right\} = \frac{3}{2}(v_{s\beta}i_{s\alpha} - v_{s\alpha}i_{s\beta})$$

$$P_r = \frac{3}{2}\text{Re}\left\{\vec{v}_r \vec{i}_r^*\right\} = \frac{3}{2}(v_{r\alpha}i_{r\alpha} + v_{r\beta}i_{r\beta})$$

$$Q_r = \frac{3}{2}\text{Im}\left\{\vec{v}_r \vec{i}_r^*\right\} = \frac{3}{2}(v_{r\beta}i_{r\alpha} - v_{r\alpha}i_{r\beta}) \tag{4.77}$$

Electromagnetic torque of the DFIG is derived as follows:

$$T_{em} = \frac{3}{2}n_p Im(\vec{\psi}_r \vec{i}_r^*) = \frac{3}{2}n_p(\psi_{r\beta}i_{r\alpha} - \psi_{r\alpha}i_{r\beta}) \tag{4.78}$$

where n_p is the pair of poles of the electric machine.

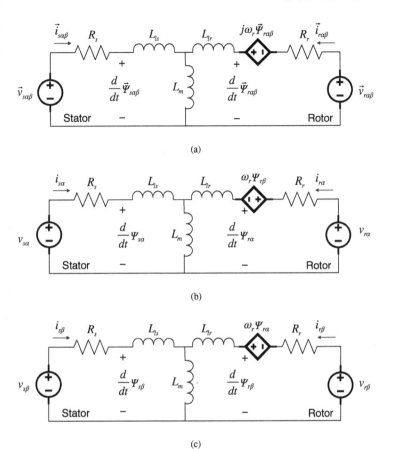

Figure 4.6   The equivalent dynamic circuit of the DFIG in $\alpha\beta$ reference frames: (a) in vector form; (b) in the α-axis; (c) in the β-axis.

With (4.75), (4.76), and (4.78), the electromagnetic torque T_{em} can be also represented as

$$T_{em} = \frac{3}{2}n_p \mathrm{Im}(\vec{\psi}_s^* \vec{i}_s) = \frac{3}{2}n_p \frac{L_m}{L_s} \mathrm{Im}(\vec{\psi}_s \vec{i}_r^*) \tag{4.79}$$

4.3.3   *dq* Model

The dynamic model of the DFIG is simplified by introducing $\alpha\beta$ reference frame as described by a group of equations from (4.73) to (4.76). If it is assumed that ω_r is constant, these equations are linear. However, electrical variables of DFIG such as the voltage, current, flux in $\alpha\beta$ reference frame are still AC components even at steady state. For the balanced three-phase system, if we observe its complex vectors

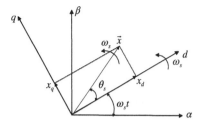

Figure 4.7   Relationship between the $\alpha\beta$ reference frame and the dq reference frame.

in a certain rotating reference frame, it is possible to further simplify them as constant vectors with their axis components having DC values. As the analysis or control design for the DC system is much simpler than that of an AC system, it is natural to transform the three-phase system into the rotating reference frame, which is known as dq rotating reference frame.

As shown in Figure 4.7, the complex vector $\vec{x}$ of the three-phase system is usually rotating in $\alpha\beta$ reference frame. Now we introduce another reference frame: dq reference frame, which is rotating with angular frequency ω_s. If the angular frequency is the same as the complex vector $\vec{x}$ rotating angular frequency, then the complex vector $\vec{x}$ becomes constant vector under the new dq reference frame. The relationship between $\alpha\beta$ reference frame and dq reference frame can be expressed as follows:

$$\vec{x}_{dq} = e^{-j\omega_s t}\vec{x} \qquad (4.80)$$

where $\vec{x}_{dq}$ is the complex vector in the rotating dq reference frame, and $\vec{x}_{dq} = x_d + jx_q$ where x_d is the d-axis component and x_q is the q-axis component.

Inverse transformation from dq reference frame to $\alpha\beta$ reference frame is

$$\vec{x} = e^{j\omega_s t}\vec{x}_{dq} \qquad (4.81)$$

Equation (4.80) can be written as

$$x_d + jx_q = (\cos \omega_s t - j \sin \omega_s t)(x_\alpha + jx_\beta) \qquad (4.82)$$

The real and virtual components of $\vec{x}_{dq}$ in the dq reference frame can be expressed as

$$x_d = x_\alpha \cos \omega_s t + x_\beta \sin \omega_s t \qquad (4.83)$$

$$x_q = x_\beta \cos \omega_s t - x_\alpha \sin \omega_s t \qquad (4.84)$$

The above two equations are often expressed in following matrix form which is known as rotating transformation from $\alpha\beta$ reference frame to rotating dq reference frame

$$\begin{bmatrix} x_d \\ x_q \end{bmatrix} = \begin{bmatrix} \cos \omega_s t & \sin \omega_s t \\ -\sin \omega_s t & \cos \omega_s t \end{bmatrix} \cdot \begin{bmatrix} x_\alpha \\ x_\beta \end{bmatrix} \qquad (4.85)$$

It is assumed that the angle between vector $\vec{x}$ and d-axis is θ_s as shown in Figure 4.7. Complex vector $\vec{x}$ in $\alpha\beta$ reference frame can be expressed as

$$\vec{x} = \|\vec{x}\|\, e^{j(\omega_s t + \theta_s)} \tag{4.86}$$

By substituting above equation (4.86) in equation (4.80), vector $\vec{x}_{dq}$ in the dq reference frame is derived as

$$\vec{x}_{dq} = e^{-j\omega_s t}\, \|\vec{x}\|\, e^{j(\omega_s t + \theta_s)} = \|\vec{x}\|\, e^{j\theta_s} \tag{4.87}$$

The real and virtual components of $\vec{x}_{dq}$ in the dq reference frame can be expressed as follows:

$$x_d = \|\vec{x}\| \cos\theta_s \tag{4.88}$$

$$x_q = \|\vec{x}\| \sin\theta_s \tag{4.89}$$

where $\|\vec{x}\|$ is the amplitude of $\vec{x}_{dq}$. Since dq reference frame rotates with the same angular frequency as that of the vector $\vec{x}$ in $\alpha\beta$ reference frame, $\vec{x}$ becomes a constant vector in the dq reference frame. Both its d-axis component x_d and q-axis component x_q are all DC values. Here θ_s is constant.

For the DFIG, if we introduce dq rotating transformation and substitute all variables in equations (4.73)–(4.76) by equations $\vec{v}_s = e^{j\omega_s t}\vec{v}_{sdq}$, $\vec{v}_r = e^{j\omega_s t}\vec{v}_{rdq}$, $\vec{i}_s = e^{j\omega_s t}\vec{i}_{sdq}$, $\vec{i}_r = e^{j\omega_s t}\vec{i}_{rdq}$, $\vec{\psi}_s = e^{j\omega_s t}\vec{\psi}_{sdq}$, and $\vec{\psi}_r = e^{j\omega_s t}\vec{\psi}_{rdq}$, we obtain

$$e^{j\omega_s t}\vec{v}_{sdq} = R_s e^{j\omega_s t}\vec{i}_{sdq} + p\left(e^{j\omega_s t}\vec{\psi}_{sdq}\right) \tag{4.90}$$

$$e^{j\omega_s t}\vec{v}_{rdq} = R_s e^{j\omega_s t}\vec{i}_{rdq} + p\left(e^{j\omega_s t}\vec{\psi}_{rdq}\right) - j\omega_r e^{j\omega_s t}\vec{\psi}_{rdq} \tag{4.91}$$

$$e^{j\omega_s t}\vec{\psi}_{sdq} = L_s e^{j\omega_s t}\vec{i}_{sdq} + L_m e^{j\omega_s t}\vec{i}_{rdq} \tag{4.92}$$

$$e^{j\omega_s t}\vec{\psi}_{rdq} = L_m e^{j\omega_s t}\vec{i}_{sdq} + L_r e^{j\omega_s t}\vec{i}_{rdq} \tag{4.93}$$

Through simplification, dq model of the DFIG is derived as

$$\vec{v}_{sdq} = R_s\vec{i}_{sdq} + p\vec{\psi}_{sdq} + j\omega_s\vec{\psi}_{sdq} \tag{4.94}$$

$$\vec{v}_{rdq} = R_r\vec{i}_{rdq} + p\vec{\psi}_{rdq} + j\left(\omega_s - \omega_r\right)\vec{\psi}_{rdq} \tag{4.95}$$

$$\vec{\psi}_{sdq} = L_s\vec{i}_{sdq} + L_m\vec{i}_{rdq} \tag{4.96}$$

$$\vec{\psi}_{rdq} = L_m\vec{i}_{sdq} + L_r\vec{i}_{rdq} \tag{4.97}$$

where $\vec{v}_{sdq} = v_{sd} + jv_{sq}$, $\vec{v}_{rdq} = v_{rd} + jv_{rq}$ are the stator voltage and the rotor voltage, respectively, in the dq reference frames. $\vec{i}_{sdq} = i_{sd} + ji_{sq}$, $\vec{i}_{rdq} = i_{rd} + ji_{rq}$, and $\vec{\psi}_{sdq} = \psi_{sd} + j\psi_{sq}$, $\vec{\psi}_{rdq} = \psi_{rd} + j\psi_{rq}$ are the stator current, rotor current, and flux, respectively, in the dq reference frame.

The equivalent dynamic circuit in the dq reference frame is derived from (4.94) to (4.97), as shown in Figure 4.8.

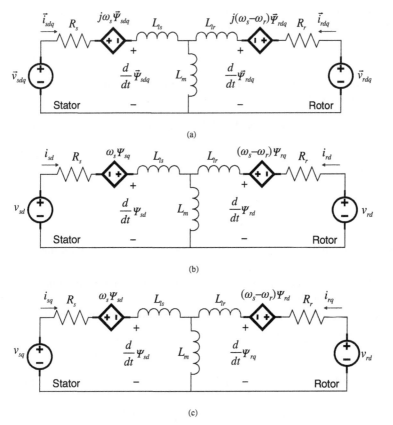

Figure 4.8   Equivalent dynamic circuit of a DFIG in dq reference frame: (a) in vector form; (b) in d-axis; (c) in q-axis.

The active power and reactive power on the stator side or on the rotor side can be calculated in the dq reference frame as follows:

$$P_s = \frac{3}{2}\text{Re}\left\{\vec{v}_{sdq}\vec{i}^*_{sdq}\right\} = \frac{3}{2}(v_{sd}i_{sd} + v_{sq}i_{sq})$$

$$Q_s = \frac{3}{2}\text{Im}\left\{\vec{v}_{sdq}\vec{i}^*_{sdq}\right\} = \frac{3}{2}(v_{sq}i_{sd} - v_{sd}i_{sq})$$

$$P_r = \frac{3}{2}\text{Re}\left\{\vec{v}_{rdq}\vec{i}^*_{rdq}\right\} = \frac{3}{2}(v_{rd}i_{rd} + v_{rq}i_{rq})$$

$$Q_r = \frac{3}{2}\text{Im}\left\{\vec{v}_{rdq}\vec{i}^*_{rdq}\right\} = \frac{3}{2}(v_{rq}i_{rd} - v_{rd}i_{rq}) \tag{4.98}$$

For the DFIG, the d-axis in the dq reference frame is normally selected to be with the same direction as the grid voltage vector $\vec{v}_s$, and the dq reference frame rotation speed is the same as the synchronous speed ω_s. In this case, the d-axis component of the grid voltage v_{sd} is the grid voltage amplitude V_s, and the q-axis component of the grid voltage v_{sq} is always zero for the ideal grid condition. Then, the stator output active and reactive power equations can be further simplified as follows:

$$P_s = \frac{3}{2} V_s i_{sd}$$

$$Q_s = -\frac{3}{2} V_s i_{sq} \tag{4.99}$$

It is observed that in this case, the stator active power will be only related to the stator d-axis current i_{sd} while the stator reactive power is only related to the stator q-axis current i_{sq}. It realizes control decoupling between the active power and reactive power. This is an important feature for the control design of the DFIG. Similarly to $\alpha\beta$ model, the electromagnetic torque in the dq reference frame can be expressed by the following equation:

$$T_{em} = \frac{3}{2} n_p \mathrm{Im}(\vec{\psi}_{rdq} \vec{i}_{rdq}^*) = \frac{3}{2} n_p (\psi_{rq} i_{rd} - \psi_{rd} i_{rq}) \tag{4.100}$$

The electromagnetic torque in the dq reference frame can be also expressed as

$$T_{em} = \frac{3}{2} n_p \mathrm{Im}(\vec{\psi}_{sdq}^* \vec{i}_{sdq}) = \frac{3}{2} n_p \frac{L_m}{L_s} \mathrm{Im}(\vec{\psi}_{sdq} \vec{i}_{rdq}^*) \tag{4.101}$$

4.4 MODELING OF THE CONVERTER

4.4.1 Steady-State Equivalent Circuit of the Converter

The typical topologies of three-phase converters with L and LCL filters are shown in Figures 4.9a and 4.9b, respectively.

The steady-state equivalent circuits of the converters are shown in Figures 4.10a and 4.10b, where the voltages of the switching bridges are shown as line-to-line voltages v_{ab}, v_{bc}, and v_{ca}.

4.4.2 *abc* Model with L Filter

There are two AC–DC converters connected in the DC bus with back-to-back way in the DFIG wind power system as shown in Figure 4.11. One of the converters, called grid-side converter (GSC), is connected to the grid with its AC side. Another converter, called rotor-side converter (RSC), is connected to the rotor windings of the DFIG machine with its AC side. RSC controls the DFIG by adjusting the rotor voltage and the frequency. GSC is used to realize power flow control to the grid with its frequency synchronized with the grid [12–17]. The diagram of the three-phase converter is shown in Figure 4.12. The converter is composed of three parts: A switch bridge, L filters, and three-phase sources. The switch bridge is composed of phase-a, phase-b, and phase-c switch legs. It is used to convert DC-side voltage into PWM waveform with the amplitude and frequency of its fundamental frequency output adjustable.

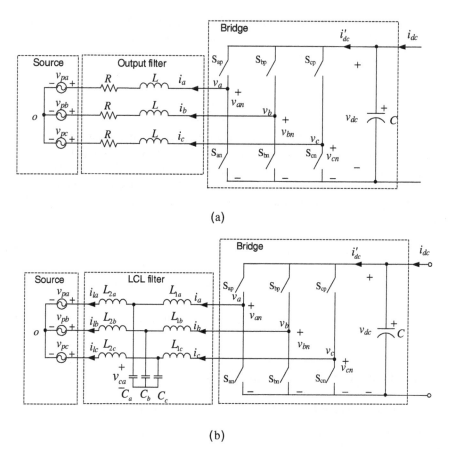

(a)

(b)

Figure 4.9   Typical topologies of three-phase converters: (a) with L filters; (b) with LCL filters.

Switches S_{ap}, S_{bp}, S_{cp} are connected between the positive pole of the DC bus and phase a, b, and c filters. Switches S_{an}, S_{bn}, S_{cn} are connected between the negative pole of the DC bus and phase a, b, and c filters. v_{dc} is the DC bus voltage, i'_{dc} is the DC bus current.

Three-phase reference voltages v_a^{ref}, v_b^{ref}, and v_c^{ref}, which are modulation signals of the PWM modulator, are generated by the front controller. They enter into the PWM modulator to produce PWM pulses to turn on or off the switches in the converter so that the expected three-phase voltages v_a, v_b, and v_c are obtained. The output filter is used to filter high frequency switching harmonics and select the fundamental-frequency component from outputting PWM voltage v_a, v_b, and v_c. v_{pa}, v_{pb}, and v_{pc} are the phase voltages of the three-phase source. For the RSC, the source voltages are expressed as v_{ra}, v_{rb}, and v_{rc}. For GSC which is connected to the grid and stator of the machine, the source voltages are expressed as v_{sa}, v_{sb}, and v_{sc}.

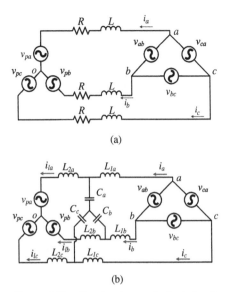

(a)

(b)

Figure 4.10   Steady-state equivalent circuits of the converters: (a) with L filters; (b) with LCL filters.

The switch bridge ac-side outputs depend on the DC bus voltage v_{dc} and the switching states of the switch bridge [18–24]. Three switching functions s_a, s_b, s_c are introduced to describe each phase switch leg state of the switch bridge. Here we take phase-a switch leg as an example. If $s_a = 1$, it indicates the upper switch S_{ap} is on and the lower switch S_{an} is off, and the voltage on the switch S_{an} is $v_{an} = v_{dc}$. If $s_a = 0$, it indicates upper switch S_{ap} is off and lower switch S_{an} is on, and the voltage across the phase-a switch S_{an} is $v_{an} = 0$. Actually, v_{an}, the voltage across phase-a switch S_{an}, is phase-a output voltage of the switch bridge to the output filter. It can be expressed with the switching function

$$v_{an} = v_a - v_n = v_{dc} s_a \tag{4.102}$$

where v_n is the voltage level of the negative pole of the DC bus.

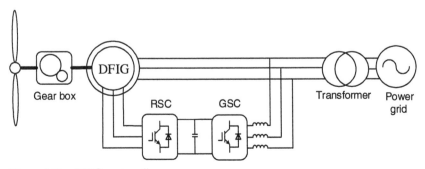

Figure 4.11   DFIG system diagram.

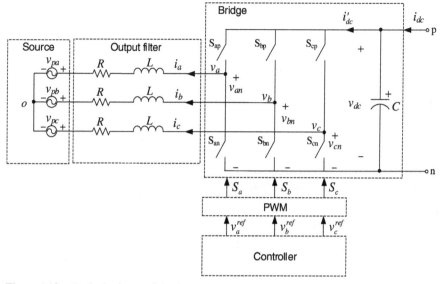

Figure 4.12   Typical scheme of the three-phase converter with L filters.

Since switching functions s_a, s_b, s_c are discontinuous in the time domain, switching period averaging is introduced to simplify the analysis. By using switching period averaging to equation (4.102), phase-a output voltage of the switch bridge is changed to

$$v_{an} \approx v_{dc} d_a \qquad (4.103)$$

where d_a is the switching period average value of s_a and is also called the duty cycle of the phase-a switch leg.

Similarly, we can derive phase-b and phase-c outputs of the switch bridge:

$$v_{bn} \approx v_{dc} d_b \qquad (4.104)$$

$$v_{cn} \approx v_{dc} d_c \qquad (4.105)$$

The AC-side model of the converter can be easily obtained from Figure 4.12 as follows:

$$v_{pa} = -L\frac{di_a}{dt} - Ri_a + v_{an} + v_{no}$$

$$v_{pb} = -L\frac{di_b}{dt} - Ri_b + v_{bn} + v_{no}$$

$$v_{pc} = -L\frac{di_c}{dt} - Ri_c + v_{cn} + v_{no} \qquad (4.106)$$

where v_{no} is the voltage between negative pole "n" of the DC bus and the central point "o" of the three-phase voltage sources. i_a, i_b, i_c are the currents through three filters $L_a, L_b,$ and $L_c,$ respectively. To simplify the analysis, it is assumed that AC-side parameters are symmetric. The filter inductances, as well as the AC resistance of the three phases, are equal, that is, $L_a = L_b = L_c = L$, and $R_a = R_b = R_c = R$.

For the DC side of the converter, the following equation is formulated:

$$C\frac{dv_{dc}}{dt} = i_{dc} - i'_{dc} \tag{4.107}$$

where i'_{dc} is DC-side current of the switch bridge. i_{dc} is external current flowing into the DC bus of the converter.

4.4.3 *dq* Model with L Filter

By using *ABC* to $\alpha\beta$ transformation (4.57), equation (4.106) can be transformed into the following equation in $\alpha\beta$ reference frame:

$$\vec{v}_p = -Lp\vec{i} - R\vec{i} + \vec{v} \tag{4.108}$$

where $\vec{v}_p = v_{p\alpha} + jv_{p\beta}$ is the source voltage in $\alpha\beta$ reference frame, $\vec{v} = v_\alpha + jv_\beta$ is the voltage output of the switch bridge in the $\alpha\beta$ reference frame, and $\vec{i} = i_\alpha + ji_\beta$ is the filter current in the $\alpha\beta$ reference frame.

If we further apply the *dq* rotating transformation by substituting variables in equation (4.108) with $\vec{v}_p = e^{j\omega_s t}\vec{v}_{pdq}$, $\vec{v} = e^{j\omega_s t}\vec{v}_{dq}$, and $\vec{i} = e^{j\omega_s t}\vec{i}_{dq}$, we obtain

$$e^{j\omega_s t}\vec{v}_{pdq} = -Lp\left(e^{j\omega_s t}\vec{i}_{dq}\right) - Re^{j\omega_s t}\vec{i}_{dq} + e^{j\omega_s t}\vec{v}_{dq} \tag{4.109}$$

where $\vec{v}_{pdq} = v_{pd} + jv_{pq}$ is the source voltage in *dq* rotating reference frame, $\vec{v}_{dq} = v_d + jv_q$ is the voltage output of the switch bridge in *dq* rotating reference frame, and $\vec{i}_{dq} = i_d + ji_q$ is the filter current in *dq* rotating reference frame. ω_s is the angular speed of *dq* rotating reference frame.

After simplification, the above equation is rewritten as follows

$$\vec{v}_{pdq} = -Lp\vec{i}_{dq} - R\vec{i}_{dq} - j\omega_s L\vec{i}_{dq} + \vec{v}_{dq} \tag{4.110}$$

If we explode above complex-value equation into real-value equations, two real-value equations are obtained, which relate to *d*-axis and *q*-axis, respectively.

$$v_{pd} = -L\frac{di_d}{dt} - Ri_d + \omega_s Li_q + v_d$$

$$v_{pq} = -L\frac{di_q}{dt} - Ri_q - \omega_s Li_d + v_q \tag{4.111}$$

where v_{pd} and v_{pq} are the source voltages in *d*-axis and *q*-axis, respectively, v_d and v_q are the switch bridge output voltages in *d*-axis and *q*-axis, respectively, and i_d and i_q are the filter currents in *d*-axis and *q*-axis, respectively. With equation (4.111), the equivalent circuits for the AC side of the converter in the *dq* reference frame are derived as shown in Figure 4.13.

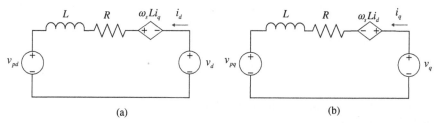

(a)   (b)

Figure 4.13   Equivalent circuit for the AC side of the converter in the dq reference frame: (a) on d-axis; (b) on q-axis.

Similarly, three-phase output voltages of the switch bridge shown in equations (4.103), (4.104), and (4.105) can be transformed to the dq reference frame:

$$\vec{v}_{dq} = v_{dc}\vec{d}_{dq} \tag{4.112}$$

where $\vec{d}_{dq}$ is the complex duty cycle in dq rotating reference frame which is transformed from the duty cycle vector $\begin{bmatrix} d_a & d_b & d_c \end{bmatrix}^{\mathrm{T}}$ in ABC reference frame

The complex output of the switch bridge $\vec{v}_{dq}$ can be expressed with d-axis component and q-axis component, respectively.

$$v_d = v_{dc}d_d$$
$$v_q = v_{dc}d_q \tag{4.113}$$

d_d and d_q are the d-axis duty cycle and the q-axis duty cycle, respectively.

The active power and reactive power transferred to the three-phase sources can be expressed by

$$P = \frac{3}{2}\mathrm{Re}\left\{\vec{v}_{pdq}\vec{i}_{dq}^{*}\right\} = \frac{3}{2}(v_{pd}i_d + v_{pq}i_q)$$

$$Q = \frac{3}{2}\mathrm{Im}\left\{\vec{v}_{pdq}\vec{i}_{dq}^{*}\right\} = \frac{3}{2}(v_{pq}i_d - v_{pd}i_q) \tag{4.114}$$

If the converter is connected to the grid and the dq reference frame is synchronized to the grid frequency with its d-axis aligned with the grid voltage complex vector, then the source voltage q-axis component is always zero while the source voltage d-axis component is constant, that is,

$$v_{pd} = V_s$$
$$v_{pq} = 0 \tag{4.115}$$

By substituting (4.115) in (4.114), the power equations mentioned above are simplified as

$$P = \frac{3}{2}V_s i_d$$

$$Q = -\frac{3}{2}V_s i_q \tag{4.116}$$

It is found that the active power and reactive power can be controlled separately. The active power is controlled by i_d while the reactive power is controlled by i_q. In other words, the active power only depends on the d-axis filter current while the reactive power only depends on the q-axis filter current.

The output active power of the switch bridge can be expressed as

$$P = \frac{3}{2}\text{Re}\left\{\vec{v}_{dq}\vec{i}_{dq}^{*}\right\} = \frac{3}{2}(v_d i_d + v_q i_q) \tag{4.117}$$

If we ignore the power loss in the converter power devices, the following power conservation equation between the AC side and DC side of the switch bridge is obtained:

$$\frac{3}{2}(v_d i_d + v_q i_q) = v_{dc} i'_{dc} \tag{4.118}$$

Then, the DC-side current can be derived:

$$i'_{dc} = \frac{3}{2}\frac{1}{v_{dc}}(v_d i_d + v_q i_q) \tag{4.119}$$

By substituting equation (4.119) in (4.107), the DC-side equation can be expressed as

$$C\frac{dv_{dc}}{dt} = i_{dc} - \frac{3}{2}\frac{1}{v_{dc}}(v_d i_d + v_q i_q) \tag{4.120}$$

By using equation (4.117), the above equation is changed into the following form:

$$C\frac{dv_{dc}}{dt} = i_{dc} - \frac{P}{v_{dc}} \tag{4.121}$$

It is seen that DC bus voltage can be controlled by the active power. Besides, DC bus voltage is affected by the external current i_{dc} flowing into the DC bus of the converter.

4.4.4   *dq* Model with LCL Filter

The AC/DC converter with LCL filter is shown in Figure 4.14. LCL filter in each phase is the third-order filter, which is composed of inductor L_1, inductor L_2, and capacitor C. For simplifying the analysis, it is assumed that the three AC-side LCL filters are identical, that is, $L_{1a} = L_{1b} = L_{1c} = L_1$, $L_{2a} = L_{2b} = L_{2c} = L_2$, and $C_a = C_b = C_c = C$.

Since it has a smaller size than L filter or LC filter, some GSCs in the DFIG wind power system use LCL filter.

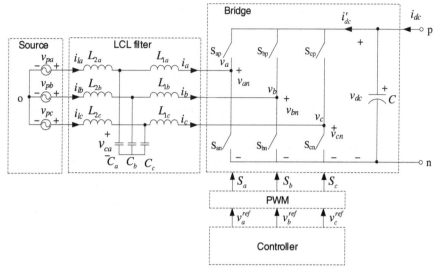

Figure 4.14   Typical scheme of a three-phase converter with LCL filter.

If the ac-side parasitic resistance is neglected, complex vector equations describing the AC side in the $\alpha\beta$ reference frame are derived:

$$L_2 p\vec{i}_l = \vec{v}_p - \vec{v}_c$$

$$Cp\vec{v}_c = \vec{i}_l - \vec{i}$$

$$L_1 p\vec{i} = \vec{v}_c - \vec{v} \tag{4.122}$$

where $\vec{v}$ is the output switch bridge voltage vector, $\vec{v}_p$ is the source voltage vector, $\vec{i}_l$ and $\vec{i}$ are the inductor L_2 current and the inductor L_1 current vector, respectively. $\vec{v}_c$ is the voltage vector on the filter capacitor.

By transforming (4.122) into the dq reference frame, the following equations are obtained:

$$L_2 p\vec{i}_{ldq} = \vec{v}_{pdq} - \vec{v}_{cdq} - j\omega_s L_2 \vec{i}_{ldq}$$

$$Cp\vec{v}_{cdq} = \vec{i}_{ldq} - \vec{i}_{dq} - j\omega_s C\vec{v}_{cdq}$$

$$L_1 p\vec{i}_{dq} = \vec{v}_{cdq} - \vec{v}_{dq} - j\omega_s L_1 \vec{i}_{dq} \tag{4.123}$$

where $\vec{v}_{dq} = v_d + jv_q$ is the output voltage vector of the switch bridge in dq reference frame, $\vec{v}_{pdq}$ is the source voltage vector, $\vec{i}_{ldq}$ and $\vec{i}_{dq}$ are the inductor L_2 and the inductor L_1 current vectors in the dq reference frame. $\vec{v}_{cdq}$ is the voltage vector in the filter capacitor in the dq reference frame. The equivalent circuit of the converter with LCL filter can then be derived from (4.123), as shown in Figure 4.15.

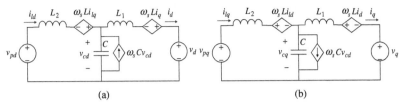

Figure 4.15   Equivalent circuit of the converter with LCL filter: (a) on d-axis; (b) on q-axis.

4.4.5   Model of the PWM Modulator

Figure 4.16 shows the PWM wave generation by comparing phase-a modulation wave v_{am} with the triangle carrier waveform. It is easy to derive the duty cycle of the phase switch leg according to the geometry relationship in the figure.

$$d_a = \frac{1}{V_{tri}} v_{am} \qquad (4.124)$$

where V_{tri} is the peak-to-peak value of the carrier wave. The range of the modulation signal v_{am} is $[0, V_{tri}]$. The range of the duty cycle d_a is $[0, 1]$.

If $K_{pwm} = 1/V_{tri}$ is defined to be the PWM coefficient, the phase-a duty cycle of the switch bridge is expressed by

$$d_a = K_{pwm} v_{am} \qquad (4.125)$$

Similarly, we can derive the expressions for phases b and c. The duty cycles for the three-phase switch bridge with PWM control are summarized as follows:

$$d_a = K_{pwm} v_{am}$$
$$d_b = K_{pwm} v_{bm}$$
$$d_c = K_{pwm} v_{cm} \qquad (4.126)$$

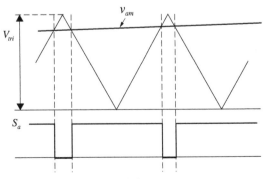

Figure 4.16   PWM modulation.

The three-phase duty cycles of the switch bridge can be transformed to dq rotating reference frame:

$$\vec{d}_{dq} = K_{pwm}\vec{v}_{mdq} \tag{4.127}$$

where $\vec{d}_{dq}$ is complex duty cycle in the dq rotating reference frame, and $\vec{v}_{mdq}$ is complex modulation wave in the dq rotating reference frame.

Equation (4.127) can be expressed using d-axis component and q-axis component:

$$d_d = K_{pwm}v_{md}$$
$$d_q = K_{pwm}v_{mq} \tag{4.128}$$

where d_d is d-axis duty cycle, and d_q is the q-axis duty cycle. v_{md} is the d-axis component of the modulation signal, and v_{mq} is the q-axis component of the modulation signal.

4.4.6   Per-Unit System

The per-unit system is the expression of system quantities as a fraction of a defined base unit quantity in power system analysis. The main idea of the per-unit system is to absorb the large difference in absolute values into a base relationship. As a result, representation of elements in the system with per-unit values become more uniform. The electrical variables expressed in the per-unit system automatically provide information on how far it is from the base or rated values. It eases the comparison between different systems. The per-unit system is widely used for analysis of the DFIG wind power system.

In the per-unit system, any variable of a system is normalized by dividing it by a base value. For the DFIG, rated phase voltage V_N and power capacity S_N provided by the manufacturers are used as the base voltage V_{base} and base power S_{base}.

$$V_{base} = V_N$$
$$S_{base} = S_N \tag{4.129}$$

Then base current I_{base} is defined by the following equation

$$I_{base} = \frac{S_{base}}{3V_N} \tag{4.130}$$

Base angular frequency ω_{base} is defined as the synchronous angular frequency ω_s

$$\omega_{base} = \omega_s \tag{4.131}$$

With equations (4.132)–(4.135), the base flux linkage ψ_{base}, base impedance, base inductance, and base capacitance can be obtained as follows:

$$\psi_{base} = \frac{V_{base}}{\omega_{base}} \tag{4.132}$$

$$Z_{base} = \frac{V_{base}}{I_{base}} \tag{4.133}$$

$$L_{base} = \frac{Z_{bace}}{\omega_{base}} \tag{4.134}$$

$$C_{base} = \frac{1}{\omega_{base}Z_{bace}} \tag{4.135}$$

To convert an electrical variable or parameter into the per-unit value, we can do it by dividing the electrical variable or parameter by its base value. For example, we can convert the stator voltage of DFIG V_s into its per-unit value V_{s_pu} by the following equation:

$$V_{s\_pu} = \frac{V_s}{V_{bace}} \tag{4.136}$$

If all the variables and parameters in the dq dynamic model of DFIG are expressed in per unit, the corresponding per-unit dq dynamic model is obtained:

$$\vec{v}_{sdq\_pu} = R_{s\_pu}\vec{i}_{sdq\_pu} + p\vec{\psi}_{sdq\_pu} + j\omega_{s\_pu}\vec{\psi}_{sdq\_pu} \tag{4.137}$$

$$\vec{v}_{rdq\_pu} = R_{r\_pu}\vec{i}_{rdq\_pu} + p\vec{\psi}_{rdq\_pu} + j\omega_{sl\_pu}\vec{\psi}_{rdq\_pu} \tag{4.138}$$

$$\vec{\psi}_{sdq\_pu} = L_{s\_pu}\vec{i}_{sdq\_pu} + L_{m\_pu}\vec{i}_{rdq\_pu} \tag{4.139}$$

$$\vec{\psi}_{rdq\_pu} = L_{m\_pu}\vec{i}_{sdq\_pu} + L_{r\_pu}\vec{i}_{rdq\_pu} \tag{4.140}$$

where $\omega_{sl\_pu} = \frac{\omega_s - \omega_r}{\omega_{base}}$, which is the slip angular frequency in the per-unit system.

4.5 SUMMARY

In this chapter, according to the steady-state model, the power relationship between the rotor and stator is derived. The rotor power P_r is a fraction of the stator power P_s. The rotor power P_r is less than 1/3 of the stator power P_s in the wind power system where the slip S_l ranges from -0.3 to 0.3. It means that the stator power can be controlled by the rotor side supplied by a power converter with 30% power rating. In the super-synchronous mode, both stator and rotor deliver power to the grid, the total power to grid $P = (1 + |S_l|)P_s$. In synchronous mode, DC current flows into the rotor winding. Only the stator delivers power to the grid. The total power to grid $P = P_s$. In the sub-synchronous mode, the stator delivers power to the grid while the rotor receives power from the grid. The total power to grid $P = (1 - |S_l|)P_s$.

The dynamic model of the DFIG in different reference frames such as *ABC* (*abc*) reference frame, $\alpha\beta$ reference frame, and dq reference frame are introduced. In *ABC* (*abc*) frame, flux expressions are time variant while flux expressions become linear in $\alpha\beta$, or dq reference frames. Besides, the three-phase system is simplified into the two-phase system in either $\alpha\beta$, or dq reference frames. The balanced three-phase system analysis can be further simplified as the DC system by using dq reference frame. The dq model is very convenient to control design.

In addition, this chapter introduces dynamic models of the converter with L filter or with LCL. Finally, the per-unit system concept is explained.

REFERENCES

[1] W. Leonhard, *Control of Electrical Drives*. Springer, 1985.

[2] M. P. Kazmierkowski, R. Krishnan, and F. Blaabjerg, *Control in Power Electronics: Selected Problems*. Academic Press, 2002.

[3] A. Veltman, D. W. J. Pulle, and R. W. DeDoncker, *Fundamentals of Electric Drives*. Springer, 2007.

[4] B. K. Bose, *Power Electronics and Drives*. Elsevier, 2006.

[5] S. J. Chapman, *Electric Machinery Fundamentals*. McGraw-Hill, 2005.

[6] B. K. Bose, *Modern Power Electronics and AC Drives*. Princeton Hall, 2002.

[7] G. Abad, J. Lopez, M. A. Rodriguez. L. Marroyo, and G. Iwanski, *Doubly Fed Induction Machine: Modeling and Control for Wind Energy Generation*. Wiley-IEEE Press, 2011.

[8] P. Kunder, *Power System Stability and Control*. McGraw-Hill, 1994.

[9] B. Wu, Y. Lang, N. Zargari, and S. Kouro. *Power Conversion and Control of Wind Energy Systems*. Wiley-IEEE Press, 2011.

[10] R. Teodorescu, M. Liserre, and P. Rodriguez, *Grid Converters for Photovoltaic and Wind Power Systems*. John Wiley & Sons, 2011.

[11] Y. He, J. Hu, and L. Xu, *Operation Control of Grid Connected Doubly Fed Induction Generator*. Electric Power Press, 2012.

[12] J. Xu, "Research on converter control strategy of doubly fed induction generator system for wind power," Ph.D. thesis, Zhejiang University, Hangzhou, China, 2011.

[13] C. Liu, "Resonant control of DFIG wind power converters for adapting to the grid environment," Ph.D. thesis, Zhejiang University, Hangzhou, China, 2012.

[14] R. Pena, J. C. Clare, and G. M. Asher, "Doubly fed induction generator using back-to-back PWM converters and its application to variable-speed wind-energy generation," *Proc. IEE Proc. Electr. Power Appl.*, vol. 143, no. 3, pp. 231–241, May 1996.

[15] S. Muller, M. Deicke, and R. W. De Doncker, "Doubly fed induction generator systems for wind turbines," *IEEE Ind. Appl. Mag.*, vol. 8, no. 3, pp. 26–33, 2002.

[16] J. G. Slootweg, S. W. H. de Haan, H. Polinder, and W. L. Kling, "General model for representing variable speed wind turbines in power system dynamics simulations," *IEEE Trans. Power Syst.*, vol. 18, no. 1, pp. 144–151, 2003.

[17] A. Tapia, G. Tapia, J. X. Ostolaza, and J. R. Saenz, "Modeling and control of a wind turbine driven doubly fed induction generator," *IEEE Trans. Energy Convers.*, vol. 18, no. 2, pp. 194–204, 2003.

[18] J. B. Ekanayake, L. Holdsworth, XueGuang Wu, and N. Jenkins, "Dynamic modeling of doubly fed induction generator wind turbines," *IEEE Trans Power Syst.*, vol. 18, no. 2, pp. 803–809, 2003.

[19] Y. Lei, A. Mullane, G. Lightbody, and R. Yacamini, "Modeling of the wind turbine with a doubly fed induction generator for grid integration studies," *IEEE Trans. Energy Convers.*, vol. 21, no. 1, pp. 257–264, 2006.

[20] D. Xiang, L. Ran, P. J. Tavner, and S. Yang, "Control of a doubly fed induction generator in a wind turbine during grid fault ride-through," *IEEE Trans. Energy Convers.*, vol. 21, no. 3, pp. 652–662, 2006.

[21] E. S. Abdin and W. Xu, "Control design and dynamic performance analysis of a wind turbine-induction generator unit," *IEEE Trans. Energy Convers.*, vol. 15, no. 1, pp. 91–96, 2000.

[22] Z. Chen and E. Spooner, "Grid power quality with variable speed wind turbines," *IEEE Trans. Energy Convers.*, vol. 16, no. 2, pp. 148–154, 2001.

[23] J. G. Slootweg, H. Polinder, and W. L. Kling, "Representing wind turbine electrical generating systems in fundamental frequency simulations," *IEEE Trans. Energy Convers.*, vol. 18, no. 4, pp. 516–524, 2003.

[24] A. Yazdani and R. Iravani, "A unified dynamic model and control for the voltage-sourced converter under unbalanced grid conditions," *IEEE Trans Power Del*, vol. 21, no. 3, pp. 1620–1629, 2006.

[25] G. Abad and G. Iwanski, *"Properties and control of a doubly fed induction machine,"* in *Power Electronics for Renewable Energy Systems, Transportation and Industrial Application*. John Wiley & Sons, 2014.

CHAPTER 5

CONTROL OF DFIG POWER CONVERTERS

Based on the dynamic model established in Chapter 4, control of the DFIG power converter will be introduced in this chapter. At first, the control of grid-side converter (GSC) will be explained. It keeps a constant DC-bus voltage for the rotor-side converter (RSC) and can also provide reactive power support to the grid. Then the controls of the RSC in starting mode, power-control mode, and speed-control mode are introduced. The switching mode for the RSC is discussed at the end of the chapter. This chapter is expected to help the reader understand the fundamentals of the control for the DFIG under ideal grid condition.

5.1 INTRODUCTION

It has been introduced in Chapter 4 that by using dq reference frame, the AC electrical variables in the three-phase balanced system can be transformed into DC components. With regards to DC linear systems, it is easy to realize better control design by using mature classic control theory. The dq model of the DFIG machine and power converters have been established in Chapter 4. Based on these dynamic models, control can be designed for both the grid-side converter (GSC) and the rotor-side converter (RSC). In this chapter, the conventional vector control for the GSC is introduced at first, the GSC is controlled to provide a DC-bus voltage for the RSC and to provide reactive power support to the grid if necessary. The vector control schemes for the RSC in starting mode, power (or torque) control mode and speed-control mode are then introduced in detail, respectively. The RSC controls DFIG to achieve different control targets in different operation modes. The mode switching for the RSC is evaluated at last. Corresponding test waveforms are also included in this chapter [1–9].

5.2 START-UP PROCESS OF THE DFIG SYSTEM

The GSC provides constant DC-bus voltage for the RSC. The RSC controls the DFIG with three control modes, starting mode, power-control mode, and speed-control mode [10–13].

Advanced Control of Doubly Fed Induction Generator for Wind Power Systems, First Edition.
Dehong Xu, Frede Blaabjerg, Wenjie Chen, and Nan Zhu.
© 2018 by The Institute of Electrical and Electronics Engineers, Inc. Published 2018 by John Wiley & Sons, Inc.

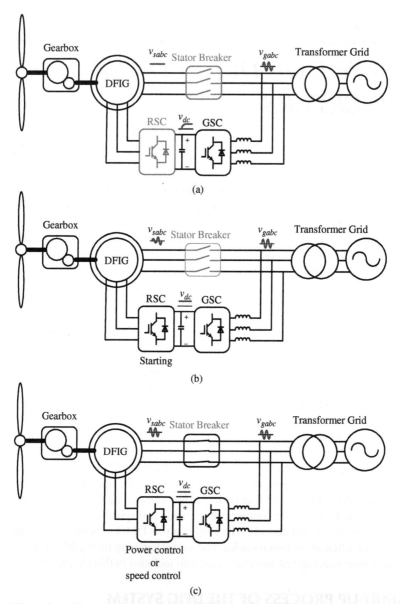

Figure 5.1 The start-up process of the DFIG system: (a) GSC builds up DC-bus voltage; (b) RSC operates in starting mode; (c) stator breaker closed, RSC operates in power-control or speed-control mode.

The start-up process of the DFIG system goes in the following steps:

Step 1: The GSC starts first and builds up the DC-bus voltage V_{dc}, as shown in Figure 5.1a.

Step 2: After the DC-bus voltage reaches the nominal value, the RSC starts to operate in starting mode. The RSC controls the DFIG to build up the stator

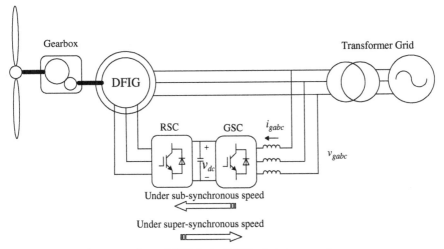

Figure 5.2   Active power flow of GSC in the DFIG wind power system.

voltage v_{sabc} to synchronize the amplitude, frequency, and phase with those of the grid voltage v_{gabc}, as shown in Figure 5.1b.

Step 3: After the stator voltage has the same amplitude, frequency, and phase as the grid, the stator breaker is closed to connect the stator to the grid.

Step 4: The stator breaker normally needs dozens of milliseconds to close and during this period, the DFIG is still running in starting mode. After the stator breaker is finally closed, the controller will receive a confirm signal from the breaker.

Step 5: As soon as the controller receives the confirm signal of the stator breaker, the control scheme is changed from starting mode to power-control or speed-control mode, as shown in Figure 5.1c.

5.3   GRID-SIDE CONVERTER

5.3.1   Control Target

As introduced in Chapter 2, the GSC provides constant DC-bus voltage for the RSC, and it can also provide reactive support to the grid if required. The diagram of the DFIG wind power system is shown again in Figure 5.2. The GSC regulates the DC-bus voltage, and this DC bus is shared with the RSC. The AC side of the GSC is connected to the power grid. GSC may receive or deliver active power according to the DFIG operating mode. Therefore, GSC is responsible for both the controlling of the power flow between the GSC and the grid and the adjusting of DC-bus voltage for the RSC [14–17].

The power flow of GSC is shown in Figure 5.2. Under sub-synchronous mode: $\omega_r < \omega_s$, the rotor side of the DFIG receives active power from the grid through

Figure 5.3   Schematic diagram of the PLL.

the BTB converter. In this case, the GSC receives power from the grid and operates as a rectifier while RSC receives power from GSC and inverts it into AC for the rotor winding. Under super-synchronous mode: $\omega_r > \omega_s$, the rotor of the DFIG delivers active power to the grid through the BTB converter. In this case, RSC receives power from the rotor while the GSC delivers power to the grid and operates as an inverter.

According to the grid code, the DFIG wind power system is sometimes required to provide reactive power support to the grid. The reactive power can be generated from the stator side of the DFIG through control of the RSC. Besides, it can also be provided by the GSC working as static var generator when the grid voltage dips and the RSC is not able to control the DFIG. However, because the capacity of the GSC is normally only 30% of the DFIG's capacity, its reactive power output is limited.

5.3.2   Grid Synchronization

For any grid-connected converter, the grid synchronization is essential as the phase difference between the grid-connected converter and the grid voltage will normally introduce undesirable surge current. For the GSC of the DFIG, if the vector control in dq reference frame is used, the grid voltage vector angle θ_s and the grid synchronous angular frequency ω_s are needed for the Park transformation. As the grid frequency and phase may vary, the GSC is required to track the grid frequency and phase dynamically. Therefore, grid synchronization is the premise of applying the vector control for the GSC.

The phase-locked loop (PLL) is successfully used to synchronize a given signal which is usually merged with a noise. Here the given signal is usually a grid voltage signal. Although it is sinusoidal in theory, it is often distorted due to nonlinear loads in the grid or disturbed by the transient process of utility systems. PLL is a closed-loop control system which can track the phase angle of the given signal even it is distorted or interfaced with noise. Therefore, it has been widely used in the industry. The diagram of a PLL is shown in Figure 5.3, which normally consists of three parts:

- The phase detector (PD) outputs an error signal e_p proportional to the phase difference between the input signal v and the output single v'. e_p is the feedback error of this close loop system.
- The loop filter (LP) is a low pass filter to extract the control signal v_{lf} from phase error signal e_p. It also works as a controller in the closed-loop system.

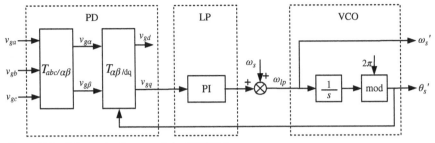

Figure 5.4   Schematic diagram of the SRF-PLL.

- The voltage-controlled oscillator (VCO) generates AC output single v', whose frequency is proportional to the LP output v_{lf}. It is seen as the plant in the closed-loop system.

By using the PLL, the output v' is able to track input signal v. In other words, the output signal not only follows the input signal frequency but also keeps its phase angle the same as that of the input signal. Both the phase and frequency information of the input signal can be extracted accurately even the input signal is distorted or immersed with the noise. Since the PLL loop has certain inertia because of the loop filter, PLL output can keep delivering periodical AC signal even the PLL input signal disappears for a certain duration of time due to the outage of the grid. It is an important characteristic of PLL for practical applications.

For the three-phase power system, three-phase grid voltage signals v_{ga}, v_{gb}, and v_{gc} are used as the PLL inputs in phase detector instead of one-phase grid voltage signal. PD is realized by the *abc* static frame to *dq* rotating frame transformation. This PLL is known as synchronous reference frame (SRF) PLL as shown in Figure 5.4. SRF-PLL output θ_s' will track phase angle θ_s of phase-a grid voltage v_{ga} while SRF-PLL output ω_s' will track the grid synchronous angular frequency ω_s.

The SRF-PLL is composed of PD, PI regulator, add junction, and VCO. We will discuss PD in detail later. PI regulator is used as the controller of the PLL. The add junction has two inputs. One input comes from PI regulator while another input is set to be a constant value ω_s, which is the nominal frequency of the grid. It is used to set up PLL output angular frequency at ω_s during PLL starting process so that it will not deviate significantly from the grid frequency. The VCO is composed of an integrator and mod operator, which produces a saw-tooth waveform with its peak value equal to 2π and its frequency equal to ω_s'.

Now let us see how PD works. In Figure 5.4, PD is Park transformation from the *abc* static frame to the *dq* rotating frame, which is composed of two sub-transformations. The first sub-transform is Clark transformation from the *abc* reference frame to the $\alpha\beta$ reference frame, which is expressed as

$$\vec{v}_g = v_{g\alpha} + j v_{g\beta} = \frac{2}{3}\left(v_{ga} + \alpha v_{gb} + \alpha^2 v_{gc}\right) \qquad (5.1)$$

where v_{ga}, v_{gb}, v_{gc} are the three-phase grid voltages, and $v_{ga} + jv_{g\beta}$ is a complex vector in $\alpha\beta$ reference frame. It is assumed that the three-phase grid voltages v_{ga}, v_{gb}, v_{gc} are sinusoidal and balanced as

$$v_{ga} = V_g \cos\theta_s = \frac{V_g}{2}(e^{j\theta_s} + e^{-j\theta_s})$$

$$v_{gb} = V_g \cos\left(\theta_s - \frac{2\pi}{3}\right) = \frac{V_g}{2}\left[e^{j\left(\theta_s - \frac{2\pi}{3}\right)} + e^{-j\left(\theta_s - \frac{2\pi}{3}\right)}\right]$$

$$v_{gc} = V_g \cos\left(\theta_s + \frac{2\pi}{3}\right) = \frac{V_g}{2}\left[e^{j\left(\theta_s + \frac{2\pi}{3}\right)} + e^{-j\left(\theta_s + \frac{2\pi}{3}\right)}\right] \tag{5.2}$$

By substituting equation (5.2) in equation (5.1), the following equation can be obtained:

$$v_{ga} + jv_{g\beta} = \frac{V_g}{3}\left[e^{j\theta_s}(1 + \alpha e^{-j\frac{2\pi}{3}} + \alpha^2 e^{j\frac{2\pi}{3}}) + e^{-j\theta_s}(1 + \alpha e^{j\frac{2\pi}{3}} + \alpha^2 e^{-j\frac{2\pi}{3}})\right] \tag{5.3}$$

Since $1 + \alpha e^{-j\frac{2\pi}{3}} + \alpha^2 e^{j\frac{2\pi}{3}} = 3$ and $1 + \alpha e^{j\frac{2\pi}{3}} + \alpha^2 e^{-j\frac{2\pi}{3}} = 0$, the above equation can be simplified as

$$v_{ga} + jv_{g\beta} = V_g e^{j\theta_s} \tag{5.4}$$

It is observed that the track of the complex vector $v_{ga} + jv_{g\beta}$ is a circle with a radius of V_g. The complex vector components in $\alpha\beta$ reference frame can also be derived through following matrix calculation:

$$\begin{bmatrix} v_{ga} \\ v_{g\beta} \end{bmatrix} = T_{abc/\alpha\beta} \begin{bmatrix} v_{ga} \\ v_{gb} \\ v_{gc} \end{bmatrix} = \frac{2}{3}\begin{bmatrix} 1 & -\frac{1}{2} & -\frac{1}{2} \\ 0 & \frac{\sqrt{3}}{2} & -\frac{\sqrt{3}}{2} \end{bmatrix} \cdot \begin{bmatrix} V_g \cos\theta_s \\ V_g \cos\left(\theta_s - \frac{2\pi}{3}\right) \\ V_g \cos\left(\theta_s + \frac{2\pi}{3}\right) \end{bmatrix} \tag{5.5}$$

The second sub-transform in PD is the rotating transformation from the $\alpha\beta$ reference frame to the rotating dq reference frame. The vector in dq reference frame is expressed as

$$v_{gd} + jv_{gq} = e^{-j\theta_s'}(v_{ga} + jv_{g\beta}) \tag{5.6}$$

where phase angle θ_s' is feedback signal from the PLL output.

By substituting equation (5.4) in equation (5.6), the grid voltage vector in the dq reference frame can be obtained:

$$v_{gd} + jv_{gq} = V_g e^{j(\theta_s - \theta_s')} \tag{5.7}$$

The above equation can also can be rewritten as

$$v_{gd} + jv_{gq} = V_g \cos(\theta_s - \theta_s') + jV_g \sin(\theta_s - \theta_s') \tag{5.8}$$

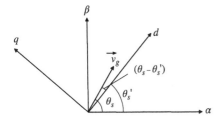

Figure 5.5   Grid voltage vector in the SRF-PLL.

The d-axis component and q-axis component of the grid voltage vector in dq reference frame are derived:

$$v_{gd} = V_g \cos(\theta_s - \theta_s') \tag{5.9}$$

$$v_{gq} = V_g \sin(\theta_s - \theta_s') \tag{5.10}$$

where $(\theta_s - \theta_s')$ is the phase angle difference between the grid phase-a voltage and PLL output phase angle.

The dq components of the grid voltage vector in dq reference frame can also be derived through the following matrix calculation:

$$\begin{bmatrix} v_{gd} \\ v_{gq} \end{bmatrix} = T_{\alpha\beta/dq} \cdot \begin{bmatrix} v_{g\alpha} \\ v_{g\beta} \end{bmatrix} = \begin{bmatrix} \cos\theta_s' & -\sin\theta_s' \\ \sin\theta_s' & \cos\theta_s' \end{bmatrix} \cdot \begin{bmatrix} v_{g\alpha} \\ v_{g\beta} \end{bmatrix} \tag{5.11}$$

It is seen from equation (5.10) that the q-axis component of the grid voltage vector in dq reference frame v_{gq} is proportional to $\sin(\theta_s - \theta_s')$. If the phase error $(\theta_s - \theta_s')$ is very small, the q-axis component v_{gq} can be approximated as

$$v_{gq} \approx V_g(\theta_s - \theta_s') \tag{5.12}$$

v_{gq} is proportional to the phase error between θ_s and θ_s'. Therefore, the Park transformation in SRF-PLL functions as a phase detector. PI controller is the loop filter in SRF-PLL. The output of the add junction is angular speed ω_s' while PLL output phase angle θ_s' is generated by integrating the angular frequency ω_s'. To limit the PLL output phase angle θ_s' range inside $[0, 2\pi]$, the mod operator is inserted in the loop. Once the PLL reaches the steady state, PD output $v_{gq} = 0$ while PLL output phase angle θ_s' is equal to the phase angle of the grid phase-a voltage θ_s. In this case, the d-axis component of the grid voltage vector in dq reference frame v_{gd} becomes constant with a value equal to the amplitude of the grid phase voltage V_g.

In SRF-PLL, as long as the q-axis component of the grid voltage vector v_{gq} is controlled to be zero, the PLL output θ_s' will track the grid phase-a angle θ_s, which is the phase angle of the grid phase-a voltage. The experiment result of the SRF-PLL in DFIG wind power system is shown in Figure 5.6. The upper two sinusoidal waveforms are the grid phase-a and phase-b voltages. The middle saw-tooth waveform is the PLL output, which has the same frequency with the grid voltage in steady state with its value equal to the instant phase angle of the grid phase-a voltage. The lower waveform shows the changing grid frequency. The grid frequency is 50 Hz initially

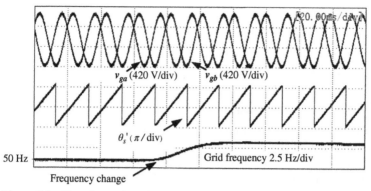

Figure 5.6   Experimental result of the SRF-PLL in the DFIG wind power system.

and later, the grid frequency increases to 52.5 Hz. PLL output can track the grid angle again in about 20 ms. The PLL is able to synchronize with the grid voltages properly under ideal grid voltage condition.

However, when the three-phase grid voltages are not ideal and there is distortion or unbalance in the three-phase grid voltages, the performance of the SRF-PLL will be affected, which will be discussed in detail in Chapter 12.

5.3.3   Control Scheme

The dynamic model of the GSC with L filter has been introduced in Chapter 4 and is rewritten as follows:

$$v_{gd} = -L\frac{di_{gd}}{dt} - Ri_{gd} + \omega_s Li_{gq} + v_d$$

$$v_{gq} = -L\frac{di_{gq}}{dt} - Ri_{gq} - \omega_s Li_{gd} + v_q \tag{5.13}$$

The active power and reactive power from the grid to the GSC are described by

$$P_g = -\frac{3}{2}\left(v_{gd}i_{gd} + v_{gq}i_{gq}\right) \tag{5.14}$$

$$Q_g = -\frac{3}{2}\left(v_{gq}i_{gd} - v_{gd}i_{gq}\right) \tag{5.15}$$

It is assumed that dq rotating reference frame is oriented with the grid voltage by aligning d-axis with the grid voltage complex vector $\vec{v}_g$. It can be realized by SRF-PLL. Once SRF-PLL is synchronized with the grid, the d-axis component of the grid voltage vector in dq reference frame $v_{gd} = V_s$ and the q-axis component of the grid voltage vector in dq reference frame $v_{gq} = 0$. Then the active power equation (5.14) and reactive power equation (5.15) are simplified to

$$P_g = -\frac{3}{2}v_{gd}i_{gd} = -\frac{3}{2}V_s i_{gd} \tag{5.16}$$

$$Q_g = \frac{3}{2}v_{gd}i_{gq} = \frac{3}{2}V_s i_{gq} \tag{5.17}$$

From equations (5.16) and (5.17), it is observed that the active power P_g depends on the d-axis component of the grid current i_{gd} while the reactive power Q_g is proportional to the q-axis component of the grid current i_{gq}. Therefore, the d-axis current i_{gd} can be used to control the active power P_g of the converter. The q-axis current i_{gq} is used to control the reactive power Q_g of the converter.

The converter dynamic model (5.13) can be expressed as

$$V_s = -L\frac{di_{gd}}{dt} - Ri_{gd} + \omega_s Li_{gq} + v_d$$

$$0 = -L\frac{di_{gq}}{dt} - Ri_{gq} - \omega_s Li_{gd} + v_q \qquad (5.18)$$

If the coupling terms $\omega_s Li_{gq}$ and $\omega_s Li_{gd}$ are ignored in equation (5.18), the first line in equation (5.18) is the first-order system deciding the relationship between the d-axis current i_{gd} and the d-axis control voltage of the converter v_d. The second line in equation (5.18) is also the first-order system deciding the relationship between the q-axis current i_{gq} and the q-axis control voltage of the converter v_q. Therefore, the d-axis current i_{gd} and the q-axis current i_{gq} can be controlled by the d-axis control voltage of the converter v_d and the q-axis control voltage of the converter v_q, respectively. In other words, the active power and reactive power from the GSC to the grid can be controlled by the d-axis control voltage of the converter v_d and the q-axis control voltage of the converter v_q, respectively.

The DC-side equation of the GSC with L filter is rewritten here from equation (4.120) by considering the given dq frame orientation:

$$C\frac{dv_{dc}}{dt} = i_{dc} - \frac{3}{2}\frac{V_s}{v_{dc}} \cdot i_{gd} \qquad (5.19)$$

where i_{dc} is the external current flowing into the DC bus of the converter. It is observed that DC-bus voltage v_{dc} can be controlled through the d-axis currents i_{gd}.

According to the analysis mentioned above, a control scheme for the GSC is drawn in Figure 5.7. Three-phase grid voltage $\vec{v}_{gabc}$, current $\vec{i}_{gabc}$, and the DC-bus voltage v_{dc} are measured. Through SRF-PLL with the measured three-phase grid voltage $\vec{v}_{gabc}$, the grid angle θ_s and angular frequency ω_s are obtained. The control scheme has two current-control loops: one is the d-axis current i_{gd} control loop, and the other is the q-axis current i_{gq} control loop [18–21].

The d-axis current i_{gd} control loop is an active power-control loop. The given DC-bus voltage reference V_{dc}^{ref} is compared with sampled DC-bus voltage v_{dc}. Their error is amplified by a PI regulator and its output is taken as d-axis current reference i_{gd}^{ref}. Then, the d-axis current i_{gd} is controlled to track the d-axis current reference i_{gd}^{ref} with PI regulator. The output of this d-axis current PI regulator is taken as the d-axis reference voltage of the converter v_{md}.

The q-axis current i_{gq} is a reactive power-control loop. According to reactive power command Q_g^{ref}, the q-axis current reference i_{gq}^{ref} is calculated using equation (5.17). Then q-axis current i_{gq} is controlled to track the q-axis current reference i_{gq}^{ref}

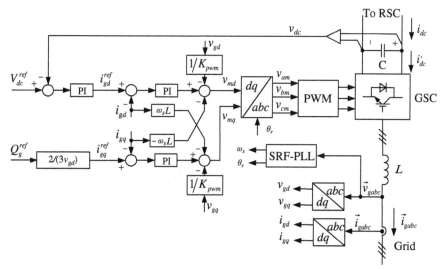

Figure 5.7   Control scheme of the GSC.

with PI regulator. The output of this q-axis current PI regulator is taken as the q-axis reference voltage of the converter v_{mq}.

In the control scheme shown in Figure 5.7, decoupling is introduced to eliminate the interference between d-axis and q-axis current controls. Once we obtain d-axis and q-axis reference voltages of the converter v_{md} and v_{mq}, respectively, PWM pulses can be generated either by carrier-based PWM or SVM. Here carrier-based PWM is shown. d-axis and q-axis reference voltages v_{md} and v_{mq}, respectively, are transformed into three-phase modulation waves, which enter PWM modulator to generate switching signals for the converter.

5.3.4   Simplified Control Model in s-Domain

For the system analysis, the frequency domain analysis is widely used. The model of the grid converter described by the differential equation (5.13) can be changed into the following equation in the s-domain:

$$v_d = (Ls + R)i_{gd} - \omega_s L i_{gq} + v_{gd}$$
$$v_q = (Ls + R)i_{gq} + \omega_s L i_{gd} + v_{gq} \tag{5.20}$$

By considering GSC, equation (4.120) can be expressed as follows:

$$C\frac{dv_{dc}}{dt} = i_{dc} - \frac{3}{2}\frac{1}{v_{dc}}(v_{gd} \cdot i_{gd} + v_{gq} \cdot i_{gq}) \tag{5.21}$$

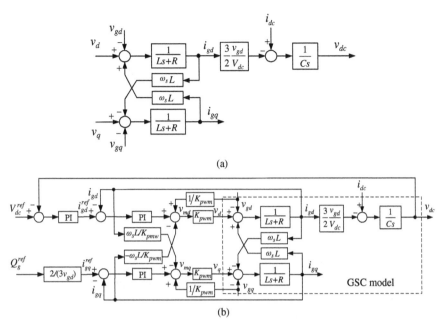

Figure 5.8 Control diagrams for the GSC: (a) the model of GSC; (b) the GSC model and control system.

Once the PLL is synchronized with the grid, the d-axis component of the grid voltage vector in dq reference frame $v_{gd} = V_s$ and the q-axis component of the grid voltage vector in dq reference frame $v_{gq} = 0$. By transferring equation (5.21) into the s-domain, the following s-domain model of the DC side can be obtained:

$$sCv_{dc} = i_{dc} - \frac{3}{2}\frac{v_{gd}}{V_{dc}}i_{gd} \tag{5.22}$$

The diagram for GSC model is drawn according to the equations (5.20) and (5.22) as shown in Figure 5.8(a). The grid voltage v_{gd}, v_{gq} and the DC-bus current i_{dc} are regarded as disturbances to the system. By replacing the GSC converter in Figure 5.7 with the diagram of GSC model in Figure 5.8(a), a diagram of the whole system is obtained as shown in Figure 5.8(b). Here PWM model in Chapter 4 is used. The relationship between the output voltage of the GSC (v_d, v_q) and d-axis and q-axis reference voltages v_{md} and v_{mq} are expressed as

$$v_d = K_{pwm}v_{md}$$

$$v_q = K_{pwm}v_{mq} \tag{5.23}$$

The d-axis loop controls the DC-bus voltage, and the q-axis loop controls the reactive power. Since these two loops have been decoupled by introducing decouple terms in the control system, the d-axis current loop and the q-axis current loop are

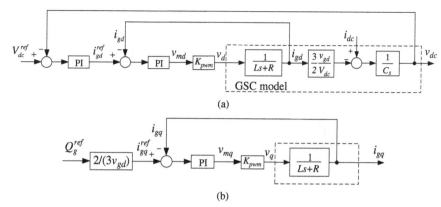

Figure 5.9   Decoupled diagram of the GSC system: (a) the d-axis control diagram; (b) the q-axis control diagram.

independent of each other. The diagram of the d-axis loop is shown in Figure 5.9a, while the diagram of the q-axis loop is shown in Figure 5.9b. With regard to the current loops, the d-axis current loop is the same as the q-axis loop. Here grid disturbance terms are eliminated by introducing the forward control.

Open-loop transfer function from the voltage reference v_{md} or v_{mq} to the grid current i_{gd} or i_{gq} are derived:

$$G_{vi}(s) = K_{pwm}\frac{1}{Ls + R} \tag{5.24}$$

Compensated open-loop transfer function $G_{ii}(s)$ for the current loop is described by

$$G_{ii}(s) = G_{PIi}(s)G_{vi}(s) = G_{PIi}(s)K_{pwm}\frac{1}{Ls + R} \tag{5.25}$$

where $G_{PIi}(s)$ is the transfer function of the PI controller in the current loop.

$$G_{PIi}(s) = k_{pi} + \frac{k_{ii}}{s} \tag{5.26}$$

The closed-loop transfer function of the current loop is

$$H_{ii}(s) = \frac{G_{ii}(s)}{1 + G_{ii}(s)} \tag{5.27}$$

With regards to the d-axis control diagram Figure 5.9a, the open-loop transfer function $G_{iv}(s)$ from i_{gd}^* to v_{dc} can be expressed as

$$G_{iv}(s) = -H_{ii}(s)\frac{3}{2}\frac{v_{gd}}{V_{dc}}\frac{1}{Cs} \tag{5.28}$$

TABLE 5.1   **Parameters of the GSC with 660 kW rated power**

Rated power P_R (kW)	660
Rated voltage V_R (V_{rms})	690
Inductance L (mH)	0.5
Parasitic resistance R (mΩ)	1.8
DC-bus capacitance C (mF)	20
DC-bus voltage V_{dc} (V)	1150
Switching frequency f_s (Hz)	2000

In the case of designing the controller for outer voltage loop, the closed-loop transfer function of the inner current loop $H_{ii}(s)$ in equation (5.28) can be approximated to 1. Then equation (5.28) can be simplified as

$$G_{iv}(s) \approx -\frac{3}{2}\frac{V_{gd}}{V_{dc}}\frac{1}{Cs} \tag{5.29}$$

Compensated open-loop transfer function $G_{vv}(s)$ is expressed as

$$G_{vv}(s) = G_{PIv}(s)G_{iv}(s) \tag{5.30}$$

where $G_{PIv}(s)$ is the transfer function of the PI controller for DC-bus voltage outer loop, which is described by

$$G_{PIv}(s) = k_{pv} + \frac{k_{iv}}{s} \tag{5.31}$$

5.3.5   Controller Design

With the GSC model in the s-domain (5.24)–(5.31), the PI controller parameters can be designed using the Bode diagram. A 660 kW power converter for 1.5 MW DFIG wind power system is used as an example. GSC parameters are listed in Table 5.1.

The PI controller of the current loop is designed firstly. The open-loop transfer function $G_{vi}(s)$ is expressed as

$$G_{vi}(s) = K_{pwm}\frac{1}{Ls + R} \tag{5.32}$$

where $K_{pwm} = V_{dc}/\sqrt{3}$ is selected, which means SVM is used.

Bode diagrams of the open-loop transfer function $G_{vi}(s)$ with the parameters in Table 5.1 are shown in Figure 5.10. It is a first-order system, the low-frequency asymptote is a horizontal line with an amplitude of 111 dB, while the high-frequency asymptote is a straight line with slope -20 dB/decade. The intersection of these two lines is at 0.57 Hz. The crossover frequency is about 200 kHz. Its phase delay is 90° at high frequency.

It is expected that the compensated open-loop transfer function has a bandwidth as wide as possible to improve the system dynamics. The crossover frequency f_c of the compensated open-loop transfer function is selected to be 1/10 of the switching frequency f_s. In this case, the switching frequency $f_s = 2000$ Hz, so the crossover

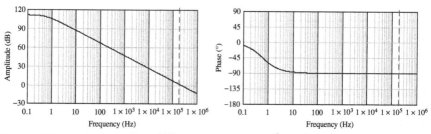

Figure 5.10   Bode diagrams of the open-loop transfer function $G_{vi}(s)$.

frequency $f_c = 200$ Hz is selected. For the compensated open-loop transfer function $G_{ii}(s)$ and the transfer function of the PI controller $G_{PIi}(s)$, the following equation can be derived:

$$|G_{ii}(s)|_{j2\pi f_c} = |G_{PIi}(s)G_{vi}(s)|_{j2\pi f_c} = 1 \tag{5.33}$$

The corner frequency of the PI controller f_i, which is related to the zero of the PI controller, is selected to be the corner frequency of the open-loop transfer function.

$$G_{PIi}(s)|_{j2\pi f_i} = k_{pi} + \frac{k_{ii}}{j2\pi f_i} = 0 \tag{5.34}$$

where $f_i = 0.57$ Hz. Finally, the PI parameters can be derived with the following two equations:

$$|G_{ii}(j2\pi f_c)| = \left|\left(k_{pi} + \frac{k_{ii}}{j2\pi f_c}\right) K_{pwm} \frac{1}{j2\pi f_c L + R}\right| = 1 \tag{5.35}$$

$$2\pi f_i = \frac{k_{ii}}{k_{pi}} \tag{5.36}$$

The PI parameters $k_{pi} = 9.46 \times 10^{-4}$, $k_{ii} = 3.4 \times 10^{-3}$ are calculated. Bode diagrams of the compensated system $G_{ii}(s)$ are shown together with Bode diagrams of the uncompensated system $G_{ui}(s)$ and PI regulator $G_{PIi}(s)$ in Figure 5.11. The crossover frequency of the compensated system is about 200 Hz, and 90° phase margin is obtained.

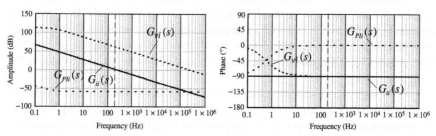

Figure 5.11   Bode diagrams for the inner current loop design.

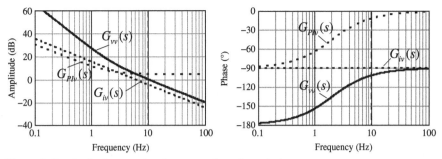

Figure 5.12   Bode diagrams for the outer voltage loop design.

As shown in Figure 5.9, the d-axis control diagram is composed of two cascaded feedback loops. After designing the inner current loop, we can design the controller for the outer voltage loop. To ensure the stability of the cascade control, the outer loop is deliberately designed to be far slower than the inner loop. Here the crossover frequency f_c of the outer loop is selected to be 1/20 of the crossover frequency of the inner loop, that is, 10 Hz. According to equation (5.30), it results in the following equation:

$$\left.|G_{vv}(s)|\right|_{j2\pi f_c} = \left.|G_{PIv}(s)G_{iv}(s)|\right|_{j2\pi f_c} = 1 \qquad (5.37)$$

Since the uncompensated system described by equation (5.29) is an integral term, the corner frequency of the outer PI regulator f_i is selected to be 1/5 of the crossover frequency, that is, 2 Hz. Therefore, the PI regulator is required to satisfy the following equation:

$$\left.G_{PIv}(s)\right|_{j2\pi f_i} = 0 \qquad (5.38)$$

From equations (5.37) and (5.38), the following two equations are derived

$$|G_{vv}(j2\pi f_c)| = \left|\left(k_{pv} + \frac{k_{iv}}{j2\pi f_c}\right)\frac{3}{2}\frac{v_{gd}}{V_{dc}}\frac{1}{j2\pi f_c C}\right| = 1 \qquad (5.39)$$

$$k_{pv} = \frac{k_{iv}}{2\pi f_i} \qquad (5.40)$$

The PI parameters for outer voltage loop $k_{pv} = 1.67$ and $k_{iv} = 21$ are obtained by solving the above two equations.

The Bode diagrams of the transfer function $G_{iv}(s)$, the outer PI regulator $G_{PIv}(s)$ and the compensated open-loop transfer function $G_{vv}(s)$ are shown in Figure 5.12. The crossover frequency is about 10 Hz, and a phase margin of more than 75° is provided.

For the q-axis control diagram shown in Figure 5.9, there is only one feedback loop. The PI regulator design is the same as the inner current loop design for the d-axis as mentioned before. Its outer reactive power is controlled with an open loop as shown in the figure.

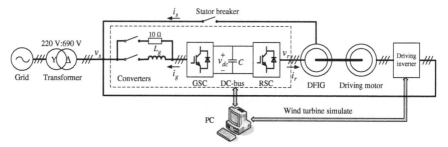

Figure 5.13   Schematic diagram of the 1.5 MW DFIG WPS test bench.

5.3.6   Test Results

Test results for the GSC in the 1.5 MW DFIG WPS are briefly presented in this section. The schematic diagram of the test bench is shown in Figure 5.13. The wind turbine is emulated by a cage rotor machine which is driven by an inverter. This machine can be operated in torque-control mode or speed-control mode to emulate different operation conditions of the DFIG. The details of the system will be introduced in Chapter 14.

The steady-state results of the GSC with light load are shown in Figure 5.14a, the DC-bus voltage is controlled to be 1150 V and it is stable under steady state. The experiment results when a step change of the output reactive power from about 0–0.25 pu is shown in Figure 5.14b. The dynamic process is finished in about 5 ms, as the crossover frequency of the inner loop is selected to be about 200 Hz. The DC-bus voltage keeps stable during this dynamic process. The control performance of the GSC vector control is satisfactory.

5.4   ROTOR-SIDE CONVERTER IN POWER-CONTROL MODE

5.4.1   Control Target

After the stator windings of the DFIG are connected to the grid, the DFIG starts to generate power to the grid. The DFIG is controlled through the RSC to generate active or reactive power according to the power command from the wind turbine central controller as shown in Figure 5.15. The power command comes from the wind turbine central controller to realize maximum power point tracking (MPPT) or the requirements of the transmission system operators, etc.

The grid synchronization for the RSC in power control is also needed. It uses a PLL to get the grid phase angle θ_s and the angular speed ω_s information as mentioned in the grid synchronization in Section 5.3.2, so it will not be repeated here.

5.4.2   Control Scheme

Under stator voltage orientation control, d-axis of dq reference frame is selected to be in the same direction as the grid voltage vector $\vec{v}_s$, and the dq reference frame rotating

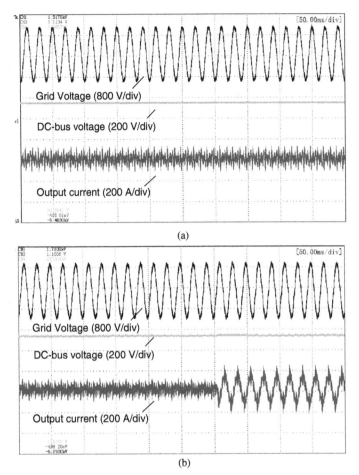

(a)

(b)

Figure 5.14　14 Experiment results of the GSC: (a) with a light load; (b) with output current step change.

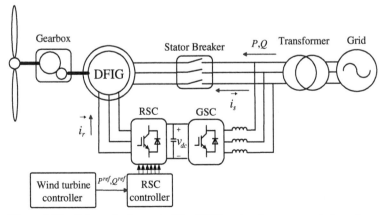

Figure 5.15　Schematic diagram of the DFIG system in power-control mode.

angular speed is the same as the synchronous speed ω_s [7, 27]. In this case, the d-axis component of the grid voltage v_{sd} is the grid phase voltage amplitude V_s while the q-axis component of the grid voltage v_{sq} is always zero. The following equation is satisfied:

$$\vec{v}_{sdq} = v_{sd} + jv_{sq} = V_s \tag{5.41}$$

The dq model of the DFIG stator equation is rewritten here:

$$\vec{v}_{sdq} = R_s \vec{i}_{sdq} + p\vec{\psi}_{sdq} + j\omega_s \vec{\psi}_{sdq} \tag{5.42}$$

If it is approximated that the stator resistance is neglected and the transient process of the flux is ignored, then equation (5.42) is simplified to

$$\vec{v}_{sdq} \approx j\omega_s \vec{\psi}_{sdq} \tag{5.43}$$

According to equations (5.41) and (5.43), the stator fluxes in the dq reference frame ψ_{sd} and ψ_{sq} are derived:

$$\psi_{sd} \approx 0$$

$$\psi_{sq} \approx -\frac{V_s}{\omega_s} \tag{5.44}$$

The stator and rotor fluxes in the dq reference frame are rewritten here:

$$\vec{\psi}_{sdq} = L_s \vec{i}_{sdq} + L_m \vec{i}_{rdq} \tag{5.45}$$

$$\vec{\psi}_{rdq} = L_m \vec{i}_{sdq} + L_r \vec{i}_{rdq} \tag{5.46}$$

According to equation (5.45), the stator current can be expressed by the rotor current

$$i_{sd} = \frac{1}{L_s}\psi_{sd} - \frac{L_m}{L_s}i_{rd}$$

$$i_{sq} = \frac{1}{L_s}\psi_{sq} - \frac{L_m}{L_s}i_{rq} \tag{5.47}$$

By substituting equation (5.44) in equation (5.47), the relationships between the stator and rotor currents under stator voltage orientated dq frame are written as

$$i_{sd} = -\frac{L_m}{L_s}i_{rd}$$

$$i_{sq} = -\frac{V_s}{\omega_s L_s} - \frac{L_m}{L_s}i_{rq} \tag{5.48}$$

With regard to equation (5.48), the stator currents i_{sd} and i_{sq} in the dq reference frame are proportional to the rotor currents i_{rd} and i_{rq}, respectively, under stator voltage orientated dq frame.

The stator output active and reactive power equations are expressed by

$$P_s = \frac{3}{2} V_s i_{sd}$$

$$Q_s = -\frac{3}{2} V_s i_{sq} \tag{5.49}$$

It is observed that the stator output active power P_s depends on the stator d-axis current i_{sd} while the stator output reactive power depends on the stator q-axis current i_{sq}.

Combining equations (5.48) and (5.49), the following equations can be derived:

$$P_s = -\frac{3}{2} \frac{L_m}{L_s} V_s i_{rd}$$

$$Q_s = \frac{3}{2} \frac{V_s^2}{\omega_s L_s} + \frac{3}{2} \frac{L_m}{L_s} V_s i_{rq} \tag{5.50}$$

It is observed that the active power and the reactive power control are decoupled. The active power P_s is controlled by the d-axis rotor current i_{rd}, while the reactive power Q_s is controlled by the q-axis rotor current i_{rq}.

According to equation (5.45), the stator current in dq reference frame is expressed by

$$\vec{i}_{sdq} = \frac{\vec{\psi}_{sdq} - L_m \vec{i}_{rdq}}{L_s} \tag{5.51}$$

By substituting equation (5.51) in (5.46), the rotor flux in dq reference frame is described by

$$\vec{\psi}_{rdq} = \frac{L_m}{L_s} \vec{\psi}_{sdq} + \sigma L_r \vec{i}_{rdq} \tag{5.52}$$

where $\sigma = 1 - \frac{L_m^2}{L_s L_r}$, which is the leakage coefficient of the machine.

The dq model of the DFIG rotor is rewritten here:

$$\vec{v}_{rdq} = R_r \vec{i}_{rdq} + p\vec{\psi}_{rdq} + j(\omega_s - \omega_r)\vec{\psi}_{rdq} \tag{5.53}$$

By substituting equation (5.52) in (5.53), the rotor voltage in the dq frame is described by

$$\vec{v}_{rdq} = \left[R_r + \sigma L_r p + j(\omega_s - \omega_r)\sigma L_r \right] \vec{i}_{rdq} + \frac{L_m}{L_s} \left[p + j(\omega_s - \omega_r) \right] \vec{\psi}_{sdq} \tag{5.54}$$

By using the approximation as equation (5.43), equation (5.54) is changed into

$$\vec{v}_{rdq} = \left[R_r + \sigma L_r p + j(\omega_s - \omega_r)\sigma L_r \right] \vec{i}_{rdq} + \frac{L_m}{L_s} \frac{(\omega_s - \omega_r)}{\omega_s} \vec{v}_{sdq} \tag{5.55}$$

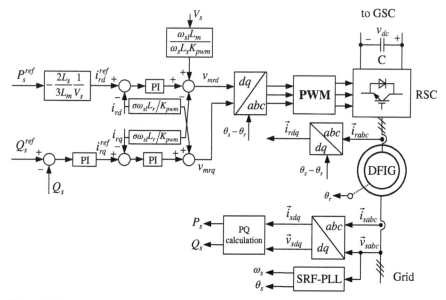

Figure 5.16   The control scheme of RSC and DFIG in power-control mode.

The above complex equation can be expressed with d-axis and q-axis real component equations as follows:

$$v_{rd} = R_r i_{rd} + \sigma L_r \frac{di_{rd}}{dt} - \omega_{sl}\sigma L_r i_{rq} + \frac{\omega_{sl}L_m}{\omega_s L_s}v_{sd}$$

$$v_{rq} = R_r i_{rq} + \sigma L_r \frac{di_{rq}}{dt} + \omega_{sl}\sigma L_r i_{rd} + \frac{\omega_{sl}L_m}{\omega_s L_s}v_{sq} \qquad (5.56)$$

where $\omega_{sl} = \omega_s - \omega_r$.

It is seen that the rotor currents i_{rd} and i_{rq} are controlled by the rotor voltages v_{rd} and v_{rq}, respectively. The rotor voltages v_{rd} and v_{rq} are output voltages of the RSC. The stator voltages v_{sd} and v_{sq} are regarded as the grid disturbance to the system.

According to the rotor equation (5.56), the control scheme of the RSC is drawn in Figure 5.16 [22–26]. The stator three-phase voltage $\vec{v}_{sabc}$, stator three-phase current $\vec{i}_{sabc}$, and rotor three-phase current $\vec{i}_{rabc}$ are sampled and transformed into dq reference frame. The active power and reactive power is controlled by the outer power loops. For the active power, open-loop control is used, while closed-loop control with a PI regulator is used for the reactive power. According to (5.50), the rotor d-axis current reference i_{rd}^{ref} is estimated from the active power command, while the rotor q-axis current reference i_{rq}^{ref} is generated by the outer reactive power control PI regulator. There are two similar current loops, that is, d-axis rotor current-control loop and q-axis rotor current-control loop. The outputs of d-axis rotor current regulator and q-axis rotor current regulator are rotor voltage references v_{mrd} and v_{mrq}. The rotor voltage references are used to generate PWM signals for the converter.

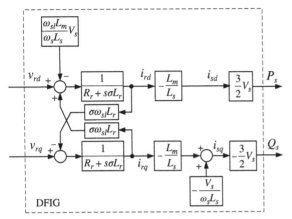

Figure 5.17   Block diagram for the model of DFIG in power-control mode.

5.4.3   Control Model in *s*-Domain

The dynamic model for the DFIG in the *s*-domain can be derived from (5.56):

$$v_{rd} = (R_r + \sigma L_r s)i_{rd} - \omega_{sl}\sigma L_r i_{rq} + \omega_{sl}\frac{L_m v_{sd}}{\omega_s L_s}$$

$$v_{rq} = (R_r + \sigma L_r s)i_{rq} + \omega_{sl}\sigma L_r i_{rd} + \omega_{sl}\frac{L_m v_{sq}}{\omega_s L_s} \qquad (5.57)$$

Combining equations (5.50) and (5.57), the block diagram for the DFIG model is derived as shown in Figure 5.17.

According to the DFIG model in Figure 5.17, the control diagram of the DFIG is illustrated in Figure 5.18. The *d*-axis rotor current loop and *q*-axis rotor current loop are decoupled by introducing decoupling terms in the current loop. Then *d*-axis and *q*-axis rotor currents i_{rd} and i_{rq}, respectively, could be controlled independently.

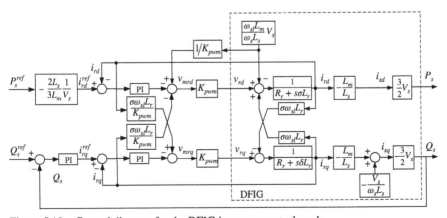

Figure 5.18   Control diagram for the DFIG in power-control mode.

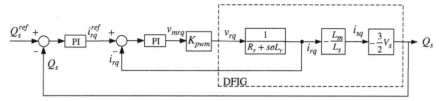

Figure 5.19   Block diagram of the q-axis loop in power-control mode.

Since two current loops are similar, we take the q-axis loop as an example of which the block diagram is drawn in Figure 5.19.

From Figure 5.19, the open-loop q-axis current loop transfer function $G_{vi}(s)$, which is from the rotor voltage reference v_{mrq} to the rotor current i_{rq}, can be written as

$$G_{vi}(s) \approx K_{pwm} \frac{1}{R_r + s\sigma L_r} \tag{5.58}$$

PI regulator for q-axis current loop is

$$G_{PIi}(s) = k_{pi} + \frac{k_{ii}}{s} \tag{5.59}$$

The compensated open-loop q-axis current loop transfer function $G_{ii}(s)$ is

$$G_{ii}(s) = G_{PIi}(s)G_{vi}(s) = \left(k_{pi} + \frac{k_{ii}}{s}\right)K_{pwm}\frac{1}{R_r + s\sigma L_r} \tag{5.60}$$

Then q-axis current closed-loop transfer function $H_{ii}(s)$ is written as

$$H_{ii}(s) = \frac{G_{ii}(s)}{1 + G_{ii}(s)} \tag{5.61}$$

Now we will deal with the outer reactive power loop. Open reactive power loop function $G_{iQ}(s)$ from the rotor current reference i_{rq}^{ref} to the reactive power can be derived as

$$G_{iQ}(s) = \frac{3}{2}\frac{L_m}{L_s}V_s H_{ii}(s) \tag{5.62}$$

PI regulator used in the outer reactive loop is

$$G_{PIQ}(s) = k_{pQ} + \frac{k_{iQ}}{s} \tag{5.63}$$

Then compensated open-loop function $G_{QQ}(s)$ for the outer reactive power loop is written as

$$G_{QQ}(s) = G_{PIQ}(s)G_{iQ}(s) = \left(k_{pQ} + \frac{k_{iQ}}{s}\right)H_{ii}(s)\frac{3}{2}\frac{L_m}{L_s}V_s \tag{5.64}$$

The design of the control parameters will be introduced in the following section.

TABLE 5.2   RSC converter parameters

Rated power (kW)	660
Rated voltage (V_{rms})	690
DC-bus capacitance (mF)	20
DC-bus voltage (V)	1150
Switching frequency (Hz)	2000

5.4.4   Controller Design

Now we will discuss the controller design of the RSC when the DFIG is in power-control mode. The parameters of the 1.5 MW DFIG wind power system is shown in the Appendix. The RSC parameters are shown in Table 5.2.

For the inner current loop design, the crossover frequency of the compensated open-loop transfer function is selected to be 1/5 of the switching frequency, which is 400 Hz. The corner frequency of the PI regulator is chosen to be the corner frequency of the uncompensated system, which is 2.4 Hz, as shown in Figure 5.20. The PI parameters are derived as $k_{pi} = 5.276 \times 10^{-4}$, $k_{ii} = 0.008$. The Bode diagrams of the uncompensated system $G_{vi}(s)$, the compensated system $G_{ii}(s)$, as well as the PI regulator $G_{PIi}(s)$ are shown in Figure 5.20. The crossover frequency of the compensated system is about 400 Hz, and 90° phase margin is obtained.

With regard to the outer reactive power loop design, the crossover frequency of the compensated open-loop transfer function is designed to be about 1/10 of the inner current loop to ensure the stability of the cascade control, which is about 40 Hz. The PI parameters of the outer loop are derived: $k_{pQ} = 2.36 \times 10^{-4}$, $k_{iQ} = 0.297$. Bode diagrams of the open-loop transfer function $G_{iQ}(s)$, the compensated system $G_{QQ}(s)$, and the PI regulator $G_{PIQ}(s)$ are shown in Figure 5.21. The crossover frequency of the compensated system is 40 Hz, and a phase margin of 90° is obtained.

5.4.5   Test Results from a 1.5 MW DFIG WPS

Test waveforms on a 1.5 MW DFIG wind power system running in power-control mode are given as an example. The steady-state waveforms for the DFIG are given first under different rotor speeds, and with different active power outputs, as shown in Figures 5.22–5.24. Notice that as the grid voltage has distortions in the real case, the stator and rotor currents also have harmonic components, which will be discussed

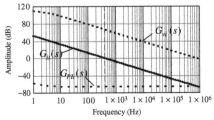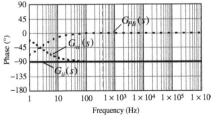

Figure 5.20   Bode diagrams of the inner current loop design in power-control mode.

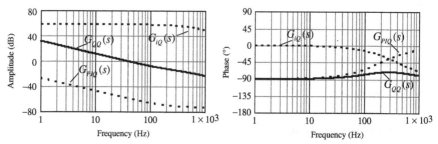

Figure 5.21 Bode diagrams of the outer reactive power loop design in power-control mode.

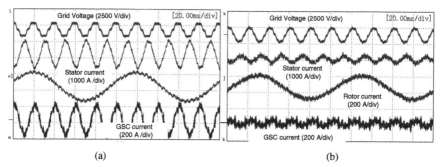

(a)　　　　　　　　　　　　　　　　(b)

Figure 5.22 Steady-state waveforms for the DFIG under the sub-synchronous speed of 1200 rpm: (a) with 0.5 pu output active power; (b) with 0.1 pu output active power.

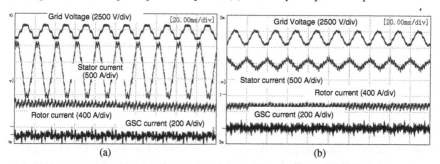

(a)　　　　　　　　　　　　　　　　(b)

Figure 5.23 Steady-state waveforms for the DFIG under the synchronous speed of 1500 rpm: (a) with 0.5 pu output active power; (b) with 0.1 pu output active power.

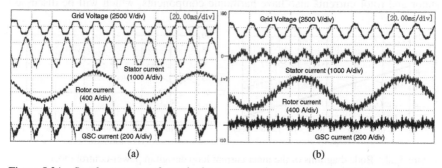

(a)　　　　　　　　　　　　　　　　(b)

Figure 5.24 Steady-state waveforms in the super-synchronous speed of 1800 rpm: (a) with 0.5 pu output active power; (b) with 0.1 pu output active power.

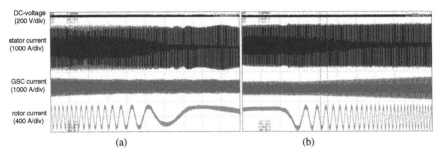

(a)   (b)

Figure 5.25   Waveforms during rotor speed change: (a) from 1200 to 1500 rpm; (b) from 1500 to 1800 rpm.

later in this chapter. The DFIG can generate active power according to the power commands from the wind turbine under different rotor speeds. It is observed that the rotor current frequency is 10 Hz when the rotor speed is 1200 or 1800 rpm. It becomes 0 Hz under 1500 rpm. The GSC current shows the direction of active power flow. The GSC power flow direction is different under rotor speeds of 1200 rpm and 1800 rpm since the rotor side absorbs power from the grid in the sub-synchronous speed of 1200 rpm while it outputs power to the grid in the super-synchronous speed of 1800 rpm.

The dynamic waveforms are shown in Figures 5.25 and 5.26. The dynamic processes of the rotor speed changing from 1200 to 1500 rpm, and from 1500 to 1800 rpm are shown in Figure 5.25. The output active power is controlled to be 0.5 pu during the speed change. The rotor current frequency and the GSC current change during the rotor speed change. The stator current remains unchanged during the dynamic process.

The dynamic process of the DFIG under a step change of active power reference from 0 to 0.5 pu is shown in Figure 5.26. The rotor speed keeps 1800 rpm during this test. The d-axis rotor current takes only about 4 ms to finish this dynamic process, which also corresponds to the crossover frequency of several hundred Hertz. The q-axis rotor current barely changes during this transient process, which indicates the d-axis and q-axis current controls are decoupled.

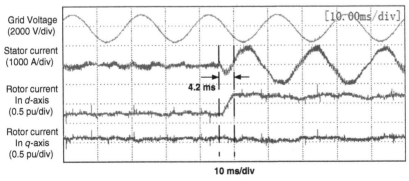

Figure 5.26   Waveforms during active power step change from 0 to 0.5 pu.

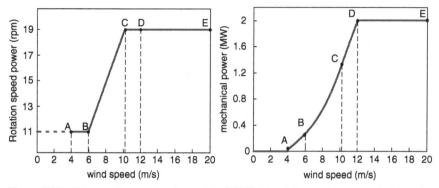

Figure 5.27   The speed-control regions of the DFIG: (a) turbine speed versus wind speed; (b) mechanical power versus wind speed.

5.5   ROTOR-SIDE CONVERTER IN SPEED-CONTROL MODE

5.5.1   Control Target [7, 9, 10]

The wind turbine has four operating modes: starting mode, MPPT mode, speed limiting mode, and power-limiting mode. In the starting mode as A–B stage shown in Figure 5.27, the DFIG runs with constant speed control when the wind speed is little higher than the cut-in speed. In the MPPT mode, as B–C stage shown in Figure 5.27, the DFIG runs at MPPT state. The rotor speed is regulated so that maximum power is harvested. When the wind speed is so large that the rotor speed has reached its mechanical limit, the rotor speed should be limited to prevent mechanical failure in the wind power system. In this case, the wind turbine is changed to speed limiting mode as C–D stage shown in Figure 5.27. Although the rotor speed is fixed, the harvested power of the wind turbine still increases with the increasing wind speed in this mode. Once the power reaches the electric limit, the wind turbine will transfer to the power-limiting mode as D–E stage shown in Figure 5.27 to avoid damaging the electric parts such as the generator and power converters.

Therefore, we can see that the DFIG rotor speed is required to be controlled through the RSC according to wind turbine operation modes.

5.5.2   Grid Synchronization

The grid synchronization is also needed for the RSC in speed control. PLL is used to get the grid phase angle θ_s and the angular speed ω_s information. The grid synchronization is similar to that in GSC control as mentioned in Section 5.3.2, so it will not be repeated here.

5.5.3   Control Scheme

The relationship among the mechanical torque, electromagnetic torque, and the rotor speed has been introduced in Chapter 2. As the response time of the pitch control

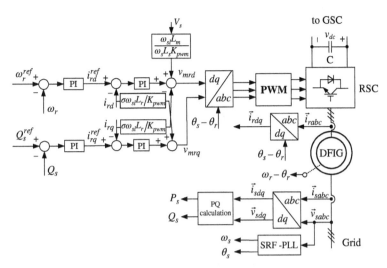

Figure 5.28   The control scheme of the DFIG in speed-control mode.

in the wind turbine is much slower than the generator controlled by the RSC, the mechanical torque can be regarded as a constant during DFIG dynamic process. If the friction coefficient is neglected, the simplified mechanical model of DFIG is represented as

$$J_G \frac{d\omega_{mech}}{dt} = T_{mech} - T_{em} \tag{5.65}$$

where J_G is equivalent inertia of the generator mechanical axis, and T_{mech} is the mechanical torque from the wind. The electromagnetic torque T_{em} is decided by the rotor d-axis current i_{rd}. The control scheme of the RSC in speed-control mode is similar to that in power-control mode, as shown in Figure 5.28.

The control loop is also composed of the inner current loop and the outer power/speed loop; the only difference from the control in power-control mode is that the outer active power control is replaced by the rotor speed control. The rotor speed reference ω_r^{ref} is from the up-level wind turbine central controller, while the actual rotor speed ω_r is measured with the encoder. The output of the rotor speed PI regulator is the rotor current reference in d-axis i_{rd}^{ref}. The rest part of the control diagram is the same as that in power-control mode.

5.5.4   Control Model in s-Domain

The PI regulator parameters for the inner current loop and reactive power loop are the same as those in power-control mode.

The electromagnetic torque in the dq reference frame can be expressed by

$$T_{em} = \frac{3}{2} n_p \frac{L_m}{L_s} \text{Im} \left(\vec{\psi}_{sdq} \vec{i}_{rdq}^* \right) = \frac{3}{2} n_p \frac{L_m}{L_s} \frac{V_s}{\omega_s} i_{rd} \tag{5.66}$$

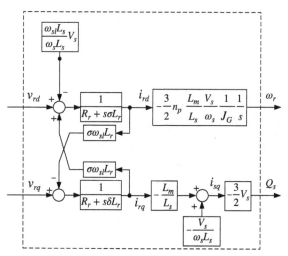

Figure 5.29   Block diagram for the model of the DFIG in speed-control mode.

where n_p is the number of pole pairs of the generator. For the outer rotor speed loop design, the open-loop transfer function from the rotor d-axis current to the rotor speed is derived by combining equations (5.65) and (5.66).

$$G_{i\omega}(s) = \frac{\omega_r(s)}{i_{rd}(s)} = -\frac{3}{2}n_p\frac{L_m}{L_s}\frac{V_s}{\omega_s}\frac{1}{J_G}\frac{1}{s} \tag{5.67}$$

The block diagram for the DFIG model is derived as shown in Figure 5.29.

According to the DFIG model in Figure 5.29, the control diagram of the DFIG is illustrated in Figure 5.30 [7, 9, 10]. The d-axis rotor current loop and q-axis rotor current loop are decoupled by introducing decoupling terms in the current loop. Then d-axis and q-axes rotor currents i_{rd} and i_{rq}, respectively, could be controlled independently.

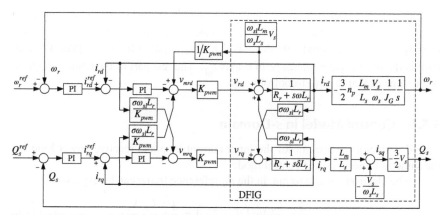

Figure 5.30   Control diagram of the DFIG in speed-control mode.

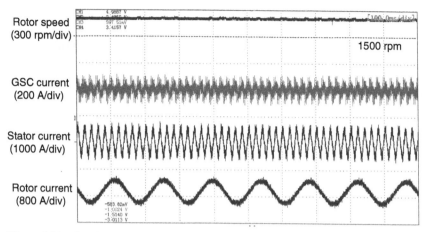

Figure 5.31   Steady performance of the DFIG to control the rotor speed to be 1680 rpm.

The PI regulator parameters for the outer rotor speed loop can then be designed with the Bode diagram.

5.5.5   Test Results

Test results from the 1.5 MW DFIG WPS test bench are shown in Figures 5.31–5.33. The driven motor here is running in torque-control mode while the speed control is taken by the DFIG. The steady-state performance of the DFIG when the rotor speed is controlled at 1680 rpm is shown in Figure 5.31. The rotor speed is stable at 1680 rpm.

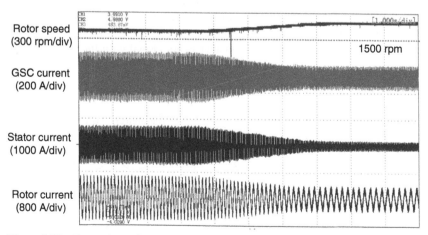

Figure 5.32   Dynamic performance waveforms during speed reference changes from 1560 to 1680 rpm.

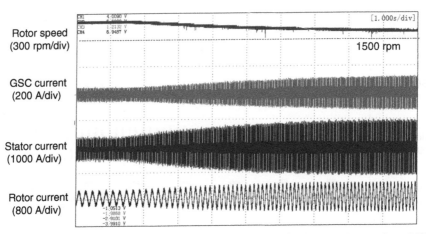

Figure 5.33   Dynamic performance waveforms during speed reference changes from 1680 to 1560 rpm.

The dynamic process of DFIG during rotor speed reference changes are shown in Figures 5.32 and 5.33. Although this situation may not happen in real wind power systems, it can be used to test the dynamic performance of the system. It can be found that the rotor speed can be controlled to smoothly vary from 1560 to 1680 rpm, and from 1560 to 1680 rpm, and no significant transient current is seen in this process.

5.6   ROTOR-SIDE CONVERTER IN STARTING MODE

5.6.1   Control Target

The starting mode is referred to the synchronous process of the DFIG stator voltage to the grid voltage before the DFIG stator side is connected to the grid, as shown in Figure 5.34. In starting mode, the amplitude of the stator voltage v_{sabc} undergoes a gradual increase process. The stator breaker is not closed until the stator voltage

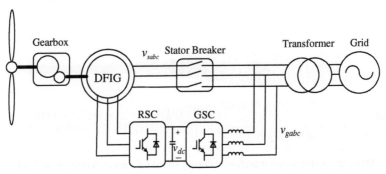

Figure 5.34   Schematic diagram of the DFIG system in starting mode.

v_{sabc} has the same amplitude, frequency, and phase angle with the grid voltage v_{gabc}. Otherwise, it may result in a large surge current when the stator breaker is closed. As a result, the control goal of the RSC in starting mode is to keep the stator voltage v_{sabc} to track the grid voltage v_{gabc} with respect to its amplitude, frequency, and phase angle so that the smooth connection of the stator windings to the grid can be achieved.

Normally, the starting mode appears once the wind speed reaches the cut-in speed. In this case, the rotor speed is usually under sub-synchronous speed. However, it is possible that the DFIG is connected to the grid with synchronous or super-synchronous speed when the wind turbine system needs to be reconnected again to the grid after short-time grid faults are cleared.

5.6.2 Grid Synchronization

The grid synchronization for the RSC in starting mode is accomplished by using a PLL to get the grid phase angle θ_s and the angular speed ω_s as mentioned before.

5.6.3 Control Scheme

The dq model of DFIG is rewritten here:

$$\vec{v}_{sdq} = R_s \vec{i}_{sdq} + p\vec{\psi}_{sdq} + j\omega_s \vec{\psi}_{sdq} \tag{5.68}$$

$$\vec{v}_{rdq} = R_r \vec{i}_{rdq} + p\vec{\psi}_{rdq} + j(\omega_s - \omega_r)\vec{\psi}_{rdq} \tag{5.69}$$

$$\vec{\psi}_{sdq} = L_s \vec{i}_{sdq} + L_m \vec{i}_{rdq} \tag{5.70}$$

$$\vec{\psi}_{rdq} = L_m \vec{i}_{sdq} + L_r \vec{i}_{rdq} \tag{5.71}$$

Since the stator breaker is not closed in the starting mode, the stator current $\vec{i}_{sdq}$ of the DFIG is zero. By substituting $\vec{i}_{sdq} = 0$ in equation (5.68), the stator circuit equation is simplified as

$$\vec{v}_{sdq} = p\vec{\psi}_{sdq} + j\omega_s \vec{\psi}_{sdq} \tag{5.72}$$

For the balanced grid voltage, both d-axis and q-axis components of $\vec{v}_{sdq}$ are DC. If the stator flux reaches steady state, the above equation can be simplified as

$$\vec{v}_{sdq} = j\omega_s \vec{\psi}_{sdq} \tag{5.73}$$

By substituting $\vec{i}_{sdq} = 0$ in equation (5.70), the stator flux is expressed as

$$\vec{\psi}_{sdq} = L_m \vec{i}_{rdq} \tag{5.74}$$

By combining equations (5.73) and (5.74), the stator voltage can be represented by the rotor current

$$\vec{v}_{sdq} = j\omega_s L_m \vec{i}_{rdq} \tag{5.75}$$

It is observed that in starting mode, the stator voltage can be controlled by the rotor current.

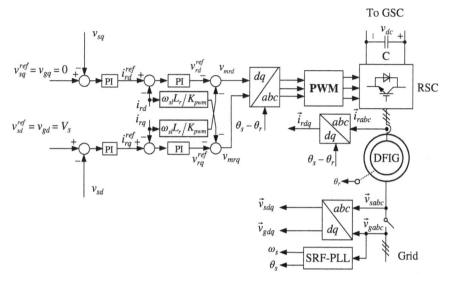

Figure 5.35   The control scheme of the RSC in starting mode.

By substituting $\vec{i}_{sdq} = 0$ in equation (5.71), the rotor flux can be expressed as

$$\vec{\psi}_{rdq} = L_r \vec{i}_{rdq} \tag{5.76}$$

By inserting the above equation to equation (5.69), the following rotor equation can be obtained

$$\vec{v}_{rdq} = R_r \vec{i}_{rdq} + L_r p \vec{i}_{rdq} + j(\omega_s - \omega_r) L_r \vec{i}_{rdq} \tag{5.77}$$

The above equation can be split into d-axis and q-axis components as follows:

$$v_{rd} = R_r i_{rd} + L_r \frac{d}{dt} i_{rd} - \omega_{sl} L_r i_{rq}$$

$$v_{rq} = R_r i_{rq} + L_r \frac{d}{dt} i_{rq} + \omega_{sl} L_r i_{rd} \tag{5.78}$$

where $\omega_{sl} = \omega_s - \omega_r$.

According to equation (5.78), the rotor currents depend on the rotor voltages, which are controlled by the AC output voltage of the RSC. The control scheme of the RSC in starting mode is shown in Figure 5.35 [13–15]. The stator voltages are controlled by the outer voltage loops. The d-axis stator voltage component v_{sd} is required to track the d-axis grid voltage component reference v_{gd}, which is the grid voltage amplitude V_s under grid voltage oriented dq frame, while the q-axis stator voltage component v_{sq} is controlled to track the d-axis grid voltage component v_{gq}, which is 0 in grid voltage oriented dq frame.

The PI regulator outputs of the outer stator voltage loops are the rotor current references i_{rq}^{ref}, i_{rd}^{ref} for the inner current loops. The rotor currents i_{rq} and i_{rd} are controlled by the inner loops. The current loops are similar to those in GSC. However, since the rotor rotates with the angular speed ω_r, the angular speed difference between

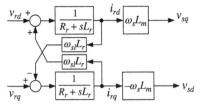

Figure 5.36   s-domain model of the DFIG in starting mode.

the synchronous dq frame and the rotor is ω_{sl} and the angle difference between the dq frame and the rotor is $\theta_s - \theta_r$. Therefore, the angle used in the dq transform for RSC is changed to $\theta_s - \theta_r$ instead of θ_s in GSC. Rotor angle θ_r is generally measured by the rotor encoder. Similarly, the d-axis and q-axis rotor current controls are decoupled as shown in Figure 5.35. The PI outputs of the inner loops are the rotor voltages reference v_{mrq} and v_{mrd}, which enters PWM modulator to realize switching control of the RSC converter.

5.6.4   Control Model in s-Domain

The dynamic model of the DFIG in the s-domain can be derived from (5.75) and (5.78):

$$v_{rd} = R_r i_{rd} + sL_r i_{rd} - \omega_{sl}L_r i_{rq}$$

$$v_{rq} = R_r i_{rq} + sL_r i_{rq} + \omega_{sl}L_r i_{rd} \tag{5.79}$$

$$v_{sd} \approx -\omega_s L_m i_{rq} \tag{5.80}$$

$$v_{sq} \approx \omega_s L_m i_{rd} \tag{5.81}$$

According to equations (5.79)–(5.81), the control diagram of the DFIG in starting mode is obtained as shown in Figure 5.36.

By replacing the DFIG in Figure 5.35 with the s-domain model in Figure 5.36, the control diagram of RSC-controlled DFIG is obtained, as shown in Figure 5.37. According to equation (5.80), q-axis rotor current i_{rq} is proportion to d-axis stator voltage component v_{sd} but with opposite polarity. To realize negative feedback

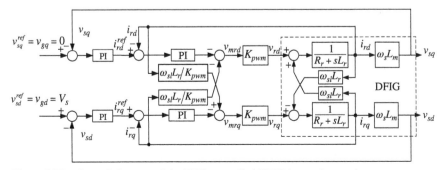

Figure 5.37   Control diagram of the RSC-controlled DFIG in starting mode.

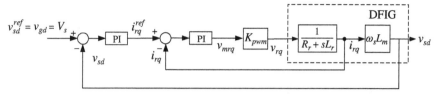

Figure 5.38   The simplified model of the RSC and the DFIG in starting mode.

control, d-axis stator voltage component v_{sd} control loop is a little revised as shown in Figure 5.37.

After introducing decoupling between the d-axis and q-axis current, the d-axis control and q-axis control are independent of each other. The d-axis loop and q-axis loop are almost the same except their difference in the voltage references. Here only q-axis control diagram is drawn in Figure 5.38. In the following control design in starting mode, the q-axis control design is taken as an example.

The open-loop transfer function $G_{vi}(s)$ for the inner current loop, which is from the rotor voltage reference v_{mrq} to the rotor current i_{rq}, is expressed as

$$G_{vi}(s) = K_{pwm} \frac{1}{R_r + sL_r} \qquad (5.82)$$

PI controller in the inner current loop is

$$G_{PIi}(s) = k_{pi} + \frac{k_{ii}}{s} \qquad (5.83)$$

The compensated open-loop function $G_{ii}(s)$ from the rotor current reference i_{rq}^{ref} to rotor current i_{rq} is described by

$$G_{ii}(s) = G_{PIi}(s)G_{vi}(s) = \left(k_{pi} + \frac{k_{ii}}{s} \right) K_{pwm} \frac{1}{R_r + sL_r} \qquad (5.84)$$

The closed-loop transfer function $H_{ii}(s)$ for the inner current loop is

$$H_{ii}(s) = \frac{G_{ii}(s)}{1 + G_{ii}(s)} \qquad (5.85)$$

For the outer voltage loop, the open-loop transfer function $G_{is}(s)$ from the rotor current reference i_{rq}^{ref} to the stator voltage v_{sd} can be derived as

$$G_{iv}(s) = H_{ii}(s)\omega_s L_m \qquad (5.86)$$

PI controller in the outer voltage loop is

$$G_{PIv}(s) = k_{pv} + \frac{k_{iv}}{s} \qquad (5.87)$$

The compensated open-loop function $G_{vv}(s)$ for the outer voltage loop is described by

$$G_{vv}(s) = G_{PIv}(s)G_{iv}(s) = \left(k_{pv} + \frac{k_{iv}}{s} \right) H_{ii}(s)\omega_s L_m \qquad (5.88)$$

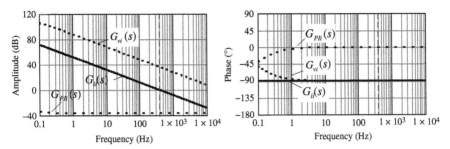

Figure 5.39   Bode diagrams of the inner current loop design.

5.6.5   Controller Design

The controller design of the RSC in power-control mode is similar to that of the RSC in starting mode. However, the inner current loop in power-control mode is different from that in starting mode. The inductance in open-loop transfer function $G_{vi}(s)$ is the leakage inductance σL_r in power-control mode while it changes to the rotor inductance L_r in starting mode.

For the inner current loop, the crossover frequency is selected to be 1/5 of the switching frequency. As the switching frequency is 2000 Hz, a crossover frequency of 400 Hz is chosen. The PI regulator of the inner current loop is designed similarly as mentioned before. The parameters of the PI regulator are $k_{pi} = 0.016$, $k_{ii} = 0.008$. Bode diagrams of the open-loop function $G_{vi}(s)$, the compensated system $G_{ii}(s)$, and PI regulator $G_{PIi}(s)$ are shown in Figure 5.39. The compensated system has a crossover frequency of 400 Hz and phase margin of 90°.

For the outer voltage loop, the crossover frequency is selected to be 1/10 of the crossover frequency of the inner current loop to ensure the stability of the cascade control, which is 40 Hz. The PI parameters of the outer voltage loop are derived such that $k_{pv} = 0.16$, $k_{iv} = 200$. Bode diagrams of the open-loop system $G_{iv}(s)$, the compensated system $G_{vv}(s)$, as well as PI regulator $G_{PIv}(s)$ are shown in Figure 5.40. For the compensated system $G_{vv}(s)$, its crossover frequency is 40 Hz and the phase margin is 90°.

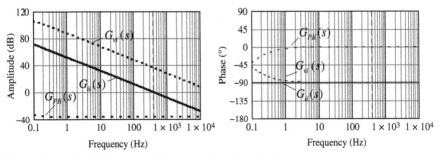

Figure 5.40   Bode diagrams of the outer voltage loop design.

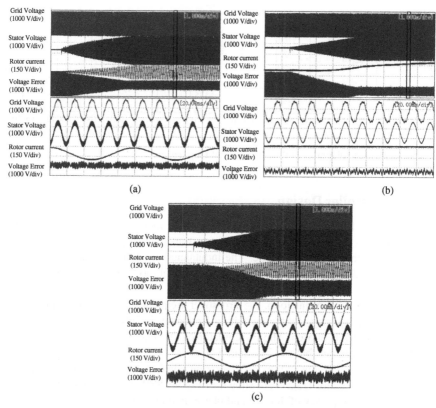

(a)   (b)

(c)

Figure 5.41   The stator voltage tracks the grid voltage under different rotor speeds in starting mode: (a) sub-synchronous speed of 1200 rpm; (b) synchronous speed of 1500 rpm; (c) super-synchronous speed of 1800 rpm.

5.6.6 Experiment Results

Experiment waveforms of a 1.5 MW DFIG WPS are shown here. During the starting mode, the stator voltage is controlled to track the grid voltage as shown in Figure 5.41. The waveforms from top to bottom in each figure in Figure 5.41 are grid voltage, stator voltage, rotor current, and the error between stator voltage and grid voltage, respectively. The rotor current is increased while the error between the grid voltage and stator voltage gradually decreases. After the soft-start process, the stator voltage has the same amplitude, frequency, and phase angle of the grid voltage. It is observed that the frequency of the rotor current depends on the rotor speed. Once the RSC finishes starting mode and the stator voltage of the DFIG is established, the stator side is ready to connect to the grid.

The grid connection of the DFIG under different rotor speeds is shown Figure 5.42. The waveforms from the top to the bottom are grid voltage, stator voltage, rotor current, and stator current. After the finishing the starting mode, the stator side is

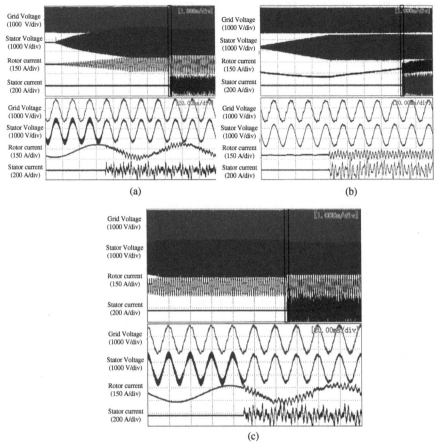

Figure 5.42   Grid connection of the DFIG under different rotor speeds: (a) sub-synchronous speed of 1200 rpm; (b) synchronous speed of 1500 rpm; (c) super-synchronous speed of 1800 rpm.

connected to the grid through the breaker. As the error between the stator voltage and the grid voltage is small before the closing of the breaker, the grid connection process is smooth and no large transient stator or rotor current is observed, as shown in Figures 5.42a–5.42c. The starting mode control of the RSC ensures the smooth connection of the DFIG stator side to the grid.

5.7   CONTROL-MODE SWITCHING

In the above sections, three control modes are introduced for the DFIG. The DFIG is operates in the starting mode before it connects to the grid. After it is connected to the grid, the DFIG begins to operate either at speed-control mode or power-control mode. As a result, the control-mode changing of the DFIG cannot be avoided.

5.7.1 From Starting Mode to Power-Control or Speed-Control Mode

If the stator voltage has tracked the grid voltage in the starting mode, the stator breaker is closed and the DFIG is smoothly connected to the grid. As the control model of the DFIG in starting mode and power control (or speed control mode) are different, the PI regulator parameters in these two modes should be changed. Since the PI regulator parameters for the inner current loop in the starting mode are much larger than that in the power-control or speed-control mode, the PI parameters need to be reduced as soon as possible once the stator breaker is closed; otherwise, the system may be unstable. On the other hand, the magnetizing current still needs to be provided in the power-control or speed-control mode, which means the q-axis rotor current reference is the same as that in the starting mode even when the reactive command is zero. The switching process from the starting mode to the power or speed-control mode normally goes in the following steps:

Step 1: After the stator voltage has the same amplitude, frequency, and phase as the grid, the stator breaker is closed to connect the stator to the grid.

Step 2: The stator breaker normally needs dozens of milliseconds to close and during this period, the DFIG is still running in the starting mode. After the stator breaker is finally closed, the controller will receive a confirm signal from the breaker.

Step 3: As soon as the controller receives the confirm signal of the stator breaker, the control scheme is changed from the starting mode to the power-control or speed-control mode. At the same time, the PI parameters of the inner current loop are refreshed to accommodate the power-control or speed-control mode. The initial values of the PI regulator of the inner current loop in starting mode are used by the updated PI regulator for the power-control or speed-control mode.

Smooth mode switching can be achieved when the DFIG is connecting to the grid, and the corresponding waveforms can be found in Figure 5.42.

5.7.2 Between Power-Control Mode and Speed-Control Mode

As the inner current loop is the same for power-control mode and speed-control mode, the switching between these two modes is easier. When the mode switching is needed, the rotor d-axis current reference swaps from the output of the active power loop to the rotor speed loop or vice versa. As the PI regulator may be saturated during the mode switching, the initialization of the integer of the PI regulator may be necessary.

5.8 SUMMARY

In this chapter, the fundamentals of the converter controls for the DFIG are introduced. The control of GSC is analyzed firstly. GSC is controlled to provide constant

DC-bus voltage for RSC and to realize bidirectional power flow. Then the control schemes of the RSC are explained. The RSC has three operation modes, that is, starting mode, power-control mode, and speed-control mode. In the starting mode, the RSC is used to realize a smooth grid connection of the DFIG stator windings to the grid. In power-control mode, the output active power and reactive power of the DFIG are controlled by the RSC. In the speed-control mode, the rotor speed is controlled by the RSC. Both the power and speed-control modes are used to accommodate the requirement of the wind turbine system. The mode switching for the RSC is also explained. Corresponding test waveforms are included to explain the control result. This chapter tries to help the readers to understand how to control the DFIG, which is the basis for the following chapters.

REFERENCES

[1] W. Leonhard, *Control of Electrical Drives*. Springer, 1985.
[2] M. P. Kazmierkowski, R. Krishnan, and F. Blaabjerg, *Control in Power Electronics: Selected Problems*. Academic Press, 2002.
[3] A. Veltman, D. W. J. Pulle, and R. W. DeDoncker, *Fundamentals of Electric Drives*. Springer, 2007.
[4] B. K. Bose, *Power Electronics and Drives*. Elsevier, 2006.
[5] S. J. Chapman, *Electric Machinery Fundamentals*. McGraw-Hill, 2005.
[6] B. K. Bose, *Modern Power Electronics and AC Drives*. Princeton Hall, 2002.
[7] G. Abad, J. Lopez, M. A. Rodriguez. L. Marroyo, and G. Iwanski, *Doubly Fed Induction Machine: Modeling and Control for Wind Energy Generation*. Wiley-IEEE Press, 2011.
[8] P. Kunder, *Power System Stability and Control*. McGraw-Hill, 1994.
[9] B. Wu, Y. Lang, N. Zargari, and S. Kouro. *Power Conversion and Control of Wind Energy Systems*. Wiley-IEEE Press, 2011.
[10] Y. He, J. Hu, and L. Xu, *Operation Control of Grid Connected Doubly Fed Induction Generator*. China: Electric Power Press, 2012.
[11] F. Blaabjerg, R. Teodorescu, M. Liserre, and A. V. Timbus, "Overview of control and grid synchronization for distributed power generation systems," *IEEE Trans. Ind. Electron.*, vol. 53, no. 5, pp. 1398–1409, Oct. 2006.
[12] Z. Chen. J. M. Guerrero, and F. Blaabjerg, "A review of the state of the art of power electronics for wind turbines," *IEEE Trans. Power Electron.*, vol. 24, no. 8, pp. 1859–1875, 2009.
[13] R. Pena, J. C. Clare, and G. M. Asher, "Doubly fed induction generator using back-to-back PWM converters and its application to variable speed wind-energy generation," *Proc. IEE Elect. Power Appl.*, vol. 143, no. 3, pp. 231–241, 1996.
[14] M. Yamamoto and O. Motoyoshi, "Active and reactive power control for doubly-fed wound rotor induction generator," *IEEE Trans. Power Electron.*, vol. 6, no. 4, pp. 624–629, 1991.
[15] D. Forchetti, G. Garcia, and M. I. Valla, "Vector control strategy for a doubly-fed stand-alone induction generator," in *Proc. IEEE Ann. Conf. Ind. Electron. Soc.*, Sevilla, Spain, 2002, pp. 991–995.
[16] L. Peng, Y. Li, J. Chai, and G. F. Yuan, "Vector control of a doubly fed induction generator for stand-alone ship shaft generator systems," in *Proc. Int. Conf. Electr. Mach. Syst.*, 2007, pp. 1033–1036.
[17] L. Xu and P. Cartwright, "Direct active and reactive power control of DFIG for wind energy generation," *IEEE Trans. Energy Convers.*, vol. 21, no. 3, pp. 750–758.
[18] L. Xu and W. Cheng, "Torque and reactive power control of a doubly fed induction machine by position sensorless scheme," *IEEE Trans. Ind. Appl.*, vol. 31, no. 3, pp. 636–642, 1995.
[19] O. A. Mohammed, Z. Liu, and S. Liu, "A novel sensorless control strategy of doubly fed induction motor and its examination with the physical modeling of machines," *IEEE Trans. Magn.*, vol. 41, no. 5, pp. 1852–1855, 2005.

[20] M. Tazil, V. Kumar, R. C. Bansal, S. Kong, Z. Y. Dong, W. Freitas, and H. D. Mathur, "Three-phase doubly fed induction generators: An overview," *IET Electric Power Appl.*, vol. 4, no. 2, pp. 75–89, 2010.

[21] E. Tremblay, S. Atayde, and A. Chandra, "Comparative study of control strategies for the doubly fed induction generator in wind energy conversion systems a DSP-based implementation approach," *IEEE Trans. Sustain. Energy*, vol. 2, no. 3, pp. 288–299, 2011.

[22] S. Muller, M. Deicke, and R. W. De Doncker, "Doubly fed induction generator systems for wind turbines," *IEEE Ind. Appl. Mag.*, vol. 8, no. 3, pp. 26–33, 2002.

[23] J. Hu, Y. He, L. Xu, and B. W. Williams, "Improved control of DFIG systems during network unbalance using PI-R current regulators," *IEEE Trans. Ind. Electron.*, vol. 56, no. 2, pp. 439–451, 2009.

[24] Z. Yi, P. Bauer, J. A. Ferreira, and J. Pierik, "operation of grid-connected DFIG under unbalanced grid voltage condition," *IEEE Trans. Energy Convers.*, vol. 24, no. 1, pp. 240–246, 2009.

[25] J. Xu, "Research on converter control strategy of doubly fed induction generator system for wind power," Ph.D. thesis, Zhejiang University, Hangzhou, China, 2011.

[26] C. Liu, "Resonant control of DFIG wind power converters for adapting to the grid environment," Ph.D. thesis, Zhejiang University, Hangzhou, China, 2012.

[27] G. Abad and G. Iwanski, "Properties and control of a doubly fed induction machine," in *Power Electronics for Renewable Energy Systems, Transportation and Industrial Application.* John Wiley & Sons, 2014.

OPERATION OF DFIG UNDER DISTORTED GRID VOLTAGE

CHAPTER 6

ANALYSIS OF DFIG UNDER DISTORTED GRID VOLTAGE

The harmonic voltages in a power system is generated by the nonlinear voltage source or loads. For example, the magnetic saturation in generators and transformers will introduce 3rd-order harmonic voltage, and nonlinear loads will bring negative 5th-, positive 7th-, 11th-, and 13th-order harmonic voltages to the grid. Normally, if the total harmonic distortion (THD) of the grid voltage is lower than 2–5%, regarding different voltage levels, the grid voltage is considered as normal. Then the harmonic voltage even in "normal grid" will influence the output power quality of the renewable energy system. Especially for wind power systems, as they are often located in remote areas, the weak grid may lead to higher THD at the wind turbine terminals. In this chapter, the performance of the DFIG wind power system under harmonic distorted grid is analyzed. The influence of the grid harmonic voltage on the grid-side converter (GSC) current, stator current, rotor current, DC-bus voltage, active/reactive power, as well as the torque fluctuations are analyzed. The challenge for the DFIG wind power system operating under harmonic distorted grid voltage is evaluated.

6.1 INTRODUCTION

In Chapter 5, the conventional vector control for the DFIG is introduced under normal grid operation, which means the grid voltage is three-phase ideal sinusoidal AC voltage with constant amplitude and frequency. Of course, this kind of ideal grid voltage does not exist. Grid harmonic distortion, unbalance, and grid fault are common cases in a real power system, and the grid code demands the wind power system should keep working in these situations. This chapter starts with the behavior of the DFIG WPS under grid harmonic distortion. The harmonic voltage in a power system is generated by a nonlinear voltage source or loads. For example, the magnetic saturation in generators and transformers will introduce 3rd-order harmonic voltage, and the nonlinear load will bring negative 5th-, positive 7th-, 11th-, and 13th-order harmonic voltages into the grid. Normally, if the total harmonic distortion (THD) of the grid voltage is lower than 2–5% [1], at different voltage levels, the grid voltage is regarded to be normal, and the DFIG should provide active power with the required

Advanced Control of Doubly Fed Induction Generator for Wind Power Systems, First Edition.
Dehong Xu, Frede Blaabjerg, Wenjie Chen, and Nan Zhu.

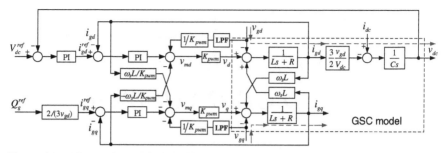

Figure 6.1   1 Dynamic model of the GSC with vector control.

power quality, following the related grid codes introduced in Chapter 3. For wind power systems, as they are often located in remote areas, the weak grid may lead to higher THD in wind turbine terminal, and make it even harder to fulfill the related grid codes. In this chapter, the performance of the DFIG wind power system under harmonic distorted grid is analyzed [2]. The influence of the grid harmonic voltage on the grid-side converter (GSC) current, stator current, rotor current, DC-bus voltage, active/reactive power, as well as the torque fluctuations are analyzed. Therefore, the challenge for the DFIG wind power system to operate under harmonic distorted grid voltage can be evaluated [3–4].

6.2   INFLUENCE ON GSC

The stator side of the DFIG is directly connected to the grid, while the rotor side is connected to the grid by the back-to-back connected rotor-side converter (RSC) and GSC, as has been introduced in Chapter 2. Consequently, the influence of the harmonic distorted grid voltage on the DFIG will be shown in two aspects: their influence on the DFIG (together with the RSC) and their influence on the GSC. However, as the RSC and GSC are connected to a common DC bus, the DC-bus voltage may be influenced by both the RSC side and the GSC side. The analysis will start with the GSC first.

6.2.1   Model of GSC under Distorted Grid Voltage

The dynamic model of the GSC with vector control has been introduced in Chapter 5, and it is represented here as shown in Figure 6.1. The grid harmonic voltage can be regarded as the disturbance in grid voltage v_{gd} and v_{gq}, and it can be found from Figure 6.1 that since the harmonic components in the decoupling terms are filtered by the low-pass filters (LPF), the harmonics in the grid voltage will influence the GSC grid current i_{gd} and i_{gq}, as well as the DC-bus voltage v_{dc} [5].

The relationship between the GSC grid current, grid voltage, and GSC output voltage in the s-domain is rewritten in (6.1).

$$v_d = (Ls + R)i_{gd} - \omega_s L i_{gq} + v_{gd}$$
$$v_q = (Ls + R)i_{gq} + \omega_s L i_{gd} + v_{gq} \tag{6.1}$$

The GSC output voltage v_d and v_q are calculated from the control loop, as shown in (6.2).

$$v_d = K_{pwm}G_{pii}(s)(i_{gd}^{ref} - i_{gd}) + \omega_s L i_{gq}$$

$$v_q = K_{pwm}G_{pii}(s)(i_{gq}^{ref} - i_{gq}) + \omega_s L i_{gd} \qquad (6.2)$$

Together with (6.1), the GSC output voltage v_d and v_q can be eliminated and the relationship between grid voltage disturbance v_{gd}, v_{gq}, and the GSC grid currents i_{gd}, i_{gq} can then be found as

$$(Ls + R)i_{gd} + v_{gd} = K_{pwm}G_{pii}(s)(i_{gd}^{ref} - i_{gd})$$

$$(Ls + R)i_{gq} + v_{gq} = K_{pwm}G_{pii}(s)(i_{gq}^{ref} - i_{gq}) \qquad (6.3)$$

As the grid voltage contains harmonic components, $v_{gd} = V_s$ and $v_{gq} = 0$ does not hold anymore. The GSC grid current references i_{gd}^{ref}, i_{gq}^{ref} are coming from the outer DC voltage/reactive power loop. We focus, in this chapter, on the harmonic voltage disturbance, normally negative 5th, positive 7th, 11th, and 13th harmonic grid voltages and with the frequency higher than 300 Hz in the dq reference frame. On the other hand, the crossover frequency of the outer loop is less than 50 Hz for large-scale DFIG WPS [6], so the harmonic components are small enough to be neglected in i_{gd}^{ref} and i_{gq}^{ref}. With $i_{gd}^{ref} = 0$ and $i_{gq}^{ref} = 0$, (6.3) can be rewritten as

$$(Ls + R)i_{gd} + v_{gd} = -K_{pwm}G_{pii}(s)i_{gd}$$

$$(Ls + R)i_{gq} + v_{gq} = -K_{pwm}G_{pii}(s)i_{gq} \qquad (6.4)$$

The transfer function $G_{vi}(s)$ from grid voltage disturbance to the GSC output current can then be derived as (6.5).

$$G_{vi}(s) = \frac{i_{gd}}{v_{gd}} = \frac{i_{gq}}{v_{gq}} = -\frac{1}{Ls + R + K_{pwm}G_{pii}(s)} \qquad (6.5)$$

This transfer function describes the influence of the grid voltage disturbance on the GSC grid current in dq reference frames, and with conventional vector control. $G_{pii}(s) = k_{pi} + k_{ii}/s$ is the transfer function of the PI controller in the inner current loop. As the grid voltage disturbance and the GSC grid current hold a linear relationship, according to the superposition principle, the influence of grid voltage disturbance on the GSC grid current with different frequencies can be evaluated separately. Particularly, as the negative fifth and positive seventh grid voltage harmonics are all having a frequency of $6f_s$ in the dq reference frame as

$$G_{vi}\left(j2\pi 6f_s\right) = \frac{1}{j2\pi 6f_s L + R + K_{pwm}G_{pii}\left(j2\pi 6f_s\right)} \qquad (6.6)$$

Equation (6.6) represents the influence of the grid negative fifth and positive seventh harmonic voltages on the GSC output current. The same results can be derived

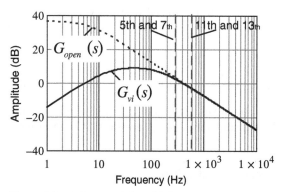

Figure 6.2 Influence of the grid disturbance on the GSC grid current. $G_{vi}(s)$, transfer function from the grid voltage disturbances to the GSC grid current using vector control; $G_{open}(s)$, transfer function from the grid voltage disturbances to the GSC grid current with open loop control.

for the 11th and 13th grid harmonic voltages, as they are all having a frequency of $12 f_s$ in the *dq* reference frame:

$$G_{vi}\left(j2\pi 12 f_s\right) = \frac{1}{j2\pi 12 f_s L + R + K_{pwm}G_{pii}\left(j2\pi 12 f_s\right)} \tag{6.7}$$

For the output active power P_g and reactive power Q_g, they can be expressed as

$$P_g = -\frac{3}{2}(v_{gd}i_{gd} + v_{gq}i_{gq}) = -\frac{3}{2}\left[v_{gd}G_{vi}(s)v_{gd} + v_{gq}G_{vi}(s)v_{gq}\right]$$

$$Q_g = -\frac{3}{2}(v_{gd}i_{gq} + v_{gq}i_{gd}) = -\frac{3}{2}\left[v_{gd}G_{vi}(s)v_{gq} + v_{gq}G_{vi}(s)v_{gd}\right] \tag{6.8}$$

As the influence of the outer loop on the harmonic component can be neglected, the DC-bus voltage can be found as v_{dc}.

$$v_{dc} = \left(i_{dc} - \frac{3}{2}\frac{v_{gd}}{V_{dc}}i_{gd}\right)\frac{1}{Cs} = \left[i_{dc} - \frac{3}{2}\frac{v_{gd}}{V_{dc}}\frac{1}{Ls + R + K_{pwm}G_{pii}(s)}\right]\frac{1}{Cs} \tag{6.9}$$

As for (6.8) and (6.9), their relationship with the grid voltage disturbance v_{gd} and v_{gq} are not linear and the superposition principle does not hold here. The influence of the grid voltage harmonic disturbance on the DC-bus voltage and output active and reactive powers cannot simply be described by one transfer function. The details will be introduced in Section 6.2.4.

6.2.2 Influence on Grid Current

The transfer function $G_{vi}(s)$ from the grid voltage disturbances to the GSC grid current has been given in (6.5). The amplitude of $G_{vi}(s)$ represents how large the amplitude will be for the GSC grid current under grid voltage disturbance. For a 1.5 MW DFIG wind power system, it is illustrated in Figure 6.2. The solid line represents the

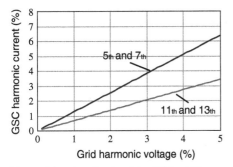

Figure 6.3   Amplitude of the GSC grid harmonic current relative to the amplitude of grid harmonic voltage.

amplitude of $G_{vi}(s)$ with different frequencies, and it can describe the influence of the grid disturbance on the GSC grid current with different frequencies. Also, the amplitude of the transfer function from the grid voltage disturbance to the GSC grid current without vector control loop (open loop) $G_{open}(s)$ is

$$G_{open}(s) = \frac{1}{Ls + R} \qquad (6.10)$$

This is also shown in Figure 6.2. It represents the influence of the grid disturbance on the GSC grid current without using the vector control loop.

It can be found from Figure 6.2 that the amplitude of $G_{vi}(s)$ is smaller than the amplitude of $G_{open}(s)$ if the frequency is smaller than 300 Hz, which means the influence of the grid disturbance on the GSC current is suppressed by the vector control loop if the frequency of the grid disturbance is smaller than 300 Hz. However, if the frequency is larger than 300 Hz, the amplitude of $G_{vi}(s)$ and $G_{open}(s)$ are about the same. The vector control cannot suppress the influence of the grid voltage disturbance above 300 Hz. As the switching frequency for large power DFIG power converter is normally a few kilohertz, the band width of the vector control is then limited to a few hundred Hz. In this case, it is only 400 Hz, so the control effect above the bandwidth is limited, and the vector control loop is not able to suppress the influence of the grid voltage disturbance on the GSC grid current above this control bandwidth.

On the other hand, the most common grid harmonic voltage disturbance in the power system is negative 5th and positive 7th harmonic voltage, as well as 11th and 13th harmonic voltage. For the negative 5th and positive 7th harmonic voltage disturbances, their frequency in the dq reference frame is 300 Hz (with a fundamental frequency of 50 Hz), and the frequency of 11th and 13th harmonic voltages in the dq reference frame is 600 Hz. It can be found from Figure 6.2 that the vector control loop is not able to suppress the grid disturbance of such a high frequency. The amplitude of the GSC grid harmonic current related to the amplitude of grid harmonic voltage is shown in Figure 6.3. As the grid harmonic voltage in the power system may sometimes reach 5% at most in remote areas, the negative fifth and positive seventh GSC grid harmonic current may reach more than 6%. It has been introduced in Chapter 4 that the largest current THD for grid-connected power converter governed by the IEEE is 5%. So, the GSC may fail to fulfill the related power quality requirement. Five percent 11th and 13th harmonic voltages will introduce around 3.5% grid harmonic current at most, and it will also reduce the power quality of the GSC.

6.2.3   Influence on Output Active and Reactive Powers

The relationship between the active and reactive powers P_g, Q_g, respectively, and the grid voltages v_{gd}, v_{gq} has been presented in (6.8). As this is not a linear relationship, a simple transfer function cannot be used to analyze it, as it was for the GSC grid current. For the GSC grid current $\vec{i}_g$, if the grid harmonic voltage disturbance is with the frequency of hf_s in the dq reference frame ($h = 6$ for negative 5th and positive 7th harmonics, and $h = 12$ for 11th and 13th harmonics), the grid current in dq reference frame can be represented as

$$\vec{i}_{gdq} = \vec{i}_{gdq0} + \vec{i}_{gdqh}e^{jh\omega_s t} \tag{6.11}$$

where $\vec{i}_{gdq0}$ is the fundamental grid current in vector form in the complex plane, and $\vec{i}_{gdqh}$ is the grid harmonic current written in vector form in the complex plane, in the dq reference frame. $\vec{i}_{gdq0} = i_{gd0} + ji_{gq0}$, $\vec{i}_{gdqh} = i_{gdh} + ji_{gqh}$. The grid voltage $\vec{v}_g$ can also be written as

$$\vec{v}_{gdq} = \vec{v}_{gdq0} + \vec{v}_{gdqh}e^{jh\omega_s t} \tag{6.12}$$

where $\vec{v}_{gdq0}$ is the fundamental grid current in vector form in the complex plane, and $\vec{v}_{gdqh}$ is the grid harmonic current written in vector form in the complex plane, in the dq reference frame. $\vec{v}_{gdq0} = v_{gd0} + jv_{gq0}$, $\vec{v}_{gdqh} = v_{gdh} + jv_{gqh}$. The apparent power S_g of the GSC to the grid can be represented as

$$S_g = -\frac{3}{2}\left(\vec{v}_{gdq}\vec{i}_{gdq}^*\right) = -\frac{3}{2}\left(\vec{v}_{gdq0} + \vec{v}_{gdqh}e^{jh\omega_s t}\right)\left(\vec{i}_{gdq0} + \vec{i}_{gdqh}e^{-jh\omega_s t}\right) \tag{6.13}$$

Then, the output active power P_g and reactive power Q_g can be expressed as

$$P_g = \mathrm{Re}\left(S_g\right) = P_{g0} + P_{g\,\cosh}\cos(h\omega_s t) + P_{g\,\sinh}\sin(h\omega_s t) \tag{6.14}$$

$$Q_g = \mathrm{Im}\left(S_g\right) = Q_{g0} + Q_{g\,\cosh}\cos(h\omega_s t) + Q_{g\,\sinh}\sin(h\omega_s t) \tag{6.15}$$

where P_{g0}, Q_{g0} are the DC component in the active and reactive power outputs, respectively. $P_{g\,\cosh}, P_{g\,\sinh}, Q_{g\,\cosh}, Q_{g\,\sinh}$ are the AC fluctuation components in the active and reactive power outputs, respectively. They can be expressed as

$$
\begin{bmatrix} P_{g0} \\ Q_{g0} \\ P_{g\,\cosh} \\ P_{g\,\sinh} \\ Q_{g\,\cosh} \\ Q_{g\,\sinh} \end{bmatrix} = \begin{bmatrix} v_{gd0} & v_{gq0} & v_{gdh} & v_{gqh} \\ v_{gq0} & -v_{gd0} & v_{gqh} & -v_{gdh} \\ v_{gdh} & v_{gqh} & v_{gd0} & v_{gq0} \\ v_{gqh} & -v_{gdh} & -v_{gq0} & v_{gd0} \\ v_{gqh} & -v_{gdh} & v_{gq0} & -v_{gd0} \\ -v_{gdh} & -v_{gqh} & v_{gd0} & v_{gq0} \end{bmatrix} \cdot \begin{bmatrix} i_{gd0} \\ i_{gq0} \\ i_{gdh} \\ i_{gqh} \end{bmatrix}
\tag{6.16}
$$

From (6.16), it can be concluded that the active power and the reactive power outputs contains AC components with a frequency of hf_s under grid harmonic voltage, and their amplitude is related to both the amplitude of the fundamental and harmonic components in the grid voltage and the GSC grid current [7–8].

6.2.4   Influence on the DC-Bus Voltage

The dynamic model of the GSC in the dq reference frame written in vector form in the complex plane can be represented as

$$
\vec{v}_{dq} = \left(R + L\frac{d}{dt} \right)\vec{i}_{gdq} + j\omega_s L\vec{i}_{gdq} + \vec{v}_{gdq}
\tag{6.17}
$$

where $\vec{v}_{dq} = v_d + jv_q$ is the output voltage of the GSC in the dq reference frame, $\vec{i}_{gdq} = i_{gd} + ji_{gq}$ is the GSC grid current in the dq reference frame, while $\vec{v}_{gdq} = v_{gd} + jv_{gq}$ is the grid voltage in the dq reference frame. As the bandwidth of the current loop is limited, the vector control loop cannot control the harmonic component, so the harmonic component in the output voltage $\vec{v}_{dq}$ can be neglected and only the fundamental component $\vec{v}_0$ will be found in the output voltage $\vec{v}_{dq}$, as shown below.

$$
\vec{v}_{dq} = \vec{v}_{dq0}
\tag{6.18}
$$

Substituting (6.18) and (6.11) in (6.17), the relationship between the grid voltage, the GSC output voltage, and the GSC grid current under grid voltage harmonic distortion can be found as

$$
\begin{aligned}
\vec{v}_{gdq} &= -\left(R + L\frac{d}{dt} \right)\vec{i}_{gdq} + j\omega_s L\vec{i}_{gdq} + \vec{v}_{dq} \\
&\approx -\left(L\frac{d}{dt} + j\omega_s L \right)\left(\vec{i}_{gdq0} + \vec{i}_{gdqh}e^{jh\omega_s t} \right) + \vec{v}_{dq0} \\
&\approx -j\omega_s L\vec{i}_{gdq0} - j(h+1)\omega_s L\vec{i}_{gdqh}e^{jh\omega_s t} + \vec{v}_{dq0}
\end{aligned}
\tag{6.19}
$$

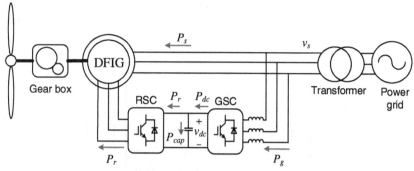

Figure 6.4   Power flow on the DC bus of the GSC in a DFIG system.

So, the fundamental and harmonic components in the grid voltage $\vec{v}_{g0}$ and $\vec{v}_{gh}$, respectively, can be expressed as

$$\vec{v}_{gdq0} \approx -j\omega_s L\vec{i}_{gdq0} + \vec{v}_{dq0} \tag{6.20}$$

$$\vec{v}_{gdqh} \approx -j(h+1)\omega_s L\vec{i}_{gdqh} = -X_{vi}(h\omega_s)\vec{i}_{gdqh} \tag{6.21}$$

If the power losses on the switches are neglected, the power flowing into the DC-bus P_{dc} can be represented as (6.22):

$$P_{dc} = -\frac{3}{2}\mathrm{Re}(\vec{v}_{dq}\vec{i}_{gdq}^*) = -\frac{3}{2}\mathrm{Re}\left(\vec{v}_{dq0}\vec{i}_{gdq0} + \vec{v}_0\vec{i}_{gdqh}e^{-jh\omega_s t}\right)$$

$$= P_{dc0} + P_{dch} \tag{6.22}$$

where $P_{dc0} = -\frac{3}{2}\mathrm{Re}(\vec{v}_0\vec{i}_{g0})$ is the DC component while $P_{dch} = -\frac{3}{2}\mathrm{Re}(\vec{v}_0\vec{i}_{gh}e^{-jh\omega_s t})$ is the AC fluctuation component with a frequency of hf_s. However, the power P_{cap} on the DC-bus capacitance C is not only related to the GSC side, it also influences the rotor-side active power P_r on the RSC side, as shown in Figure 6.4. P_{cap} can be expressed as

$$P_{cap} = P_{dc} - P_r \tag{6.23}$$

In the next section, it will be introduced that the rotor-side active power P_r also contains the AC fluctuation with a frequency of hf_s and it is expressed as P_{rh}. So, the AC-fluctuation components P_{caph} on the DC-bus voltage will be derived as

$$P_{caph} = P_{dch} - P_{rh} = -\frac{3}{2}\mathrm{Re}\left(\vec{v}_0\vec{i}_{gh}e^{-jh\omega_s t}\right) - P_{rh} \tag{6.24}$$

So, P_{caph} is related to the active power fluctuations on both the GSC and the RSC side. The precise expression will be given in the next section. With P_{caph}, the current fluctuation on the DC bus i_{dch} with the frequency of hf_s can be found as

$$i_{dch} = \frac{P_{caph}}{V_{dc}} \tag{6.25}$$

The DC-bus voltage fluctuation v_{dch} can be derived as

$$v_{dch} = i_{dch} \frac{1}{jh\omega_s C} \tag{6.26}$$

Note that the current fluctuation on the DC bus i_{dch} and the DC-bus voltage fluctuation v_{dch} will not generate active power, as they are on a DC-bus capacitor C [9].

6.2.5 Example of a 1.5 MW DFIG WPS

Simulations on a 1.5 MW DFIG WPS is given as an example, and the parameters of the GSC has been given in Chapter 5. The grid voltage contains 3% negative fifth-order harmonic and 2% positive seventh-order harmonic, which makes the THD of the grid voltage to be about 3.6%. The DFIG is working with rated active power under the rotor speed of 1800 rpm (1.2 pu), so about 0.2 pu active power flows from the GSC to the grid. The grid voltage, GSC grid current, the active and reactive powers of the GSC, as well as the DC-bus voltage under the distorted grid is shown in Figure 6.5. The GSC grid current is heavily distorted under grid voltage harmonics and the FFT analysis in Figure 6.6 shows that 3% negative fifth-order harmonic voltage introduced more than 5% negative fifth-order harmonic GSC current, while 2% positive seventh-order harmonic voltage introduced about 3% positive seventh harmonic GSC current, which makes the THD of the GSC grid current reach 6.99%, larger than the THD of the grid voltage, and it cannot meet most of the related grid codes. The grid harmonic voltage introduces sixth-order fluctuations on the active power and reactive power outputs of the GSC. The fluctuations can be found on the DC-bus voltage as well. The simulation results are provided to verify the analysis in the above sections.

6.3 INFLUENCE ON DFIG AND RSC

6.3.1 Model of DFIG and RSC under Distorted Grid Voltage

The grid harmonic distortion will directly influence the stator side of the DFIG, as the stator side of the DFIG is connected to the grid. The dynamic model of the DFIG with vector control has been introduced in Chapter 5. However, the grid voltage disturbance is not considered in that model as it is for the normal operation under ideal grid voltage. If the grid voltage disturbance is taken into consideration, the dynamic model of the DFIG in the dq reference frame written in vector form in the complex plane can be represented as (6.27) and (6.28).

$$\vec{v}_{sdq} = R_s \vec{i}_{sdq} + \frac{d}{dt} \vec{\psi}_{sdq} + j\omega_s \vec{\psi}_{sdq}$$

$$\vec{v}_{rdq} = R_r \vec{i}_{rdq} + \frac{d}{dt} \vec{\psi}_{rdq} + j\omega_{sl} \vec{\psi}_{sdq} \tag{6.27}$$

$$\vec{\psi}_{sdq} = L_s \vec{i}_{sdq} + L_m \vec{i}_{rdq}$$

$$\vec{\psi}_{rdq} = L_r \vec{i}_{rdq} + L_m \vec{i}_{sdq} \tag{6.28}$$

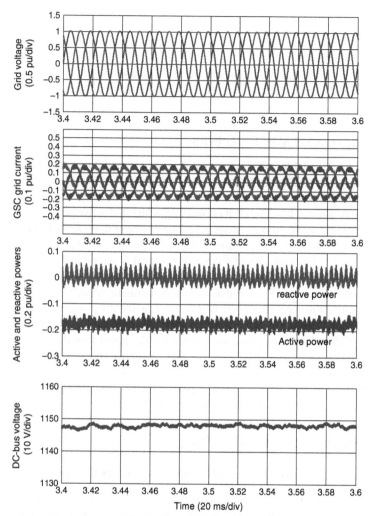

Figure 6.5   Simulation of GSC under distorted grid voltage containing 3% negative fifth-order harmonic and 2% positive seventh-order harmonics.

where $\vec{v}_{sdq} = v_{sd} + jv_{sq}$, $\vec{v}_{rdq} = v_{rd} + jv_{rq}$ are the stator and rotor voltage in the dq reference frame, respectively, $\vec{i}_{sdq} = i_{sd} + ji_{sq}$ and $\vec{i}_{rdq} = i_{rd} + ji_{rq}$ are the stator and rotor current in the dq reference frame, respectively. $\vec{\psi}_{sdq} = \psi_{sd} + j\psi_{sq}$, $\vec{\psi}_{rdq} = \psi_{rd} + j\psi_{rq}$ are the stator and rotor flux linkages in the dq reference frame, respectively. The stator current and rotor flux can be derived from (6.28), as shown in (6.29) and (6.30).

$$\vec{i}_{sdq} = \frac{\vec{\psi}_{sdq} - L_m \vec{i}_{rdq}}{L_s} \tag{6.29}$$

$$\vec{\psi}_{rdq} = L_m \frac{\vec{\psi}_{sdq} - L_m \vec{i}_{rdq}}{L_s} + L_r \vec{i}_{rdq} = \frac{L_m}{L_s} \vec{\psi}_{sdq} + \sigma L_r \vec{i}_{rdq} \tag{6.30}$$

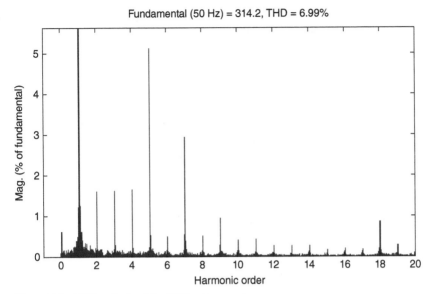

Figure 6.6   FFT analysis of the GSC grid current.

By substituting (6.30) in (6.27), the relationship between the rotor voltage $\vec{v}_{rdq}$ and the stator flux linkage $\vec{\psi}_{sdq}$ can be found as (6.31).

$$\vec{v}_{rdq} = \left(R_r + \sigma L_r \frac{d}{dt} + j\omega_{sl}\sigma L_r \right) \vec{i}_{rdq} + \frac{L_m}{L_s} \left(\frac{d}{dt} + j\omega_{sl} \right) \vec{\psi}_{sdq} \qquad (6.31)$$

If the stator resistance R_s is neglected, the stator flux linkage $\vec{\psi}_{sdq}$ can be expressed as

$$\vec{\psi}_{sdq} = \frac{\vec{v}_{sdq}}{\frac{d}{dt} + j\omega_{sl}} \qquad (6.32)$$

With (6.29), (6.31), and (6.32), the equivalent block diagram of the DFIG model in the s-domain, and in the dq reference frame can be derived and it is shown in Figure 6.7, where the grid voltage $\vec{v}_{sdq}$ is considered to be a disturbance input.

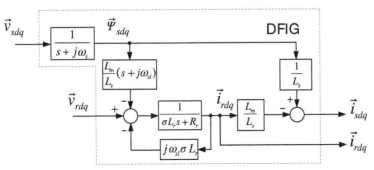

Figure 6.7   Model of the DFIG considering the grid voltage disturbance.

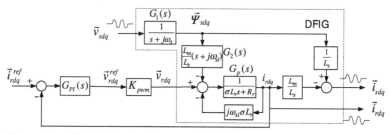

Figure 6.8 Model of the DFIG and the RSC when using conventional vector control.

The open loop transfer function $G_{viro}(s)$ from the grid voltage disturbance to the rotor current in the dq reference frame can be found as

$$G_{viro}(s) = \frac{\vec{i}_{rdq}}{\vec{v}_{sdq}} = \frac{1}{s+j\omega_s}\frac{L_m}{L_s}(s+j\omega_{sl})\frac{1}{\sigma L_r s + R_r} \tag{6.33}$$

The open loop transfer function $G_{viso}(s)$ from the grid voltage disturbance to the stator current in the dq reference frame is derived as

$$G_{viso}(s) = \frac{\vec{i}_{sdq}}{\vec{v}_{sdq}} = \frac{1}{L_s}\frac{1}{s+j\omega_s} - \frac{1}{s+j\omega_s}(s+j\omega_{sl})\frac{1}{\sigma L_r s + R_r} \tag{6.34}$$

These two transfer functions describe the influence of the grid voltage disturbance on the rotor current and stator current when the suppression effects of the vector control loop are not considered.

If the vector control loop is taken into account, the model of DFIG and RSC with conventional vector control can be derived, as shown in Figure 6.8. As the crossover frequency of the outer loop is normally no more than 50 Hz for large-scale DFIG wind power system, much smaller than the harmonic frequency (>300 Hz), only the inner current loop is considered in Figure 6.8. $G_1(s) = \frac{1}{s+j\omega_s}$, $G_2(s) = \frac{L_m}{L_s}(s+j\omega_{sl})$, and $G_p(s) = \frac{1}{\sigma L_r s + R_r}$ are defined to be the transfer function in the three blocks as shown in Figure 6.8.

With Figure 6.8, the stator and rotor currents under distorted grid voltage can be derived. Then, the active power and reactive power, as well as the torque fluctuations, can be calculated [10].

6.3.2 Influence on Rotor Current

From Figure 6.8, the stator and rotor currents in the s-domain can be expressed as (6.35).

$$\begin{cases} \vec{i}_{sdq}(s) = -\left[\left(\vec{i}_{rdq}^{ref}(s) - \vec{i}_{rdq}(s)\right)G_{PI}(s)K_{pwm} - \vec{v}_{sdq}(s)G_1(s)G_2(s)\right] \\ \qquad \times G_p(s)\frac{L_m}{L_s} + \vec{v}_{sdq}(s)G_1(s)\frac{1}{L_s} \\ \vec{i}_{rdq}(s) = \vec{v}_{sdq}(s)G_1(s)\frac{1}{L_m} - \frac{L_s}{L_m}\vec{i}_{sdq}(s) \end{cases} \tag{6.35}$$

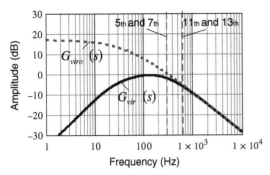

Figure 6.9   Amplitude of transfer function $G_{vir}(s)$ and $G_{viro}(s)$.

With (6.35), by eliminating the stator current $\vec{i}_{sdq}$, the rotor current $\vec{i}_{rdq}$ can be expressed as

$$\vec{i}_{rdq}(s) = \frac{G_{PI}(s)G_p(s)K_{pwm}}{1 + G_{PI}(s)G_p(s)K_{pwm}}\vec{i}_{rdq}^{ref}(s) + \frac{G_1(s)G_2(s)G_p(s)}{1 + G_{PI}(s)G_p(s)K_{pwm}}\vec{v}_{sdq}(s) \quad (6.36)$$

Define the transfer function from the rotor current reference to the rotor current as

$$G_{iir}(s) = \frac{\vec{i}_{rdq}}{\vec{i}_{rdq}^{ref}} = \frac{G_{PI}(s)G_p(s)K_{pwm}}{1 + G_{PI}(s)G_p(s)K_{pwm}} \quad (6.37)$$

Define the transfer function from the grid voltage disturbance to the rotor current as

$$G_{vir}(s) = \frac{\vec{i}_{rdq}}{\vec{v}_{sdq}} = \frac{G_1(s)G_2(s)G_p(s)}{1 + G_{PI}(s)G_p(s)K_{pwm}} \quad (6.38)$$

Then, (6.36) can be expressed as

$$\vec{i}_{rdq}(s) = G_{iir}(s)\vec{i}_{rdq}^{ref}(s) + G_{uir}(s)\vec{v}_{sdq}(s) \quad (6.39)$$

The rotor current is not only related to the current reference $\vec{i}_{rdq}^{ref}(s)$ but also related to the grid voltage disturbance $\vec{v}_{sdq}^{ref}(s)$. The influence of the grid voltage disturbance $\vec{v}_{sdq}^{ref}(s)$ on the rotor current $\vec{i}_{rdq}(s)$ can be described by the transfer function $G_{vir}(s)$. The smaller amplitude of $G_{vir}(s)$ indicates the lighter influence of the grid voltage disturbance on the rotor current [11].

Take again the 1.5 MW DFIG as an example, the amplitude of transfer function $G_{vir}(s)$ from grid voltage disturbance to the rotor current at different frequencies are shown in Figure 6.9 (solid line), together with the open loop transfer function $G_{viro}(s)$ (dotted line). The amplitude of $G_{vir}(s)$ in the low frequency range is much lower than the amplitude of $G_{viro}(s)$, as the vector control loop is able to suppress the grid disturbance with a frequency lower than the crossover frequency of the control loop. However, as in the large-scale DFIG wind power system, the switching frequency is limited, the crossover frequency of the inner loop is also limited to

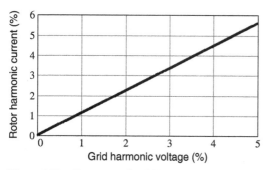

Figure 6.10 Rotor negative fifth and positive seventh harmonic current (the same line) relative to the grid harmonic voltage.

hundreds of Hz, in this case, it is 300 Hz. Consequently, the suppression effect of the vector control loop on the grid disturbance larger than 300 Hz is limited, as shown in Figure 6.9. The amplitude of $G_{viro}(s)$ and $G_{vir}(s)$ are about the same above 300 Hz. On the other hand, the most common harmonic voltage disturbance is negative 5th and positive 7th-order harmonic voltage, which is with a frequency of 300 Hz in the dq reference frame, as well as the 11th and 13th, which is with a frequency of 600 Hz in the dq reference frame. The suppression effect of the vector control is rather limited, the grid harmonic voltage disturbance may introduce large harmonic rotor current with the same frequency in dq reference frame. The negative fifth and the positive seventh harmonic current related to the grid harmonic voltage in the 1.5 MW DFIG WPS is shown in Figure 6.10. About 5% negative fifth-order grid harmonic voltage will introduce nearly 6% rotor harmonic current, with the rated power output, while 5% positive seventh-order harmonic voltage may introduce about 4% positive seventh-order rotor harmonic current, with the rated power output.

It should be noticed the negative fifth and the positive seventh order is referred to the grid or synchronous frequency in the dq reference frame, as the frequency of the rotor current varies from 15 to -15 Hz with the slip frequency, the negative fifth-order rotor harmonic current does not have any frequency which is five times of the rotor fundamental frequency [12].

6.3.3 Influence on Stator Current

Similar to the stator current, if the rotor current $\vec{i}_{rdq}$ in (6.35) is eliminated, the stator current $\vec{i}_{sdq}$ can be expressed as (6.40).

$$\vec{i}_{sdq}(s) = -\frac{L_m}{L_s}\frac{G_{PI}(s)K_{pwm}G_p(s)}{1+G_{PI}(s)K_{pwm}G_p(s)}\vec{i}_{rdq}^{ref}(s)$$

$$+\frac{G_1(s)/L_s - G_1(s)G_{PI}(s)K_{pwm}G_p(s)/L_s - G_1(s)G_2(s)G_p(s)L_m/L_s}{1+G_{PI}(s)K_{pwm}G_p(s)}\vec{v}_{sdq}(s)$$

$$(6.40)$$

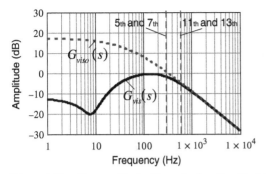

Figure 6.11   The amplitude of transfer function $G_{vis}(s)$ and $G_{viso}(s)$.

Also, defining

$$G_{iis}(s) = \frac{\vec{i}_{sdq}}{\vec{i}_{rdq}^{ref}} = -\frac{L_m}{L_s} \frac{G_{PI}(s)K_{pwm}G_p(s)}{1 + G_{PI}(s)K_{pwm}G_p(s)} \tag{6.41}$$

to obtain the transfer function from the rotor current reference to the stator current, and

$$
\begin{aligned}
G_{vis}(s) &= \frac{\vec{i}_{sdq}}{\vec{v}_{sdq}^{ref}} \\
&= \frac{G_1(s)/L_s + G_1(s)G_{PI}(s)K_{pwm}G_p(s)/L_s + G_1(s)G_2(s)G_p(s)L_m/L_s}{1 + G_{PI}(s)K_{pwm}G_p(s)}
\end{aligned} \tag{6.42}
$$

to be the transfer function from the grid voltage disturbance to the stator current, (6.40) can be represented as

$$\vec{i}_{sdq}(s) = G_{iis}(s)\vec{i}_{rdq}^{ref}(s) + G_{vis}(s)\vec{v}_{sdq}(s) \tag{6.43}$$

The stator current is also related to the current reference $\vec{i}_{rdq}^{ref}(s)$ and the grid voltage disturbance $\vec{v}_{sdq}^{ref}(s)$. The influence of the grid voltage disturbance $\vec{v}_{sdq}^{ref}(s)$ with different frequency of the rotor current $\vec{i}_{sdq}(s)$ can be described by the transfer function $G_{vis}(s)$. The smaller amplitude of $G_{vis}(s)$ indicates lighter influence of the grid voltage disturbance on the stator current. The amplitude of the transfer function $G_{vis}(s)$ and $G_{viso}(s)$ with different frequencies are shown in Figure 6.11. Similar with that for rotor harmonic current, the amplitude of $G_{vis}(s)$ is smaller than the amplitude of $G_{viso}(s)$ in low frequency range, as the vector control loop is able to suppress the influence of the grid voltage disturbance on the stator current. However, for grid negative 5th and positive 7th harmonic, as well as 11th and 13th harmonics, as their frequency is higher than the crossover frequency of the inner loop, the suppression effect is rather weak. The amplitude of $G_{vis}(s)$ and $G_{viso}(s)$ are about the same for the negative 5th and positive 7th, as well as 11th and 13th harmonic voltage frequencies. So, the

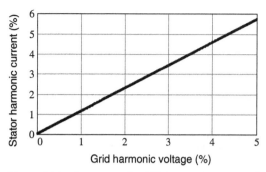

Figure 6.12   Stator negative fifth harmonic current relative to the grid harmonic voltage.

harmonic grid voltage may introduce relatively larger corresponding harmonic stator current. Taking negative fifth-order harmonic as an example, the stator harmonic current related to the grid harmonic voltage is shown in Figure 6.12, when the DFIG is generating the rated active power. It can be found that 5% negative fifth-order grid harmonic voltage will introduce more than 5% stator harmonic current, and this is the best situation as the fundamental stator current reaches the maximum value. The harmonic component may increase if the DFIG is not working at full load [13].

6.3.4   Influence on Active and Reactive Powers

With the stator harmonic current, the active power P_s and reactive power Q_s can be derived using the same method as shown in (6.13)–(6.16). Assuming only one frequency component exists in the grid voltage, in this case, the stator active power P_s and reactive power Q_s can be derived as shown in (6.44) and (6.45).

$$P_s = P_{s0} + P_{s\,\cosh} \cos(h\omega_s t) + P_{s\,\sinh} \sin(h\omega_s t) \tag{6.44}$$

$$Q_s = Q_{s0} + Q_{s\,\cosh} \cos(h\omega_s t) + Q_{s\,\sinh} \sin(h\omega_s t) \tag{6.45}$$

where P_{s0}, Q_{s0} are the DC components in the active and reactive power outputs, respectively. $P_{s\,\cosh}, P_{s\,\sinh}, Q_{s\,\cosh}, Q_{s\,\sinh}$ are the AC fluctuations components in the stator active and reactive power outputs, respectively. They can be expressed as

$$
\begin{bmatrix} P_{s0} \\ Q_{s0} \\ P_{s\,\cosh} \\ P_{s\,\sinh} \\ Q_{s\,\cosh} \\ Q_{s\,\sinh} \end{bmatrix} =
\begin{bmatrix}
v_{sd0} & v_{sq0} & v_{sdh} & v_{sqh} \\
v_{sq0} & -v_{sd0} & v_{sqh} & -v_{sdh} \\
v_{sdh} & v_{sqh} & v_{sd0} & v_{sq0} \\
v_{sqh} & -v_{sdh} & -v_{sq0} & v_{sd0} \\
v_{sqh} & -v_{sdh} & v_{sq0} & -v_{sd0} \\
-v_{sdh} & -v_{sqh} & v_{sd0} & v_{sq0}
\end{bmatrix} \cdot
\begin{bmatrix} i_{sd0} \\ i_{sq0} \\ i_{sdh} \\ i_{sqh} \end{bmatrix} \tag{6.46}
$$

where v_{sd0}, v_{sq0} are the fundamental components in d-axis and q-axis stator voltage (grid voltage), respectively, i_{sd0}, i_{sq0} are the fundamental components in d-axis and q-axis stator current, respectively. v_{sdh}, v_{sqh}, i_{sdh}, i_{sqh} are the hth-order harmonic components in d-axis and q-axis stator voltage and current, respectively.

If more than one frequency harmonic components exist in the grid voltage, the situations will be much more complex. Assuming h_1-order and h_2-order harmonic voltages ($\vec{v}_{sdqh_1}$ and $\vec{v}_{sdqh_2}$, respectively) exist at the same time in the stator voltage, the stator voltage in dq reference frame $\vec{v}_{sdq}$ can be expressed as

$$\vec{v}_{sdq} = \vec{v}_{sdq0} + \vec{v}_{sdqh_1} e^{j(h_1-1)\omega_s t} + \vec{v}_{sdqh_2} e^{j(h_2-1)\omega_s t} \tag{6.47}$$

The h_1-order and h_2-order harmonic voltages will introduce corresponding stator harmonic currents ($\vec{i}_{sdqh_1}$ and $\vec{i}_{sdqh_2}$, respectively), and the stator current $\vec{i}_{sdq}$ can be expressed as

$$\vec{i}_{sdq} = \vec{i}_{sdq0} + \vec{i}_{sdqh_1} e^{j(h_1-1)\omega_s t} + \vec{i}_{sdqh_2} e^{j(h_2-1)\omega_s t} \tag{6.48}$$

as the apparent power in the stator side can be expressed as (6.49).

$$
\begin{aligned}
S_s &= \vec{v}_{sdq}\vec{i}_{sdq}^* \\
&= \left(\vec{v}_{sdq0} + \vec{v}_{sdqh_1} e^{j(h_1-1)\omega_s t} + \vec{v}_{sdqh_2} e^{j(h_2-1)\omega_s t} \right) \\
&\quad \times \left(\vec{i}_{sdq0}^* + \vec{i}_{sdqh1}^* e^{-j(h_1-1)\omega_s t} + \vec{i}_{sdqh2}^* e^{-j(h_2-1)\omega_s t} \right)
\end{aligned} \tag{6.49}
$$

The stator active power P_s and the reactive power Q_s can be expressed as

$$
\begin{aligned}
P_s &= \mathrm{Re}\left(S_s \right) \\
&= \mathrm{Re}\left[\vec{v}_{sdq0}\vec{i}_{sdq0}^* + \vec{v}_{sdqh_1}\vec{i}_{sdqh1}^* + \vec{v}_{sdqh_2}\vec{i}_{sdqh2}^* \right] \\
&\quad + \mathrm{Re}\left[\vec{v}_{sdq0}\vec{i}_{sdqh1}^* e^{-j(h_1-1)\omega_s t} + \vec{v}_{sdqh_1}\vec{i}_{sdq0}^* e^{j(h_1-1)\omega_s t} \right] \\
&\quad + \mathrm{Re}\left[\vec{v}_{sdq0}\vec{i}_{sdqh2}^* e^{-j(h_2-1)\omega_s t} + \vec{v}_{sdqh_2}\vec{i}_{sdq0}^* e^{j(h_2-1)\omega_s t} \right] \\
&\quad + \mathrm{Re}\left[\vec{v}_{sdqh_1}\vec{i}_{sdqh2}^* e^{j(h_1-1-h_2+1)\omega_s t} + \vec{v}_{sdqh_2}\vec{i}_{sdqh1}^* e^{j(-h_1+1+h_2-1)\omega_s t} \right]
\end{aligned} \tag{6.50}
$$

$$
\begin{aligned}
Q_s &= \mathrm{Im}\left(S_s \right) \\
&= \mathrm{Im}\left[\vec{v}_{sdq0}\vec{i}_{sdq0}^* + \vec{v}_{sdqh_1}\vec{i}_{sdqh1}^* + \vec{v}_{sdqh_2}\vec{i}_{sdqh2}^* \right] \\
&\quad + \mathrm{Im}\left[\vec{v}_{sdq0}\vec{i}_{sdqh1}^* e^{-j(h_1-1)\omega_s t} + \vec{v}_{sdqh_1}\vec{i}_{sdq0}^* e^{j(h_1-1)\omega_s t} \right] \\
&\quad + \mathrm{Im}\left[\vec{v}_{sdq0}\vec{i}_{sdqh2}^* e^{-j(h_2-1)\omega_s t} + \vec{v}_{sdqh_2}\vec{i}_{sdq0}^* e^{j(h_2-1)\omega_s t} \right] \\
&\quad + \mathrm{Im}\left[\vec{v}_{sdqh_1}\vec{i}_{sdqh2}^* e^{j(h_1-1-h_2+1)\omega_s t} + \vec{v}_{sdqh_2}\vec{i}_{sdqh1}^* e^{j(-h_1+1+h_2-1)\omega_s t} \right]
\end{aligned} \tag{6.51}
$$

There will be three harmonic components in the active and reactive power outputs in this case. The $(h_1 - 1)$ order is introduced by the fundamental and h_1-order harmonic components; the $(h_2 - 1)$ order is introduced by the fundamental and h_2-order harmonic components, and the additional $(h_1 - 1 - [h_2 - 1])$ order is introduced by the h_1-order and h_2-order harmonic components [14–15].

6.3.5 Influence on Electromagnetic Torque

The electromagnetic torque of the DFIG will also be influenced by the grid harmonic voltage and the stator and rotor harmonic currents. If h-order harmonic components exist in stator voltage and current, the electromagnetic torque can be expressed as

$$
\begin{aligned}
T_{em} &= \frac{3}{2}n_p\text{Re}\left(j\vec{\psi}_{sdq}\vec{i}^*_{sdq}\right) \\
&= \frac{3}{2}n_p\text{Re}\left[\left(\frac{\vec{v}_{sdq0}}{\omega_{sl}} + \frac{\vec{v}_{sdqh}}{h\omega_{sl}}e^{j(h-1)\omega_s t}\right)\left(\vec{i}^*_{sdq0} + \vec{i}^*_{sdqh}e^{-j(h-1)\omega_s t}\right)\right] \\
&= T_{em0} + T_{emh}
\end{aligned}
\tag{6.52}
$$

There will be two frequency components in the electromagnetic torque of the DFIG in this case: the DC components T_{em0} and the AC fluctuation components T_{emh}, as shown in (6.53) and (6.54). The AC fluctuation in the electromagnetic torque is always to be avoided, as it will influence the reliability of the mechanical system, for DFIG WPS, especially the gearbox.

$$
T_{em0} = \frac{3}{2}n_p\text{Re}\left(\frac{\vec{v}_{sdq0}}{\omega_{sl}}\vec{i}^*_{sdq0} + \frac{\vec{v}_{sdqh}}{h\omega_{sl}}\vec{i}^*_{sdqh}\right)
\tag{6.53}
$$

$$
T_{emh} = \frac{3}{2}n_p\text{Re}\left(\frac{\vec{v}_{sdq0}}{\omega_s}\vec{i}^*_{sdqh}e^{-j(h-1)\omega_s t} + \frac{\vec{v}_{sdqh}}{h\omega_s}\vec{i}^*_{sdq0}e^{j(h-1)\omega_s t}\right)
\tag{6.54}
$$

If more than one frequency harmonic components exist in the stator voltage (grid voltage), for example, the h_1-order and h_2-order harmonic voltage exist at the same time in the grid voltage, the electromagnetic torque fluctuation will contains $(h_1 - 1)$-order components, $(h_2 - 1)$-order components, as well as the $(h_1 - 1 - h_2 - 1)$ components, similar to that for active and reactive powers.

6.3.6 Influence on DC-Bus Voltage

It has been discussed in Section 6.2.4 that the fluctuation in the DC bus is determined by the active power from both the GSC side and the RSC side, as shown in Figure 6.4 and equation (6.24).

Assuming h-order grid harmonic voltage exists, for the RSC side, as the suppression effect of the vector control loop on the grid harmonic voltage distortion is limited, it can be assumed that the rotor voltage does not contain harmonic components; in this case, the rotor voltage $\vec{v}_{rdq}$ can be represented as

$$
\vec{v}_{rdq} = \vec{v}_{rdq0}
\tag{6.55}
$$

where $\vec{v}_{rdq0} = v_{rd0} + jv_{rq0}$ are the fundamental components of the rotor voltage in dq reference frame. The rotor current $\vec{i}_{rdq}$ contains fundamental components $\vec{i}_{rdq0}$ and $\vec{i}_{rdqh}$, as shown in (6.56).

$$\vec{i}_{rdq} = \vec{i}_{rdq0} + \vec{i}_{rdq0}e^{jh\omega_s t} \tag{6.56}$$

So, the rotor-side active power P_r can be represented as (6.57).

$$
\begin{aligned}
P_r &= \frac{3}{2}\text{Re}\left(\vec{v}_{rdq}\vec{i}^*_{rdq}\right) = \frac{3}{2}\text{Re}\left[\vec{v}_{rdq0}\left(\vec{i}^*_{rdq0} + \vec{i}^*_{rdqh}e^{-jh\omega_s t}\right)\right] \\
&= \frac{3}{2}\text{Re}\left(\vec{v}_{rdq0}\vec{i}^*_{rdq0}\right) + \frac{3}{2}\text{Re}\left(\vec{v}_{rdq0}\vec{i}^*_{rdqh}e^{-jh\omega_s t}\right) \\
&= P_{r0} + P_{rh}
\end{aligned}
\tag{6.57}
$$

The rotor active power will contain the DC-steady component P_{r0} and the fluctuation component P_{rh} with a frequency of hf_s. If the power loss in the switches is neglected, P_{rh} is the h-order fluctuation component from the DC bus to the RSC. Then, with (6.24), the active power fluctuation in the DC bus can be found as (6.58).

$$P_{caph} = P_{dch} - P_{rh} = \frac{3}{2}\text{Re}\left(\vec{v}_0\vec{i}_{gh}e^{-jh\omega_s t}\right) - \frac{3}{2}\text{Re}\left(\vec{v}_{rdq0}\vec{i}^*_{rdqh}e^{-jh\omega_s t}\right) \tag{6.58}$$

Consequently, the DC-bus voltage fluctuation can be found by (6.25) and (6.26). The h-order fluctuations will be introduced on the DC-bus voltage, by the influence on both the GSC side and the RSC side [16].

6.3.7 Example of a 1.5 MW DFIG WPS

Simulation on a 1.5 MW DFIG WPS is given as an example, the parameters of the DFIG has been given in Chapter 5. The grid voltage contains 5% negative fifth-order harmonic and 5% positive seventh-order harmonic. The DFIG is working with rated active power under a rotor speed of 1800 rpm (1.2 pu). The grid voltage, rotor current, stator current, the active and reactive powers of the WPS, as well as the electromagnetic torque of the DFIG under distorted grid is shown in Figure 6.13. It can be found that the rotor current and stator current is heavily distorted under grid voltage harmonics; the FFT analysis in Figure 6.14 shows that 5% negative fifth-order harmonic voltage introduced more than 5% negative fifth-order harmonic GSC current, while 5% positive seventh-order harmonic voltage introduced about 4% positive seventh harmonic GSC current, which makes the THD of the GSC grid current reach 6.80%, and it cannot meet most of the related grid codes. The rotor current also contains the corresponding harmonic component. The grid harmonic voltage also introduced sixth-order fluctuations on the stator active power, reactive power, as well as the electromagnetic torque. The simulation results are able to verify the analysis in this section.

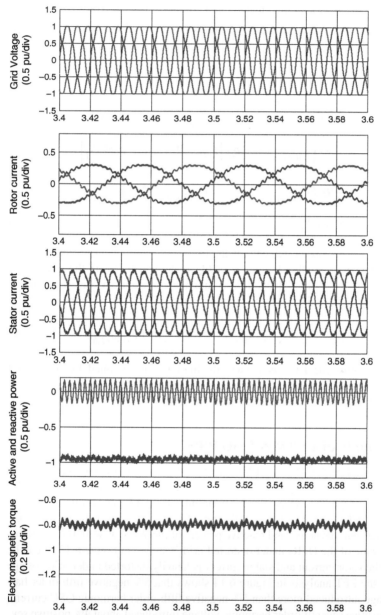

Figure 6.13 Simulation of DFIG under distorted grid voltage containing 5% negative fifth-order harmonic and 5% positive seventh-order harmonic.

Fundamental (50 Hz) = 0.9506 , THD = 6.80%

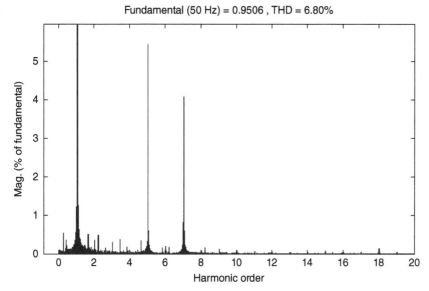

Figure 6.14   FFT analysis of the stator current.

The test results of the steady-state performance of the reduced-scale 30 kW DFIG WPS under harmonic distorted grid voltage is shown in Figure 6.15. The DFIG operatingat a sub-synchronous rotor speed of 1200 rpm, the grid voltage, stator current, rotor current of one phase, as well as the electronic torque fluctuation is shown in Figure 6.15. The grid voltage contains about 1% negative fifth-order harmonic and 1% positive seventh-order harmonic voltage. It can be found that the rotor current

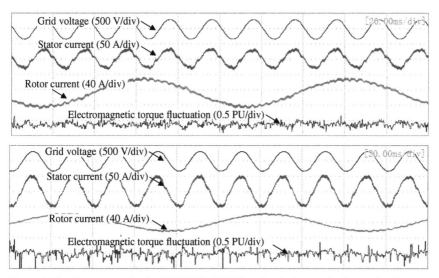

Figure 6.15   The test result of a 30kW DFIG WPS under the harmonic distorted grid.

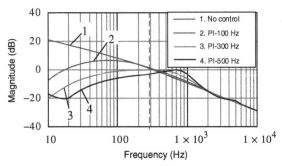

Figure 6.16   Magnitude of $G_{vis}(s)$ with different controller bandwidth.

and stator current are heavily distorted. When the output power is 0.5 pu (15 kW), the negative fifth- and positive seventh-order harmonic current and the THD of stator current are decreased from 9.10%, 4.45%, and 10.79%, respectively. When the output active power is 0.8 pu (24 kW), the negative fifth- and positive seventh-order harmonic current and the THD of stator current are decreased from 4.12%, 2.68%, and 5.41%, respectively. The AC fluctuations on the electromagnetic torque are also introduced. The test results are able to verify the analysis in this section.

6.4 DISCUSSION ON DIFFERENT CONTROLLER PARAMETERS

It has been concluded in Sections 6.2 and 6.3 that the suppression effect of the vector control loop on the influence of the grid harmonic voltage disturbance is limited in a large-scale DFIG WPS. The main reason for this limitation is that the bandwidth of the current loop is lower than the harmonic frequency. As a result, different controller parameters will lead to different controller bandwidth and may influence the performance of the DFIG WPS under harmonic distorted grid voltage [17].

Take the RSC and DFIG as an example. The transfer function $G_{vis}(s)$ in (6.42) represents the influence of the grid voltage disturbance on the stator current in the dq reference frame. In the 1.5 MW DFIG WPS, the magnitudes of $G_{vis}(s)$ with different controller bandwidths, together with the open loop control are shown in Figure 6.16. It can be concluded that basically, at the frequency of 300 Hz, which corresponds to the negative fifth and positive seventh harmonic voltages in the dq reference frame; the magnitude of $G_{vis}(s)$ is decreased with the increase of the bandwidth, as higher bandwidth provides better suppression effect on the grid harmonic voltage [18–19]. However, the difference is not so obvious. When the bandwidth is increased from 100 to 500 Hz, the magnitude of $G_{vis}(s)$ decreased about 2 dB.

If the grid voltage contains 5% negative fifth- and positive seventh-order harmonics, the corresponding stator harmonic current with different controller parameters k_{pi} are shown in Figure 6.17, and the electromagnetic torque fluctuations with different controller parameters is shown in Figure 6.18, for the 1.5 MW DFIG WPS with rated power output. $k_{pi} = 0.28$, $k_{pi} = 0.85$, and $k_{pi} = 1.4$ is corresponding to

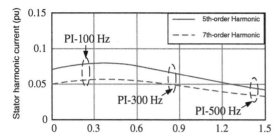

Figure 6.17   Corresponding stator harmonic current with different controller parameters k_{pi}.

bandwidths of 100, 300, and 500 Hz, respectively. In this case, when the controller bandwidth is increased from 100 to 500 Hz, the fifth-order stator harmonic current is decreased from about 0.07 to 0.05 pu, while the seventh-order harmonic current is decreased from about 0.06 pu to about 0.03 pu, respectively. The electromagnetic torque fluctuations introduced by fifth-order voltage harmonics is decreased from about 0.07 pu to about 0.04 pu, while the electromagnetic torque fluctuations introduced by seventh-order harmonic voltage is decreased from about 0.06 pu to about 0.04 pu, respectively.

As the switching frequency is only 2000 Hz in this case, the bandwidth of the controlled system is limited, 500 Hz has already been a relatively high value. However, it can be found the stator harmonic current is also large with the 500 Hz controller parameters, as well as the electromagnetic torque fluctuations.

6.5   DISCUSSION ON DIFFERENT POWER SCALES

For different power-scale DFIGs, the different influences of the grid harmonic voltage disturbance on the stator, rotor current, as well as the GSC grid current and electromagnetic torque may be caused by two reasons:

- The larger-scale DFIG normally has smaller stator and rotor resistances and sometimes smaller leakage inductance. Also, for the GSC, the grid inductance for the larger-scale DFIG WPS is also smaller. It can be found from (6.5),

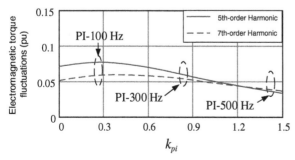

Figure 6.18   Electromagnetic torque fluctuations with different controller parameters k_{pi}.

(6.38), and (6.42) that the smaller inductance and resistance will make the corresponding transfer function from the grid harmonic voltage disturbance to the GSC grid current, rotor current, and stator current larger. So, the influence of the grid harmonic voltage disturbance on the performance of the DFIG is more obvious in larger-power-scale DFIG WPS.

- The switching frequency of the GSC and RSC is normally decreased with an increase of the power scale, as the switching loss must be limited. For MW rated DFIGs, the switching frequency of the GSC and RSC is normally only a few kHz, while more than 10 kHz is reasonable for kW rated DFIG. As the controller parameters need to ensure that the crossover frequency of the inner loop is kept at about 1/5 to 1/10 of the switching frequency, the bandwidth of the control loop for the large-scale DFIG is also limited. As discussed in the last section, the influence of the grid harmonic voltage disturbance on the performance of the DFIG is more obvious in a larger power-scale DFIG WPS, as the bandwidth of the control loop is limited [20].

6.6   SUMMARY

In this chapter, the performance of the DFIG wind power system under harmonic distorted grid is analyzed. The grid harmonic voltage will have an influence on the GSC current, stator current, rotor current, DC-bus voltage, active/reactive power, as well as the electromagnetic torque fluctuations of the DFIG. The mathematic model for the influence of grid harmonic voltage disturbance on the GSC current, stator current, rotor current, DC-bus voltage, active/reactive power, as well as the electromagnetic torque fluctuations of the DFIG, is established. The analysis indicates that as the bandwidth of the control loop is limited, the vector control used in the GSC and RSC has a limited suppression effect on the influence of the grid harmonic voltage disturbance. The corresponding harmonic current, DC-bus fluctuations, and electromagnetic torque fluctuations will still be introduced under grid harmonic disturbance. The increase of the bandwidth may lead to a smaller influence of the grid harmonic voltage distortions, but the effect is still limited as the bandwidth should be limited between about 1/5 to 1/10 of the switching frequency in order to ensure the stability of the system.

As the performance of the DFIG with normal vector control cannot meet the grid codes related to the power quality, harmonic suppression control strategies must be implied so that the DFIG WPS is able to operate normally under distorted grid voltage. This part will be presented in Chapters 7 and 8.

REFERENCES

[1] W. Chen, J. Xu, N. Zhu, C. Liu, M. Chen, and D. Xu, "Stator harmonic current suppression for DFIG wind power system under distorted grid voltage," in *IEEE Int. Symp. Power Electron. Distrib. Gener. Syst.*, pp. 307–314, 2012.

[2] H. Xu, J. Hu, and Y. He, "Operation of wind-turbine-driven DFIG systems under distorted grid voltage conditions: Analysis and experimental validations," *IEEE Trans. Power Electron.*, vol. 27, no. 5, pp. 2354–2366, 2012.

[3] C. Liu, F. Blaabjerg, W. Chen, and D. Xu, "Stator current harmonic control with resonant controller for doubly fed induction generator," *IEEE Trans. Power Electron.*, vol. 27, no. 7, pp. 3207–3220, 2012.

[4] H. Xu, J. Hu, and Y. He, "Integrated modeling and enhanced control of DFIG under unbalanced and distorted grid voltage conditions," *IEEE Trans. Energy Convers.*, vol. 27, no. 3, pp. 725–736, 2012.

[5] Y. Zhou, P. Bauer, J. A. Ferreira, and J. Pierik, "Operation of grid-connected DFIG under unbalanced grid voltage condition," *IEEE Trans. Energy Convers.*, vol. 24, no. 1, pp. 240–246, 2009.

[6] J. Hu, Y. He, and H. Nian, "Enhanced control of DFIG-used back-to-back PWM VSC under unbalanced grid voltage conditions," *J. Zhejiang Univ. Sci. A*, vol. 8, no. 8, pp. 1330–1339, 2007.

[7] J. Hu and Y. He, "Modeling and control of grid-connected voltage-sourced converters under generalized unbalanced operation conditions," *IEEE Trans. Energy Convers.*, vol. 23, no. 3, pp. 903–913, 2008.

[8] M. I. Martinez, G. Tapia, A. Susperregui, and H. Camblong, "DFIG power generation capability and feasibility regions under unbalanced grid voltage conditions," *IEEE Trans. Energy Convers.*, vol. 26, no. 4, pp. 1051–1062, 2011.

[9] Z. Ivanovic, M. Vekcic, S. Grabic, and V. Katic, "Control of multilevel converter driving variable speed wind turbine in case of grid disturbances," in *Int. Power Electron. Motion Control Conf.*, Portoroz, Slovenia, 2006, pp. 1569–1573.

[10] P. Xiao, K. A. Corzine, and G. K. Venayagamoorthy, "Multiple reference frame-based control of three-phase PWM boost rectifiers under unbalanced and distorted input conditions," *IEEE Trans. Power Electron.*, vol. 23, no. 4, pp. 2006–2017, 2008.

[11] D. G. Giaourakis, A. N. Safacas, and S. Tsotoulidis, "Wind energy conversion system equipped with doubly-fed induction generator under various grid faults," in *Proc. Eur. Conf. Power Electron. Appl.*, 2013, pp. 1–10.

[12] M. I. Maetinez, A. Susperregui, and H. Camblong, "Sliding-mode control for DFIG rotor- and grid-side converters under unbalanced and harmonically distorted grid voltage," *IEEE Trans. Energy Convers.*, vol. 27, no. 2, pp. 328–339, 2012.

[13] J. Hu, H. Nian, H. Xu, and Y. He, "Dynamic modeling and improved control of DFIG under distorted grid voltage conditions," *IEEE Trans. Energy Convers.*, vol. 26, no. 1, pp. 163–175, 2011.

[14] H. Nian and Y. Song, "Direct power control of doubly fed induction generator under distorted grid voltage," *IEEE Trans. Power Electron.*, vol. 29, no. 2, pp. 894–905, 2014.

[15] J. Hu, H. Xu, and Y. He, "Coordinated control of DFIG's RSC and GSC under generalized unbalanced and distorted grid voltage conditions," *IEEE Trans. Ind. Electron*, vol. 60, no. 7, pp. 2808–2819, 2013.

[16] L. Fan, S. Yuvarajan, and R. Kavasseri, "Harmonic analysis of a DFIG for a wind energy conversion system," *IEEE Trans. Energy Convers.*, vol. 25, no. 1, pp. 181–190, 2010.

[17] E. Tremblay, A. Chandra, and P. J. Lagace, "Grid-side converter control of DFIG wind turbines to enhance power quality of distribution network," in *Proc. IEEE Power Eng. Soc. Gen. Meeting*, Montreal, Canada, 2006.

[18] M. Kiani and W. J. Lee, "Effects of voltage unbalance and system harmonics on the performance of doubly fed induction wind generators," *IEEE Trans. Ind. Appl.*, vol. 46, no. 2, pp. 562–568, 2010.

[19] M. Lindholm and T. W. Rasmussen, "Harmonic analysis of doubly fed induction generators," in *Proc. Fifth Int. Conf. Power Electron. Drive Syst.*, Singapore, vol. 2, 2003, pp. 837–841.

[20] Y. Liao, L. Ran, G. A. Putrus, and K. S. Smith, "Evaluation of the effects of rotor harmonics in a doubly-fed induction generator with harmonic induced speed ripple," *IEEE Trans. Energy Convers.*, vol. 18, no. 4, pp. 508–515, 2003.

CHAPTER 7

MULTIPLE-LOOP CONTROL OF DFIG UNDER DISTORTED GRID VOLTAGE

In this chapter, a harmonic suppression control strategy based on multiple-loop control is introduced. The current is transferred to dq reference frames rotating at the harmonic angular speed, so the harmonic currents are DC components in these frames. In such cases, they can be controlled as DC variables using PI controllers, and a much higher controller gain can be achieved for the harmonic components. This control strategy can be used both on the GSC side and on the RSC side. Also by choosing different harmonic current references, different control targets can also be achieved. If more than one frequency of harmonic components exists in the grid voltage disturbance, more than one harmonic current loop needs to be added into the control scheme. All these control schemes also work together with the fundamental control loop, to ensure they are not disturbed by each other; the design of the controller parameters is introduced, as well as the influence of the harmonic current loop on the fundamental current loop is evaluated. Simulation and test results are given to demonstrate the effectiveness of the control strategy.

7.1 INTRODUCTION

The analysis in the last chapter indicates that in large power scale DFIG WPS, the suppression effect on the grid harmonic disturbance of the control loop is limited due to the limited bandwidth of the control loop restricted by the switching frequency [1]. The corresponding harmonic current, active/reactive power fluctuations, as well as the DC-bus voltage and torque fluctuations, will be introduced under harmonic distorted grid voltages, which may cause the DFIG WPS to fail to meet the related grid codes, and may also influence the reliability of the system [2]. As a result, harmonic suppression strategy needs to be included to ensure the operation of the DFIG under harmonic distorted grid voltage. In this chapter, a harmonic suppression control strategy based on multiple-loop control is introduced. The current is transferred to the dq

Advanced Control of Doubly Fed Induction Generator for Wind Power Systems, First Edition.
Dehong Xu, Frede Blaabjerg, Wenjie Chen, and Nan Zhu.
© 2018 by The Institute of Electrical and Electronics Engineers, Inc. Published 2018 by John Wiley & Sons, Inc.

reference frames rotating at the harmonic angular speed, so that the harmonic currents are DC components in the dq reference frames. In this case, they can be controlled as DC variables using PI controllers, thereby a much higher controller gain can be achieved for the harmonic components. This control strategy can be used on both the GSC side and the RSC side, by choosing different harmonic current references, and different control target can be achieved [3, 4].

If more than one frequency of harmonic components exists in the grid voltage disturbance, more than one harmonic current loops needs to be added to the control scheme. And all these control schemes work together with the fundamental control loop. To ensure they are not interrupted by each other, the design of the controller parameters is introduced as well in this chapter, and the influence of the harmonic current loop on the fundamental current loop is evaluated. Simulation and test results are given to prove the effectiveness of the control strategy [5].

7.2   GSC CONTROL

7.2.1   Control Target

The former analysis indicated that the grid harmonic voltage distortion will introduce GSC grid harmonic current, active/reactive power fluctuations, as well as DC-bus voltage fluctuations. An ideal thought is to suppress them all so that the GSC could operate in the same way as under ideal grid voltage conditions. However, this may not always be the case even in theory [6].

For example, under grid harmonic voltages, if the GSC grid harmonic current is controlled to be zero, then the GSC grid current $\vec{i}_{gdq}$ in dq reference frame will only contain fundamental components, as shown in (7.1).

$$\vec{i}_{gdq} = \vec{i}_{gdq0} \tag{7.1}$$

However, as the grid voltage $\vec{v}_{gdq}$ still contains harmonic components, like given as

$$\vec{v}_{gdq} = \vec{v}_{gdq0} + \vec{v}_{gdqh}e^{jh\omega_s t} \tag{7.2}$$

The output active P_g and reactive power Q_g can then be found as in (7.3) and (7.4).

$$
\begin{aligned}
P_g &= -\frac{3}{2}\mathrm{Re}\left(\vec{v}_{gdq}\vec{i}^{*}_{gdq}\right) = -\frac{3}{2}\mathrm{Re}\left[\left(\vec{v}_{gdq0} + \vec{v}_{gdqh}e^{jh\omega_s t}\right)\vec{i}^{*}_{gdq0}\right] \\
&= -\frac{3}{2}\mathrm{Re}\left(\vec{v}_{gdq0}\vec{i}^{*}_{gdq0}\right) - \frac{3}{2}\mathrm{Re}\left(\vec{v}_{gdqh}\vec{i}^{*}_{gdq0}e^{jh\omega_s t}\right) \\
&= P_{g0} + P_{g\,\mathrm{sinh}}\sin(h\omega_s t) + P_{g\,\mathrm{cosh}}\cos(h\omega_s t)
\end{aligned}
\tag{7.3}
$$

$$Q_g = -\frac{3}{2}\text{Im}\left(\vec{v}_{gdq}\vec{i}^*_{gdq}\right) = -\frac{3}{2}\text{Im}\left[\left(\vec{v}_{gdq0} + \vec{v}_{gdqh}e^{jh\omega_s t}\right)\vec{i}^*_{gdq0}\right]$$

$$= -\frac{3}{2}\text{Im}\left(\vec{v}_{gdq0}\vec{i}^*_{gdq0}\right) - \frac{3}{2}\text{Im}\left(\vec{v}_{gdqh}\vec{i}^*_{gdq0}e^{jh\omega_s t}\right)$$

$$= Q_{g0} + Q_{g\sinh}\sin(h\omega_s t) + Q_{g\cosh}\cos(h\omega_s t) \tag{7.4}$$

The hth-order AC fluctuation component still exists in the active and reactive power outputs, even when the GSC grid harmonic current is controlled to be zero, as shown in (7.5).

$$P_{g\cosh} = v_{gdh}i_{gd0} + v_{gqh}i_{gq0}$$
$$P_{g\sinh} = v_{gqh}i_{gd0} - v_{gdh}i_{gq0}$$
$$Q_{g\cosh} = v_{gqh}i_{gd0} - v_{gdh}i_{gq0}$$
$$Q_{g\sinh} = -v_{gdh}i_{gd0} - v_{gqh}i_{gq0} \tag{7.5}$$

The AC fluctuation components in the active and reactive power outputs can still be introduced by the fundamental GSC grid current and the grid harmonic voltages. Nevertheless, compared to (6.16), the AC fluctuation components in the active and reactive power outputs are still suppressed, compared to the situation when only using conventional vector control.

With respect to the DC-bus voltage fluctuation, if the GSC grid harmonic current is controlled to be zero, it can be concluded from (6.17) that the hth-order harmonic output voltage of the GSC $\vec{v}_{dqh}$ is the same as the hth-order grid harmonic voltage $\vec{v}_{gdqh}$

$$\vec{v}_{dqh} = \vec{v}_{gdqh} \tag{7.6}$$

So, if the power loss in the switching devices is neglected, the active power P_{dc} from the DC bus can be derived from (6.22), as expressed in

$$P_{dc} = \frac{3}{2}\text{Re}\left[\left(\vec{v}_{dq0} + \vec{v}_{dqh}e^{jh\omega_s t}\right)\vec{i}^*_{gdq0}\right] = \frac{3}{2}\text{Re}\left(\vec{v}_{dq0}\vec{i}^*_{gdq0} + \vec{v}_{dqh}\vec{i}^*_{gdq0}e^{-jh\omega_s t}\right)$$

$$= P_{dc0} + P_{dch} \tag{7.7}$$

Also, there are still AC fluctuation components P_{dch} which exist in the active power from the DC bus P_{dc}, even when the GSC grid harmonic current is controlled to be zero, and the corresponding DC-bus voltage fluctuation will still be seen.

On the other hand, if the AC fluctuation components in active power output is controlled to be zero, when it goes back to Chapter 6, it means that the $P_{g\cosh}$ and $P_{g\sinh}$ terms in (6.16) are controlled to be zero, which means

$$P_{g\cosh} = v_{gdh}i_{gd0} + v_{gqh}i_{gq0} + v_{gd0}i_{gdh} + v_{gq0}i_{gqh} = 0$$
$$P_{g\sinh} = v_{gqh}i_{gd0} - v_{gdh}i_{gq0} - v_{gq0}i_{gdh} + v_{gd0}i_{gqh} = 0 \tag{7.8}$$

Then the GSC harmonic grid currents in the dq reference frame i_{gdh} and i_{gqh} can be derived as

$$i_{gdh} = \frac{v_{gdh}v_{gq0} + v_{gqh}v_{gd0}}{v_{gd0}^2 + v_{gq0}^2}i_{gd0} + \frac{v_{gqh}v_{gq0} - v_{gdh}v_{gd0}}{v_{gd0}^2 + v_{gq0}^2}i_{gq0}$$

$$i_{gqh} = \frac{v_{gdh}v_{gd0} - v_{gqh}v_{gq0}}{v_{gd0}^2 + v_{gq0}^2}i_{gd0} + \frac{v_{gqh}v_{gd0} + v_{gdh}v_{gq0}}{v_{gd0}^2 + v_{gq0}^2}i_{gq0} \qquad (7.9)$$

So, if the AC fluctuation components in the active power output are controlled to be zero, there will exist GSC harmonic grid current. The same conclusion can also be drawn for the reactive power fluctuations. If the reactive power fluctuations are controlled to be zero, there will exist GSC harmonic grid current. For the DC-bus voltage fluctuation, as it is also related to the active power from the RSC side, the case is more complicated. However, if the AC fluctuation components P_{dch} in the active power from the DC bus is controlled to be zero, the GSC harmonic grid current also exists.

Consequently, there are four control targets for the GSC under grid harmonic voltages, which are [7, 8]:

Target 1: Control the GSC grid harmonic current to be zero.

Target 2: Control the AC fluctuation components in the active power output to be zero.

Target 3: Control the AC fluctuation components in the reactive power output to be zero.

Target 4: Control the AC fluctuation components in the DC-bus voltage to be zero.

Only Targets 2 and 4 can be achieved at the same time, while the other targets have a mutual exclusion and the decision must be made on which target is going to be achieved before starting the design the control scheme [9].

In this book, as the power quality of the GSC is mainly determined by the THD of the GSC grid current, fulfilling Target 1 has a better effect in increasing the power quality of the GSC. Also, when there are more than one frequency harmonic components existing in the grid voltage, the GSC harmonic grid currents in dq reference frame i_{gdh} and i_{gqh} becomes much more complicated than (7.9), if Targets 2–4 are selected. So, control Targets 2–4 are not easy to realize in practical applications. As a result, Target 1 is selected to ensure the good power quality of the GSC, which means the control target is to suppress the GSC grid harmonic current to be zero [10].

7.2.2 Control Scheme

For conventional vector control, the grid harmonic voltages are the AC disturbance components with frequency higher than the crossover frequency of the control loop, so that they cannot be controlled with good performance because of the limited controller gain. If the harmonic components can also be controlled as DC components, high-performance control can also be achieved by a PI controller for the harmonic components. The hth-order Park transform can transfer the hth-order harmonic components into DC components, and it is introduced as

$$\begin{bmatrix} x_d \\ x_q \end{bmatrix} = \frac{2}{3} \begin{bmatrix} \sin\left(h\omega_s t\right) & \sin\left(h\omega_s t - \frac{2}{3}\pi\right) & \sin\left(h\omega_s t + \frac{2}{3}\pi\right) \\ \cos\left(h\omega_s t\right) & \cos\left(h\omega_s t - \frac{2}{3}\pi\right) & \cos\left(h\omega_s t + \frac{2}{3}\pi\right) \end{bmatrix} \cdot \begin{bmatrix} x_a \\ x_b \\ x_c \end{bmatrix} \quad (7.10)$$

The hth-order Park transfer matrix $\mathbf{T}_{ph}$ is defined as

$$\mathbf{T}_{ph} = \frac{2}{3} \begin{bmatrix} \sin\left(h\omega_s t\right) & \sin\left(h\omega_s t - \frac{2}{3}\pi\right) & \sin\left(h\omega_s t + \frac{2}{3}\pi\right) \\ \cos\left(h\omega_s t\right) & \cos\left(h\omega_s t - \frac{2}{3}\pi\right) & \cos\left(h\omega_s t + \frac{2}{3}\pi\right) \end{bmatrix} \quad (7.11)$$

With the hth-order Park transform, the three-phase hth-order components can be transferred to DC component in the hth-order dq reference frame, as this frame is rotating with an angular speed of $h\omega_s$, which is h times the synchronous speed. The fundamental components will be transferred to $(h-1)$-order components in this hth-order dq reference frame. Practically when $h = 1$, this is the synchronous dq reference frame.

After the hth order, harmonic components have been transferred to the DC components, it can be controlled with a PI controller. If there are more than one harmonic component, each of the harmonic components can be controlled in the control loop based on an individual dq reference frame. It is therefore called multiple-loop control. The multiple-loop control scheme of the GSC under grid harmonic voltage disturbance is shown in Figure 7.1, where the negative fifth- and positive seventh-order GSC harmonic grid currents are suppressed.

The fifth- and seventh-order GSC grid harmonic currents $\vec{i}_{gdq5}$ and $\vec{i}_{gdq7}$ are extracted from the GSC grid current $\vec{i}_{gdq}$. The $\vec{i}_{gdq}$ is directly measured and transferred in negative fifth- and positive seventh-order dq reference frames, where the negative fifth and positive seventh harmonic components are DC components, while the fundamental component in the GSC grid current has a frequency of $6f_s$. The low pass filter (LPF) with the cross frequency of about f_s is used to separate fifth and seventh harmonic components in GSC grid current. After the LPF, the fifth and seventh harmonic components $\vec{i}_{gdq5}$ and $\vec{i}_{gdq7}$ can be extracted as DC components in fifth and seventh dq reference frames.

As Target 1 is selected to minimize the GSC harmonic grid current, the current references for the fifth and seventh GSC harmonic current are set to zero. The PI controller is used in the negative fifth and positive seventh dq reference frames to control the fifth and seventh GSC grid harmonic currents $\vec{i}_{gdq5}$ and $\vec{i}_{gdq7}$, which are already transferred to DC components, to be zero, respectively. The outputs of the PI controllers are then transferred back to the synchronous dq reference frame, and

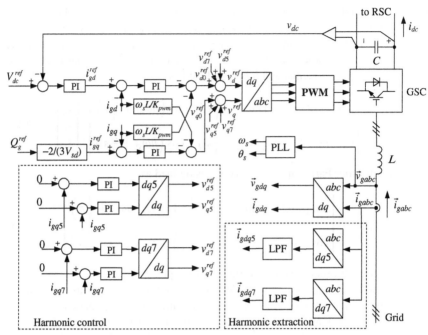

Figure 7.1   Multiple-loop control scheme of the GSC under grid harmonic voltage disturbance.

to be the fifth and seventh harmonic output voltage commands v_{d5}^{ref}, v_{q5}^{ref} and v_{d7}^{ref}, v_{q7}^{ref}. The fundamental control loop is the same as the conventional vector control loop introduced in Chapter 5. The output voltage commands from the fundamental control loop v_{d0}^{ref}, v_{q0}^{ref} are added to the fifth and seventh harmonic output voltage commands v_{d5}^{ref}, v_{q5}^{ref} and v_{d7}^{ref}, v_{q7}^{ref}, and the sum of them is final output voltage commands v_d^{ref}, v_q^{ref}. They are transferred back to *abc* reference frames and the drive signals are generated by the SPWM modulation, as illustrated in Figure 7.1.

If the harmonic components in the GSC grid current with other frequencies need to be suppressed, other control loops can be added in parallel connection to the control loops as shown in Figure 7.1. However, the most commonly seen low-order harmonic voltage in the power grid is the 5th and 7th order, as well as the 11th and the 13th harmonic grid voltages, so the suppression control loops for those components may be used more often in practice.

7.2.3   System Model with Harmonic Suppression Loop

It has been introduced in Chapter 6 that the model of the GSC in the *dq* reference frame can be expressed in *s*-domain as shown in (7.12).

$$(Ls + R)\vec{i}_{gdq} + \vec{v}_{gdq} = \vec{v}_{dq}^{ref} \tag{7.12}$$

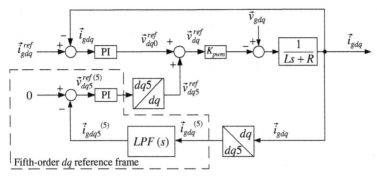

Figure 7.2   Simplified model of the GSC with multiple-loop control.

If the conventional vector control is used, the output voltage of the GSC $\vec{v}_{dq}^{ref}$ can be found as

$$\vec{v}_{dq}^{ref} = \vec{v}_{dq0}^{ref} = K_{pwm} G_{pii}(s) \left(\vec{i}_{gdq}^{ref} - \vec{i}_{gdq} \right) \tag{7.13}$$

When the fifth harmonic suppression control loop is used, the output voltage of the GSC $\vec{v}_{dq}^{ref}$ is the sum of $\vec{v}_{dq0}^{ref}$, $\vec{v}_{dq5}^{ref}$. The output voltage command of the fifth-order harmonic loop $\vec{v}_{dq5}^{ref}$ is coming from another reference frame, which may make the mathematical expression more complicated. The simplified model of the GSC in dq reference frame with multiple-loop control is shown in Figure 7.2. The transfer matrix from the dq reference frame to the negative fifth-order dq reference frame in the time domain T_{0-5} can be expressed in complex form as

$$T_{0-5} = e^{j(-6\omega_s)t} \tag{7.14}$$

The corresponding transfer matrix from the fifth-order dq reference frame to the synchronous dq reference frame T_{5-0} can be expressed as

$$T_{5-0} = e^{j(6\omega_s)t} \tag{7.15}$$

The GSC grid current in the synchronous dq reference frame $\vec{i}_{gdq}$ is transferred to the negative fifth-order dq reference frame; assuming the GSC grid current in the negative fifth-order dq reference frame is $\vec{i}_{gdq}^{(5)}$, it can be derived as

$$\vec{i}_{gdq}^{(5)}(t) = \vec{i}_{gdq}^{(5)}(t)e^{j(-6\omega_s)t} \tag{7.16}$$

Transfer to the s-domain gives

$$\vec{i}_{gdq}^{(5)}(s) = L\left[\vec{i}_{gdq}^{(5)}(t) \right] = L\left[\vec{i}_{gdq}(t)e^{j(-6\omega_s)t} \right] = \vec{i}_{gdq}(s + j6\omega_s) \tag{7.17}$$

After the LPF, the fifth-order grid harmonic current $\vec{i}_{gdq5}^{(5)}$ can be extracted, as shown in (7.18).

$$\vec{i}_{gdq5}^{(5)}(s) = \vec{i}_{gdq}^{(5)}(s)LPF(s) = \vec{i}_{gdq}(s + j6\omega_s)LPF(s) \tag{7.18}$$

where $LPF(s)$ is the transfer function of the LPF. If a second-order Butterworth LPF is used, it can be expressed as $LPF(s) = \omega_c \big/ s^2 + \sqrt{2}s\omega_c + \omega_c^2$. As the fundamental component has a frequency of $6f_s$ in the negative fifth-order dq reference frame, the cutoff frequency $\omega_c/2\pi$ is chosen to be 50 Hz. After the PI controller, the output voltage reference of the fifth-order harmonic loop in the negative fifth-order dq reference frame $\vec{v}_{dq5}^{ref\,(5)}(s)$ can be expressed as

$$\vec{v}_{dq5}^{ref\,(5)}(s) = -G_{pi5}(s)\vec{i}_{gdq5}^{ref\,(5)}(s) = -\vec{i}_{gdq}(s+j6\omega_s)LPF(s)G_{pi5}(s) \qquad (7.19)$$

where $G_{pi5}(s) = k_{p5} + k_{i5}/s$ is the transfer function of the PI controller in the harmonic current loop. Then, $\vec{v}_{dq5}^{ref\,(5)}$ needs to be transferred to the synchronous dq reference frame, as shown in

$$\vec{v}_{dq5}^{ref\,(5)}(t) = \vec{v}_{dq5}^{ref\,(5)}(t)\,e^{j(6\omega_s)t} \qquad (7.20)$$

In the s-domain, the output voltage reference of the fifth-order harmonic loop in synchronous the dq reference frame $\vec{v}_{dq5}^{ref\,(5)}(s)$ can be represented as

$$
\begin{aligned}
\vec{v}_{dq5}^{ref\,(5)}(s) &= L\left[\vec{v}_{dq5}^{ref\,(5)}(t)\right] = L\left[\vec{v}_{dq5}^{ref\,(5)}(t)e^{j(6\omega_s)t}\right] \\
&= \vec{v}_{dq5}^{ref\,(5)}(s - j6\omega_s) = -\vec{i}_{gdq}(s - j6\omega_s + j6\omega_s)LPF(s - j6\omega_s)G_{pi5}(s - j6\omega_s) \\
&= -\vec{i}_{gdq}(s)LPF(s - j6\omega_s)G_{pi5}(s - j6\omega_s) \qquad (7.21)
\end{aligned}
$$

Together with (7.12) and (7.13), the model of the GSC with harmonic suppression control in the s-domain can be written as

$$
\begin{aligned}
(Ls &+ R)\vec{i}_{gdq}(s) + \vec{v}_{gdq}(s) \\
&= K_{pwm}G_{pii}(s)\left[\vec{i}_{gdq}^{ref}(s) - \vec{i}_{gdq}^{ref}(s)\right] - K_{pwm}LPF(s - j6\omega_s)G_{pi5}(s - j6\omega_s)\vec{i}_{gdq}(s)
\end{aligned}
$$
$$(7.22)$$

So, the GSC current in the synchronous dq reference frame can be expressed as

$$
\begin{aligned}
\vec{i}_{gdq}(s) =\ & \frac{K_{pwm}G_{pii}(s)}{K_{pwm}[G_{pii}(s) + LPF(s - j6\omega_s)G_{pi5}(s - j6\omega_s)] + Ls + R}\vec{i}_{gdq}^{ref}(s) \\
& - \frac{1}{K_{pwm}[G_{pii}(s) + LPF(s - j6\omega_s)G_{pi5}(s - j6\omega_s)] + Ls + R}\vec{v}_{gdq}(s) \qquad (7.23)
\end{aligned}
$$

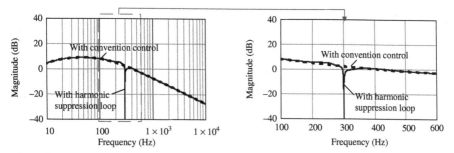

Figure 7.3   Magnitude frequency response of $G_{vi}(s)$ in synchronous dq reference frame.

7.2.4   Control Effect

It can be found that after the harmonic suppression loop is used, the transfer function $G_{vi}(s)$ from the grid voltage disturbance $\vec{v}_{gdq}(s)$ to the GSC grid current $\vec{i}_{gdq}(s)$ can become

$$G_{vi}(s) = -\frac{1}{K_{pwm}[G_{pii}(s) + LPF(s - j6\omega_s)G_{pi5}(s - j6\omega_s)] + Ls + R} \qquad (7.24)$$

Compared to the situation in the conventional vector control in (6.5), this is rewritten as (7.25)

$$G_{vi}(s) = -\frac{1}{K_{pwm}G_{pii}(s) + Ls + R} \qquad (7.25)$$

Their magnitude in a 1.5 MW DFIG WPS with different frequencies are shown in Figure 7.3. The frequency response of $G_{vi}(s)$ with the harmonic suppression loop (solid line) and with normal vector control (dotted line) is almost the same, except for a frequency range around 300 Hz, which corresponds to the fifth-order harmonic components in the synchronous dq reference frame. The magnitude of $G_{vi}(s)$ with harmonic suppression loop is decreased dramatically around 300 Hz, which means the influence of the grid harmonic distortion at the frequency of 300 Hz in the synchronous dq reference frame on the GSC grid current will be greatly suppressed by the harmonic control loop. And this 300 Hz frequency corresponds to the fifth-order harmonic components.

7.2.5   Test Results

Test results on the 30 kW reduced-scale DFIG test platform are shown here to verify the harmonic current suppression effect on the GSC grid current. The test platform will be introduced in Chapter 14. The GSC grid current without and with the introduced harmonic suppression control is shown in Figure 7.4. The DFIG is generating about 0.5 pu active power under a rotor speed of 1200 rpm and the grid voltage contains about 1% fifth-order harmonics. A fifth-order harmonic control loop is used for the harmonic suppression control. Without this control loop, the GSC current is distorted as shown in Figure 7.4a, and the FFT analysis shows that 4.7% fifth-order GSC harmonic current is introduced. After the harmonic suppression control is used, the

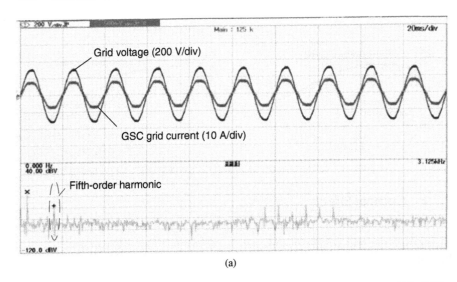

(a)

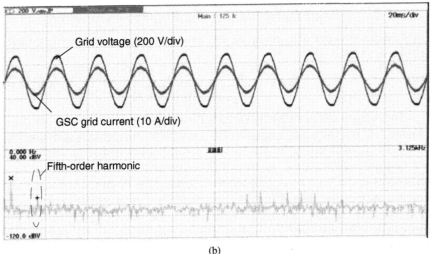

(b)

Figure 7.4   GSC grid current: (a) With conventional vector control and (b) with harmonic suppression control.

GSC grid current becomes more sinusoidal, and the fifth-order GSC harmonic current is reduced to about 1%, and the fifth-order GSC harmonic current is suppressed by the multiple-loop harmonic suppression control strategy.

7.3   DFIG AND RSC CONTROL

7.3.1   Control Target

Similar to that of the GSC operation, the grid harmonic voltage disturbance will introduce rotor harmonic current, stator harmonic current, active power/reactive power fluctuations, as well as torque fluctuations in the DFIG. Before designing the control scheme, the control target needs to be defined first—as we cannot suppress the rotor harmonic current, stator harmonic current, active power/reactive power fluctuations, and electromagnetic torque fluctuations at the same time.

For example, assuming that hth-order harmonic voltage exists in the grid voltage and the rotor harmonic current is suppressed to zero, meaning that only the fundamental components exist in the rotor harmonics. From (6.35), it can be derived that

$$\vec{i}_{sdq}(s) = \vec{v}_{sdq}(s)G_1(s)\frac{1}{L_m} - \frac{L_m}{L_s}\vec{i}_{rdq}(s) \tag{7.26}$$

For the hth-order harmonic components, as the rotor harmonic current is suppressed to zero, which means $\vec{i}_{rdq}(jh\omega_s) = \vec{i}_{rdqh} = 0$, (7.26) can be rewritten as

$$\vec{i}_{sdqh} = \vec{v}_{sdqh}\frac{1}{(h+1)j\omega_s}\frac{1}{L_m} \tag{7.27}$$

It can be concluded that even the rotor harmonic current $\vec{i}_{rdqh}$ is suppressed to zero, the stator harmonic current component $\vec{i}_{sdqh}$ will still exist, and it is proportional to the grid harmonic voltage $\vec{v}_{sdqh}$. Also, the active power and reactive power fluctuations, as well as the electromagnetic torque fluctuations will be introduced in this case, similar to the situation in the GSC.

If the hth-order stator harmonic current is controlled to be zero, which means $\vec{i}_{sdq}(jh\omega_s) = \vec{i}_{sdqh} = 0$ in (7.26), the rotor harmonic current $\vec{i}_{rdqh}$ can be expressed as

$$\vec{i}_{rdqh} = \vec{v}_{sdqh}\frac{1}{(h+1)j\omega_s}\frac{1}{L_m} \tag{7.28}$$

So, the rotor harmonic current $\vec{i}_{rdqh}$ still exists if the stator harmonic current is controlled to be zero. A similar conclusion can also be drawn for the active/reactive powers, as well as for the electromagnetic torque fluctuations—only one of these fluctuations can be suppressed to zero in theory at the same time.

As a result, one of the following control targets must be selected [11, 12]:

Target 1: Control the rotor harmonic current to be zero.

Target 2: Control the stator harmonic current to be zero.

Target 3: Control the AC fluctuation components in the stator active power output to be zero.

Target 4: Control the AC fluctuation components in the stator reactive power output to be zero.

Target 5: Control the AC fluctuation components in the electromagnetic torque fluctuations to be zero.

As for the DFIG WPS, the related grid code for power quality is mainly focused on the THD of the output current, thus in this book, control Target 2 is selected, and the control target is to control the stator harmonic current to be zero.

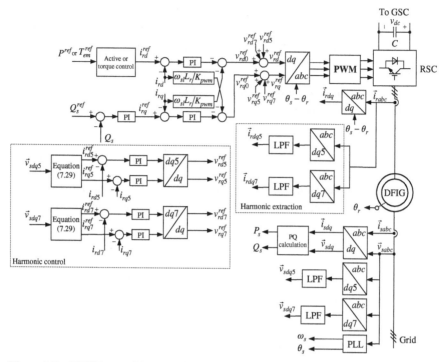

Figure 7.5   DFIG harmonic suppression control based on rotor current multiple-loop control.

7.3.2   Control Scheme

It has been introduced in Chapter 6 that the main problem of the conventional vector control is that the bandwidth is limited so that the current loop cannot suppress all the influences of the grid harmonic voltages on the stator harmonic current. This problem can be solved using an extra control loop in hth-order dq reference frame to be in parallel connection with the fundamental loop, which is similar to the GSC. The stator harmonic current is selected to be suppressed. However, as introduced in Chapter 5, the inner current loop of the RSC is the rotor current loop and the stator current is not directly controlled in this current loop. So the corresponding rotor harmonic current reference when the stator harmonic current is controlled to be zero can be calculated from (7.28), as shown in (7.29).

$$\vec{i}_{rdqh}^{ref} = \vec{v}_{sdqh} \frac{1}{(h+1)j\omega_s} \frac{1}{L_m} \tag{7.29}$$

As long as the rotor harmonic current is controlled to track the reference $\vec{i}_{rdqh}^{ref}$ as shown in (7.29), the stator harmonic current can be controlled to be zero. So, the control scheme can be designed as shown in Figure 7.5. The fifth- and seventh-order stator harmonic currents are suppressed in this case, based on the rotor current multiple-loop control. The rotor current $\vec{i}_{rabc}^{ref}$ is transferred to negative fifth- and positive seventh-order dq reference frame, respectively, where the fifth- and seventh-order rotor harmonic currents are DC components, and the fundamental components are AC components with the frequency of $6f_s$. Note that the transfer angle from the rotor ABC reference frame to the fifth and seventh dq reference frame should be $-5\theta_s - \theta_r$ and $7\theta_s - \theta_r$, respectively. After the LPF, the fifth- and seventh-order rotor harmonic currents $\vec{i}_{rdq5}$ and $\vec{i}_{rdq7}$, respectively, can be then extracted. The rotor harmonic current references $\vec{i}_{rdq5}^{ref}$ and $\vec{i}_{rdq7}^{ref}$ are calculated from (7.29), and the rotor harmonic currents $\vec{i}_{rdq5}$ and $\vec{i}_{rdq7}$ are controlled to track the rotor harmonic current references $\vec{i}_{rdq5}^{ref}$ and $\vec{i}_{rdq7}^{ref}$ so that the stator current can be controlled to be zero. PI controllers are used in fifth- and seventh dq reference frame and the output of the PI controllers are transferred back to the synchronous dq reference frame, to be the rotor harmonic voltage references v_{rd5}^{ref}, v_{rq5}^{ref} and v_{rd7}^{ref}, v_{rq7}^{ref}. They are added to the rotor voltage reference of the fundamental loop v_{rd0}^{ref}, v_{rq0}^{ref}, and then the rotor voltage references v_{rd}^{ref} and v_{rq}^{ref} are derived. After the inverse Park transformation and the SVM modulation, the RSC is driven to control the DFIG and suppress the harmonics.

By using the multiple-loop control, the harmonic components in the stator components can be suppressed. However, in this control scheme, the rotor harmonic reference is calculated from grid voltage harmonic components $\vec{v}_{sdqh}$ using (7.29), and it is also related to the mutual inductance of the DFIG L_m. When the DFIG is running, the mutual inductance of the DFIG L_m may vary at different operation conditions, causing errors in the rotor harmonic reference, and this error cannot be compensated by the control loop. Furthermore, the accuracy of the grid voltage harmonic components $\vec{v}_{sdqh}$ will also influence the control system [13].

As the stator current is normally sampled in the DFIG WPS for the power control, another multiple-loop control scheme based on the stator current control is shown in Figure 7.6. In this control scheme, the stator harmonic currents $\vec{i}_{sdq5}$ and $\vec{i}_{sdq7}$ are directly extracted from the sampled stator current $\vec{i}_{sdq}$, in the negative fifth and positive seventh dq reference frame. As the stator current $\vec{i}_{sdq}$ holds a linear relationship with the rotor current $\vec{i}_{rdq}$, $\vec{i}_{sdq}$ can be directly controlled to be zero by the PI controllers in fifth and seventh dq reference frames, as shown in Figure 7.6.

In this control scheme, the control reference does not need to be calculated from the grid voltage and the DFIG parameters, so it should be more robust against parameter changes compared to the control scheme shown in Figure 7.5. Also, in this control scheme, the grid voltage harmonic components $\vec{v}_{sdqh}$ do not need to be extracted, which makes the control scheme simpler. So, it will be a better choice to suppress the stator harmonic current under distorted grid voltage. The following analysis will be based on the control scheme shown in Figure 7.6.

7.3.3 System Model and Control Effect

With the control scheme shown in Figure 7.6, the model of the DFIG in the s-domain can then be derived, as shown in Figure 7.7. With the same method introduced in

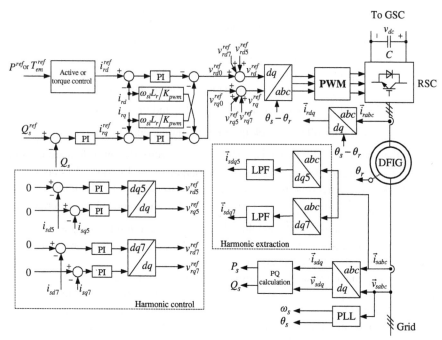

Figure 7.6   DFIG harmonic suppression control based on stator current multiple-loop control.

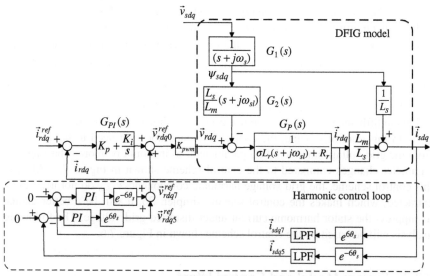

Figure 7.7   Model of DFIG in the s-domain with harmonic suppression control based on stator current multiple-loop control.

Section 7.2.3, the rotor harmonic voltage references $\vec{v}_{rdq5}^{ref}$ and $\vec{v}_{rdq7}^{ref}$ can be found as (7.30).

$$\vec{v}_{rdq5}^{ref} = G_{pi5}(s + j6\omega_s)LPF(s + j6\omega_s)\vec{i}_{sdq}$$

$$\vec{v}_{rdq7}^{ref} = G_{pi7}(s - j6\omega_s)LPF(s - j6\omega_s)\vec{i}_{sdq} \qquad (7.30)$$

where $G_{pi5}(s) = k_{p5} + k_{i5}/s$ and $G_{pi7}(s) = k_{p7} + k_{i7}/s$ are the transfer functions of the PI controllers in fifth and seventh dq reference frame. $LPF(s)$ is the transfer function of the LPF, as introduced in Section 7.2.3. Then the rotor voltage v_{rdq} can be derived as (7.31).

$$\vec{v}_{rdq} = K_{PMW}\left(\vec{v}_{rdq0}^{ref} + \vec{v}_{rdq5}^{ref} + \vec{v}_{rdq7}^{ref}\right)$$

$$= K_{PMW}[G_{pii}(s)\left(\vec{i}_{rdq}^{ref} - \vec{i}_{rdq}\right)$$

$$+ G_{pi5}(s + j6\omega_s)LPF(s + j6\omega_s)\vec{i}_{sdq} + G_{pi7}(s - j6\omega_s)LPF(s - j6\omega_s)\vec{i}_{sdq}] \qquad (7.31)$$

By substituting (7.31) in (6.29), (6.31), and (6.32), the stator current $\vec{i}_{sdq}$ of the DFIG can be expressed as

$$\vec{i}_{sdq}(s) = -\frac{L_m}{L_s}\frac{G_o(s)}{1 + G_o(s)}\vec{i}_{rdq}^{ref}(s)$$

$$+ \frac{1}{L_s}\frac{L_m\left[G_2(s) + G_{PI}(s)\right]G_1(s)G_P(s) + G_1(s)}{1 + G_o(s)}\vec{v}_{sdq}(s) \qquad (7.32)$$

where $G_o(s) = K_{pwm}G_P(s)[G_{Pii}(s) + \frac{L_m}{L_s}LPF(s + j6\omega_s)G_{pi5}(s + j6\omega_s) + \frac{L_m}{L_s}LPF(s - j6\omega_s)G_{pi7}(s - j6\omega_s)]$ is the open-loop transfer function of the system, when the harmonic suppression loop is considered. From (7.32), the transfer function from the grid voltage disturbance to the stator current $G_{vis}(s)$ can then be derived as

$$G_{vis}(s) = \frac{\vec{i}_{sdq}}{\vec{v}_{sdq}^{ref}} = \frac{1}{L_s}\frac{L_m\left[G_2(s) + G_{PI}(s)\right]G_1(s)G_P(s) + G_1(s)}{1 + G_o(s)} \qquad (7.33)$$

As introduced in Chapter 6, $G_{vis}(s)$ can describe the influence of the grid harmonic voltage disturbance on the stator current. The frequency response of the magnitude of $G_{vis}(s)$ with the harmonic suppression control loop, as shown in (7.33), and with the conventional vector control, as shown in (6.42), as well as in the case with open-loop control $G_{viso}(s)$, as expressed in (6.34), is shown in Figure 7.8. It can be found that the frequency response of the $G_{vis}(s)$ with harmonic suppression control loop and with the conventional vector control are basically the same in the large frequency range, except for the frequency range around 300 Hz, which is the frequency of negative fifth and positive seventh harmonic components in the synchronous dq reference frame. By applying the multiple-loop control scheme, large attenuation of $G_{vis}(s)$ at the frequency of 300 Hz is introduced, and therefore the influence of the

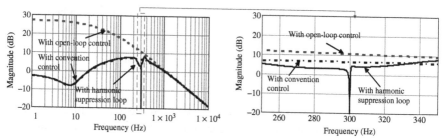

Figure 7.8 Frequency response of the transfer function from grid voltage disturbance to the stator current.

negative fifth- and positive seventh-order grid harmonic voltage on the stator current can be effectively suppressed.

7.3.4 Controller Design

After the harmonic suppression control loop is applied, the DFIG system becomes a multiple input and multiple output system, which makes the controller design for the harmonic loop much more complicated compared to that of a conventional vector control loop. However, as in the fundamental loop, the cross frequency is limited so that its control effect on the harmonic components is relatively small; on the other hand, in the harmonic suppression loop, the fundamental components and other harmonic components with different frequency has been filtered by the LPF. So, an assumption can be made that the fundamental loop and the harmonic suppression loop will not influence each other. Consequently, the controller parameters in each of the harmonic loops can be designed separately, as a single-input single-output system, with the same method used for the fundamental loop control design.

Take the fifth-order harmonic suppression loop as an example. For the fifth-order harmonic components, the dynamic model of the DFIG in the fifth-order dq reference frame can be found as (7.34) and (7.35).

$$\vec{v}_{sdq5} = R_s\vec{i}_{sdq5} + \frac{d}{dt}\vec{\psi}_{sdq5} - j5\omega_s\vec{\psi}_{sdq5}$$

$$\vec{v}_{rdq5} = R_r\vec{i}_{rdq5} + \frac{d}{dt}\vec{\psi}_{rdq5} + j(-5\omega_s - \omega_r)\vec{\psi}_{sdq5} \qquad (7.34)$$

$$\vec{\psi}_{sdq5} = L_s\vec{i}_{sdq5} + L_m\vec{i}_{rdq5}$$

$$\vec{\psi}_{rdq5} = L_r\vec{i}_{rdq5} + L_m\vec{i}_{sdq5} \qquad (7.35)$$

By eliminating the rotor current fifth-order harmonic $\vec{i}_{rdq5}$ and the stator and rotor flux linkage fifth-order harmonics $\vec{\psi}_{sdq5}$ and $\vec{\psi}_{sdq5}$, the relationship between the rotor voltage fifth-order harmonic $\vec{v}_{rdq5}$ and the stator current fifth-order harmonic

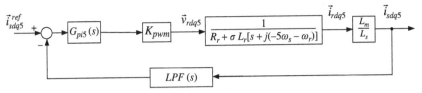

Figure 7.9   Control loop scheme of the fifth-order harmonic loop.

$\vec{i}_{sdq5}$ can be found as (7.36).

$$\vec{v}_{rdq5} = -\frac{L_s R_r}{L_m}\vec{i}_{sdq5} + \frac{L_s}{L_m}\sigma L_r[s + (-5\omega_s - \omega_r)j]\vec{i}_{sdq5}$$

$$+ \frac{(5\omega_s + \omega_r)L_r - jR_r}{5\omega_s L_m}\vec{v}_{sdq5} \tag{7.36}$$

Note that the stator resistance R_s is neglected in (7.36). Then the transfer function from the rotor voltage $\vec{v}_{rdq5}$ to the stator current $\vec{i}_{sdq5}$ can be found as $G_{rs5}(s)$.

$$G_{rs5}(s) = \frac{\vec{i}_{sdq5}}{\vec{v}_{rdq5}} = -\frac{L_m}{L_s}\frac{1}{R_r + \sigma L_r[s + j(-5\omega_s - \omega_r)]} \tag{7.37}$$

So, the control loop can be illustrated as shown in Figure 7.9 in the fifth-order dq reference frame. The open-loop transfer function $G_{rso5}(s)$ from the reference to the feedback can then be derived as

$$G_{rso5}(s) = G_{pi5}(s)LPF(s)G_{rs5}(s) \tag{7.38}$$

The parameters of the PI controller $G_{pi5}(s)$ can be designed using Bode diagram, which has been introduced already in Chapter 5. Nevertheless, for the harmonic control loop, the crossover frequency is designed to only about 10 Hz in order to ensure that the harmonic loop does not influence the performance of the fundamental loop. The low crossover frequency may lead to a poor dynamic response of the harmonic loop, but for the harmonic loop, the steady-state performance is much more important, and the slow dynamic response is acceptable [14].

The corresponding Bode diagrams in a 1.5 MW DFIG WPS can be found in Figure 7.10. The PI controller parameters are designed to be $k_{p5} = 0.17$ and $k_{i5} = 20$, the crossover frequency is about 11 Hz, and the phase margin is much larger than 45°. The stability is ensured for the harmonic suppression loop.

7.3.5   Simulation and Test Results

Simulations based on MATLAB/SIMULINK are made to verify the stator harmonic current multiple-loop control. A 1.5 MW DFIG is used, with parameters shown in Table 4.1. The steady-state performances of stator harmonic current multiple-loop control compared with vector control in the dq reference frame are shown in Figures 7.11 and 7.12. The DFIG was operating at a super-synchronous speed of 1800 rpm,

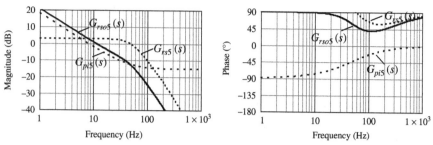

Figure 7.10   Bode diagram of the harmonic control loop.

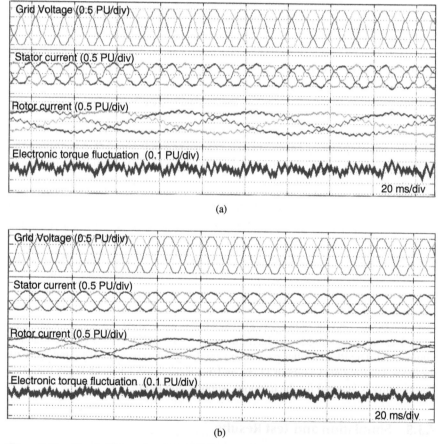

(a)

(b)

Figure 7.11   Simulation results of the DFIG under distorted grid with 0.5 pu output active power: (a) With conventional vector control and (b) with stator harmonic current multiple-loop control.

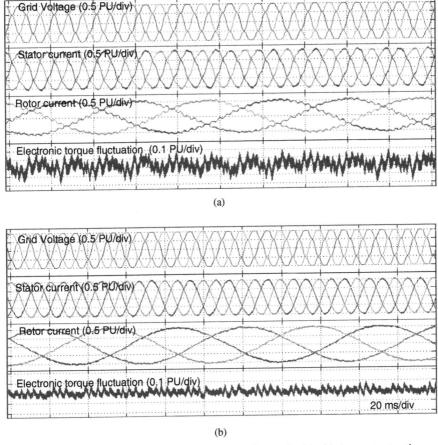

(a)

(b)

Figure 7.12   Simulation results of the DFIG under distorted grid with 1 pu output active power: (a) With conventional vector control and (b) with stator harmonic current multiple-loop control.

5% negative fifth-order and 5% positive seventh harmonics were injected into the grid voltage. The fifth- and seventh-order stator harmonic currents and the THD of the stator current were calculated using FFT and listed in Table 7.1. By using the proposed stator harmonic current control, when the output power is 0.5 pu, the fifth- and seventh-order harmonic current and the THD of stator current are decreased from 8.3%, 6.1%, and 10.8% to 1.0%, 1.6%, and 4.3%, respectively. When the output active power is 1.0 pu, the fifth- and seventh-order harmonic current and the THD of stator current are decreased from 5.15%, 3.12%, and 6.77% to 0.38%, 0.27%, and 3.18%, respectively. The fifth- and seventh-order stator harmonic current were suppressed by the proposed control method and the stator current quality is still good under distorted grid voltage. The electromagnetic torque fluctuation is also decreased after the proposed control is applied [15].

TABLE 7.1 Fifth- and seventh-order harmonic currents, the THD of stator current with vector control, and proposed stator harmonic current control under a super-synchronous speed of 1800 rpm

	Grid voltage		Stator current with vector control		Stator current with proposed stator harmonic current control	
Output power (pu)	0.5	1.0	0.5	1.0	0.5	1.0
Fifth-order harmonic (%)	5	5	8.3	5.1	1.0	0.4
Seventh-order harmonic (%)	5	5	6.1	3.1	1.6	0.3
Total THD (%)	6.85	6.85	10.8	6.7	4.3	3.2

The test results of the steady-state performance of the DFIG with proposed stator harmonic current control compared with vector control is shown in Figures 7.13 and 7.14, the 30 kW reduced-scale DFIG test system. The DFIG is operated at the sub-synchronous rotor speed of 1200 rpm; the grid voltage, stator current, rotor current of one phase, and electronic torque fluctuation is shown in Figures 7.13 and 7.14. The stator current harmonics and the THD of the stator current with proposed stator harmonic current control are compared with the vector control. The stator current harmonics and THD are measured by the power analyzer and listed in Table 7.2. It can be found the stator negative fifth and positive seventh harmonic currents are suppressed by the proposed control. By using the proposed stator harmonic current

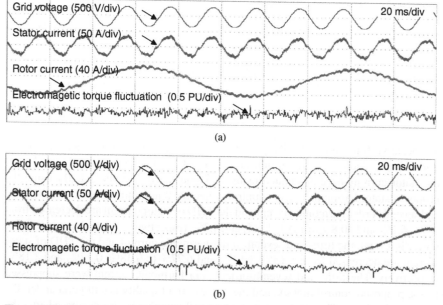

(a)

(b)

Figure 7.13   Test results of the DFIG under distorted grid with 0.5pu output active power: (a) With conventional vector control and (b) with stator harmonic current multiple-loop control.

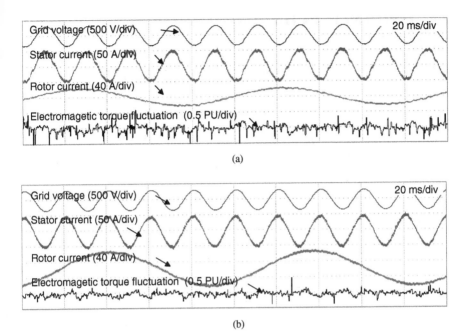

Figure 7.14   Test results of the DFIG under distorted grid with 0.8 pu output active power:
(a) With conventional vector control and (b) with stator harmonic current multiple-loop
control.

control, when the output power is 0.5 pu (15 kW), the fifth- and seventh-order harmonic currents and the THD of the stator current are decreased from 9.1%, 4.5%, and
10.8% to 1.2%, 0.2%, and 3.5%, respectively. When the output active power is 0.8 pu
(24 kW), the fifth- and seventh-order harmonic currents and the THD of stator current
are decreased from 4.1%, 2.7%, and 5.4% to 1.0%, 0.2%, and 2.4%, respectively. The
stator current quality is significantly improved. Although the control target is selected
to suppress the stator harmonic current, it can be found that the rotor harmonic current
and the electromagnetic torque fluctuation is suppressed as well.

**TABLE 7.2   Fifth- and seventh-order harmonic currents, the THD of stator current with
vector control, and proposed stator harmonic current control under a sub-synchronous
speed of 1200 rpm**

	Grid voltage		Stator current with vector control		Stator current with proposed stator harmonic current control	
Output power (pu)	0.5	1.0	0.5	1.0	0.5	1.0
Fifth-order harmonic (%)	1.15	0.99	9.10	4.12	1.19	0.96
Seventh-order harmonic (%)	0.77	1.17	4.45	2.68	0.22	0.17
Total THD (%)	1.81	2.05	10.79	5.41	3.49	2.44

7.4 INFLUENCE ON THE FUNDAMENTAL CURRENT LOOP

7.4.1 Influence on the Stability and Dynamic Response

The multiple-loop harmonic controls are in parallel with the fundamental control loop, and they will be enabled together at the same time in most cases. So, it is important to make sure that the harmonic control loop will have minimal influence on the dynamic performance of the fundamental loop, and this is also the premise of the controller design introduced in Section 7.3.4.

Take the stator harmonic current multiple-loop control as an example; the dynamic performance of the system is determined by the open-loop transfer function of the system. When the DFIG is controlled with conventional vector control given in (6.36), the open-loop transfer function $G'_o(s)$ of the system can be expressed as

$$G'_o(s) = G_{Pii}(s)G_p(s)K_{pwm} \tag{7.39}$$

When the DFIG is controlled with stator harmonic current multiple-loop control, and the negative fifth and positive seventh harmonic loop is applied, the open-loop transfer function of the system $G_o(s)$ can be derived from (7.32) and expressed as (7.40).

$$G_o(s) = K_{pwm}G_P(s)[G_{Pii}(s)$$
$$+ \frac{L_m}{L_s}LPF(s+j6\omega_s)G_{pi5}(s+j6\omega_s) + \frac{L_m}{L_s}LPF(s-j6\omega_s)G_{pi7}(s-j6\omega_s)] \tag{7.40}$$

Notice that $G_o(s)$ and $G'_o(s)$ are written in the synchronous dq reference frame. In (7.40), the fundamental loop and the harmonic control loop are all considered here. In (7.38), it has already been assumed that the fundamental loop and harmonic loop have no influence on each other: so the fundamental loop is not considered. As a result, (7.38) can only be used as a simplified method to design the controller parameters and it cannot be used in precise system analysis. In a 1.5 MW DFIG WPS with the controller parameters introduced above, the bode diagram of $G'_o(s)$ and $G_o(s)$ are compared in Figure 7.15. It can be found that the gain of $G_o(s)$ increase rapidly around 300 Hz, which corresponds to the frequency of the negative fifth and positive seventh frequency in the synchronous dq reference frame. The phase of $G_o(s)$ jumps from about 0° to 180° at the frequency of 300 Hz; besides, the Bode diagram of $G_o(s)$ and $G'_o(s)$ are almost the same.

For the DFIG with the conventional vector control, the crossover frequency is about 400 Hz while the phase margin is about 90°, as shown in Figure 7.15. For the DFIG with multiple-loop control, the Bode diagram of the system is almost the same around 400 Hz, the crossover frequency of the system is also about 400 Hz. The phase margin of the system may decrease a little, but it is also more than 80°. The stability of the system is mainly determined by the phase margin in the Bode diagram,

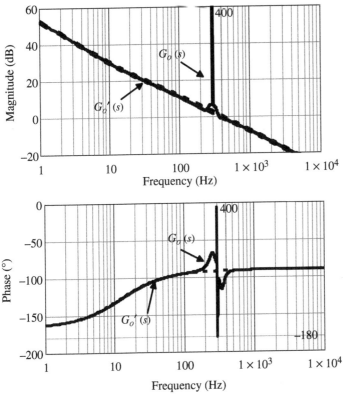

Figure 7.15   Bode diagram of DFIG with conventional vector control (dotted line) and multiple-loop control (solid line).

while the crossover frequency is normally related to the crossover frequency. As a result, a conclusion can be drawn from Figure 7.15 that after the multiple-loop control is applied, the phase margin and crossover frequency of the system are almost the same to that when conventional control is used. The influences from the multiple-loop control on the dynamic performance of the fundamental control loop are small.

7.4.2   Simulation and Test Results

Simulations were made on a 1.5 MW DFIG simulation model to verify the influence of the stator harmonic current multiple-loop control on the dynamic performance of the fundamental control loop. The dynamic performance of the DFIG with stator harmonic current multiple-loop control compared to conventional vector control is shown in Figure 7.16. The d-axis rotor current reference changes from 0 to 1 pu while the q-axis rotor current reference keeps at zero. The DFIG was operating at a super-synchronous speed of 1800 rpm, 5% negative fifth-order and 5% positive seventh-order harmonic is injected into the grid voltage. The response time of the system is

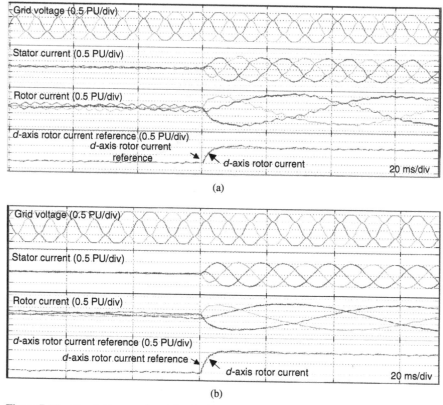

(a)

(b)

Figure 7.16 Simulation results of the dynamic performance of the DFIG with the step increase of the active power from 0 to 1 pu: (a) With conventional vector control and (b) with stator harmonic current suppression control.

still the same after the stator harmonic current multiple-loop control is applied (about 4 ms). The dynamic performance is not influenced by the stator harmonic current multiple-loop control method. More ever, as the negative fifth and positive seventh stator harmonic currents are suppressed by the proposed control method, the over-shoot in the d-axis rotor current during the transient stage is reduced.

The influence of the stator harmonic current multiple-loop control on the dynamic performance of the fundamental control loop is also verified in the 30 kW DFIG test system. The dynamic performance of the DFIG with stator harmonic current multiple-loop control compared to conventional vector control is shown in Figure 7.17. The d-axis rotor current reference changes from 0 to 0.5 pu while the d-axis rotor current reference keeps zero. The DFIG was operating at a sub-synchronous speed of 1200 rpm. The response time of the system is still the same after the stator harmonic current multiple-loop control is applied (about 5 ms). The dynamic performance is not influenced by the stator harmonic current multiple-loop control. Furthermore, as the negative fifth and positive seventh stator harmonic currents are suppressed by the

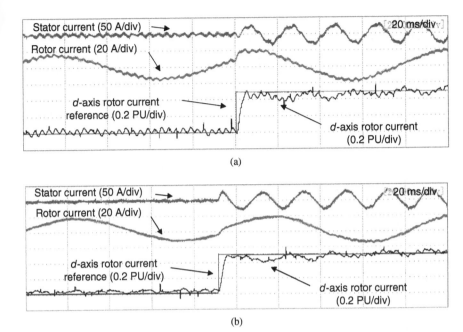

Figure 7.17   Test results of the dynamic performance of the DFIG with the step increase of the active power from 0 to 0.5 pu: (a) With conventional vector control and (b) with stator harmonic current suppression control.

proposed control method, the over shoot in the d-axis rotor current during the transient stage is reduced.

7.5   SUMMARY

In this chapter, a harmonic suppression control strategy based on multiple-loop control is introduced. The current is transferred to the dq reference frames which are rotating at harmonic angular speed, so the harmonic currents are DC components in these frames. In this case, they can be controlled as DC variables using a PI controller and a much higher controller gain can be achieved for the harmonic components. This control strategy can be used on both the GSC side and RSC side, and by choosing different harmonic current references, different controller target can be achieved. For the GSC, the control target is selected to suppress the GSC grid current, while for the RSC and DFIG, the control target is to suppress the stator harmonic current. If more than one frequency of the harmonic components exists in the grid voltage disturbance, more than one harmonic current loop needs to be added in the control scheme [16, 17]. All these control schemes work together with the fundamental control loop. A control scheme to suppress the negative fifth and positive seventh GSC grid harmonic current and DFIG stator harmonic current are introduced in this chapter. Furthermore, for the DFIG stator harmonic current control, two control schemes are introduced and compared; the direct stator harmonic current control seems to be a better choice and it is analyzed in detail [18, 19]. To ensure the control loops do not interrupt each other, the design of the controller parameters is introduced as well in this chapter, and the influence of the harmonic current loop on the fundamental current loop is evaluated. Simulation and test results are given to prove the effectiveness of the control strategy.

Besides for the multiple-loop control, there exists another control method to ensure a higher controller gain at a certain frequency, named resonant control. It can also be used for the harmonic suppression control of the DFIG. In Chapter 8, the DFIG stator harmonic current control based on the resonant controller will be discussed in detail [20, 21].

REFERENCES

[1] W. Chen, J. Xu, N. Zhu, C. Liu, M. Chen, and D. Xu, "Stator harmonic current suppression for DFIG wind power system under distorted grid voltage," in *3rd IEEE Int. Symp. Power Electron. Distributed Generation Systems (PEDG)*, Jun. 25–28, 2012, pp. 307, 314.

[2] C. Liu, F. Blaabjerg, W. Chen, and D. Xu, "Stator current harmonic control with resonant controller for doubly fed induction generator," *IEEE Trans. Power Electron.*, vol. 27, no. 7, pp. 3207–3220, Jul. 2012.

[3] H. Xu, J. Hu, and Y. He, "Operation of wind-turbine-driven DFIG systems under distorted grid voltage conditions: Analysis and experimental validations," *IEEE Trans. Power Electron.*, vol. 27, no. 5, pp. 2354–2366, May 2012.

[4] J. Xu, "Research on converter control strategy of doubly fed induction generator system for wind power," Ph.D. thesis, Zhejiang University, Hangzhou, China, 2011.

[5] Y. Song and H. Nian, "Sinusoidal output current implementation of DFIG using repetitive control under a generalized harmonic power grid with frequency deviation," *IEEE Trans. Power Electron.*, vol. 30, no. 12, pp. 6751–6762, Dec. 2015.

[6] Y. Song and H. Nian, "Enhanced grid-connected operation of DFIG using improved repetitive control under generalized harmonic power grid," *IEEE Trans. Energy Convers.*, vol. 30, no. 3, pp. 1019–1029, Sep. 2015.

[7] Y. Song and H. Nian, "Modularized control strategy and performance analysis of DFIG system under unbalanced and harmonic grid voltage," *IEEE Trans. Power Electron.*, vol. 30, no. 9, pp. 4831–4842, Sep. 2015.

[8] H. Nian and Y. Song, "Optimised parameter design of proportional integral and resonant current regulator for doubly fed induction generator during grid voltage distortion," *IET Renew. Power Gen.*, vol. 8, no. 3, pp. 299–313, Apr. 2014.

[9] L. Fan, S. Yuvarajan, and R. Kavasseri, "Harmonic analysis of a DFIG for a wind energy conversion system," *IEEE Trans. Energy Convers.*, vol. 25, no. 1, pp. 181–190, Mar. 2010.

[10] V. T. Phan and H. H. Lee, "Control strategy for harmonic elimination in stand-alone DFIG applications with nonlinear loads," *IEEE Trans. Power Electron.*, vol. 26, no. 9, pp. 2662–2675, Sep. 2011.

[11] J. Hu, H. Xu, and Y. He, "Coordinated control of DFIG's RSC and GSC under generalized unbalanced and distorted grid voltage conditions," *IEEE Trans. Ind. Electron.*, vol. 60, no. 7, pp. 2808–2819, Jul. 2013.

[12] A. E. Leon, J. M. Mauricio, and J. A. Solsona, "Fault ride-through enhancement of DFIG-based wind generation considering unbalanced and distorted conditions," *IEEE Trans. Energy Convers.*, vol. 27, no. 3, pp. 775–783, Sep. 2012.

[13] V. T. Phan and H. H. Lee, "Performance enhancement of stand-alone DFIG systems with control of rotor and load side converters using resonant controllers," *IEEE Trans. Ind. Appl.*, vol. 48, no. 1, pp. 199–210, Jan.–Feb. 2012.

[14] Y. Song and F. Blaabjerg, "Overview of DFIG-based wind power system resonances under weak networks," *IEEE Trans. Power Electron.*, vol. 32, no. 6, pp. 4370–4394.

[15] M. Kiani and W. J. Lee, "Effects of voltage unbalance and system harmonics on the performance of doubly fed induction wind generators," *IEEE Trans. Ind. Appl.*, vol. 46, no. 2, pp. 562–568, 2010.

[16] J. Hu, H. Nian, H. Xu, and Y. He, "Dynamic modeling and improved control of DFIG under distorted grid voltage conditions," *IEEE Trans. Energy Convers.*, vol. 26, no. 1, pp. 163–175, 2011.

[17] H. Nian and Y. Song, "Direct power control of doubly fed induction generator under distorted grid voltage," *IEEE Trans. Power Electron.*, vol. 29, no. 2, pp. 894–905, 2014.

[18] M. I. Martinez, G. Tapia, A. Susperregui, and H. Camblong, "Sliding-mode control for DFIG rotor- and grid-side converters under unbalanced and harmonically distorted grid voltage," *IEEE Trans. Energy Convers.*, vol. 27, no. 2, pp. 328–339, 2012.

[19] H. Xu, J. Hu, and Y. He, "Integrated modeling and enhanced control of DFIG under unbalanced and distorted grid voltage conditions," *IEEE Trans. Energy Convers.*, vol. 27, no. 3, pp. 725–736, 2012.

[20] M. Lindholm and T. W. Rasmussen, "Harmonic analysis of doubly fed induction generators," in *Proc. Fifth International Conf. Power Electron. Drive Systems*, vol. 2, 2003, pp. 837–841.

[21] A. E. Leon, J. M. Mauricio, and J. A. Solsona, "Fault ride-through enhancement of DFIG-based wind generation considering unbalanced and distorted conditions," *IEEE Trans. Energy Convers.*, vol. 27, no. 3, pp. 775–783, 2012.

CHAPTER 8

RESONANT CONTROL OF DFIG UNDER GRID VOLTAGE HARMONICS DISTORTION

8.1 INTRODUCTION

According to the analysis given in Chapter 6, under harmonically distorted grid conditions, the grid voltage harmonics, mainly the fifth and seventh order, will cause sixth-order harmonics in the electromagnetic torque and stator active and reactive currents of the DFIG. Therefore typically six times the grid frequency is the characteristic oscillation frequency of the DFIG under harmonically distorted grid. If the fluctuating components of such frequencies can be effectively suppressed without influencing the normal operation of the DFIG, the capability of the DFIG wind power system to overcome harmonically distorted grid conditions will be largely improved. Resonant controllers, also known as generalized AC integrators, achieve a large gain at its resonant frequency and damp rapidly other frequency components. Such a feature makes resonant controller suitable for the suppression of the fluctuating components that have a characteristic oscillation frequency.

This chapter discusses the improvement of current controllers for DFIG under grid voltage harmonics using resonant controllers in order to enhance the compatibility of DFIG wind power generation systems according to the grid codes.

8.2 RESONANT CONTROLLER

8.2.1 Mathematical Model of a Resonant Controller

The transfer function of a resonant controller can be expressed as [1, 2]

$$G_R(s) = \frac{2K_r\omega_c s}{s^2 + 2\omega_c s + \omega_o{}^2} \tag{8.1}$$

where K_r is the controller gain at resonant frequency ω_o, ω_c is the open-loop cut-off frequency of the resonant controller when $K_r = 1$. Due to the small value

---

Advanced Control of Doubly Fed Induction Generator for Wind Power Systems, First Edition.
Dehong Xu, Frede Blaabjerg, Wenjie Chen, and Nan Zhu.

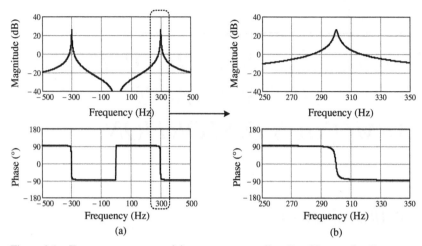

Figure 8.1 Frequency response of the resonant controller, $K_r = 20$, $\omega_c = 5$ rad/s, $\omega_o = 1884$ rad/s.

of ω_c, the gain of the resonant controller decreases rapidly at frequencies other than the resonant frequency and thereby achieving a good frequency-selection characteristic.

To eliminate the fifth and seventh harmonic currents, the resonant controller is required to have a large gain at $6\omega_s$, where ω_s is the grid frequency. Figure 8.1 shows the frequency response of the resonant controller when the resonant frequency is set as $\omega_o = 6\omega_s$. It is shown that the resonant controller achieves a very large gain at frequencies of $+300$ Hz (the frequency of the positive-sequence seventh-order harmonic in the stationary reference frame) and -300 Hz (the frequency of the negative-sequence fifth-order harmonic in the stationary reference frame). Figure 8.1b shows the frequency response of the resonant controller around the resonance frequency.

It can be seen from Figure 8.1b that the resonant controller has a very large control gain at 300 Hz, and the gain decreases rapidly at other frequencies. Such a frequency-selection characteristic is suitable for the control of AC components that have frequencies several times the grid frequency [1]. However, when the grid frequency deviates from its normal value, the fifth- and seventh-order harmonic frequency would also deviate from 300 Hz. In that way, the resonant controller tuned at 300 Hz would become ineffective in suppressing the harmonics. Therefore, the resonant controller is sensitive to the change of the grid frequency. Nevertheless, this problem can be dealt with in modern digital control systems. The resonant frequency of the resonant controller can be adjusted according to the measured grid frequency by the phase locked loop (PLL), so that the resonant frequency always stays at the harmonic frequency of the grid, despite the changes in the fundamental grid frequency. The resonant controller that has a frequency tracking capability is called an adaptive resonant controller [2, 13].

8.2.2 Resonant Controller in *dq* Frames

The resonant controller shown in equation (8.1) can be deduced by a reference frame transformation. In a synchronous rotating reference frame, the non-ideal integrator can be expressed as [6]

$$G_I(s) = \frac{K_I}{1 + s/\omega_c} \tag{8.2}$$

where K_I is the controller gain and ω_c is the cut-off frequency. When the integrators in the *dq* rotating frames synchronous to the negative fifth-order and positive seventh-order harmonics are transferred into the *dq* frame synchronous to the fundamental frequency of the grid, the following equation can be obtained:

$$G_I^{(1)}(s) = G_I^{(5)}\left(s + j\omega_o\right) + G_I^{(7)}\left(s - j\omega_o\right)$$

$$= \frac{K_I}{1 + \left(s + j\omega_o\right)/\omega_c} + \frac{K_I}{1 + \left(s - j\omega_o\right)/\omega_c} = \frac{2K_I\left(\omega_c s + \omega_c^2\right)}{s^2 + 2\omega_c s + \left(\omega_c^2 + \omega_o^2\right)} \tag{8.3}$$

where $G_I^{(5)}(s)$ and $G_I^{(7)}(s)$ are the integrators in the negative fifth-order and positive seventh-order synchronous rotating frames, respectively, $G_I^{(1)}(s)$ is their equivalence in the grid synchronous frame, ω_o is six times the grid frequency ($\omega_o = 6\omega_s$). Since the cut-off frequency $\omega_c << \omega_o$, $G_I^{(1)}(s)$ can be approximated as [13]

$$G_I^{(1)}(s) \approx \frac{2K_I\omega_c s}{s^2 + 2\omega_c s + \omega_o^2} \tag{8.4}$$

It can be seen that equation (8.4) has the same form as equation (8.1). Therefore, a resonant controller in the grid synchronous frame is equivalent to the conventional PI controllers in the *dq* rotating frames synchronous to the harmonics [14, 15].

8.3 STATOR CURRENT CONTROL USING RESONANT CONTROLLERS

8.3.1 Control Target

Similar to the control strategy introduced in Chapter 7, four control targets can be adopted for the control of a DFIG under harmonically distorted grid voltage [16, 17]:

- Elimination of rotor current oscillations.
- Suppression of stator current harmonics.
- Elimination of the fluctuations in stator active and reactive powers.
- Elimination of the fluctuations in stator reactive power and electromagnetic torque.

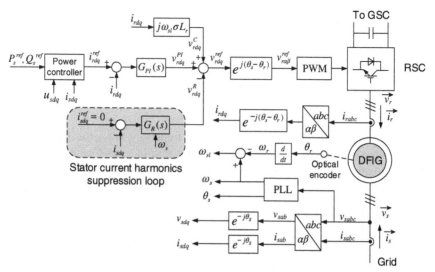

Figure 8.2   Control of the DFIG including stator current harmonics control scheme based on resonant controller.

The above-mentioned control targets can be achieved by modifying the control scheme described in Section 5.3. One possible solution is to calculate the rotor current reference for the rotor current control loop according to the generator parameters and grid voltage harmonics. However, the variations in generator parameters and the difficulty of accurate extraction of the grid voltage harmonics will affect the effectiveness of this method. On the other hand, a control scheme based on the resonant controller does not rely on the generator parameters and the extraction of grid voltage harmonics is not required, which is suitable for practical applications with better robustness [11, 12, 24, 25].

8.3.2   Control Scheme

The schematic diagram of the stator current harmonics elimination control scheme based on the resonant controller is shown in Figure 8.2. Compared to the control strategy introduced in Chapter 7, since the resonant controller is used for harmonic suppression, the harmonic extraction is no longer needed and the multiloop PI controllers are substituted by a single resonant controller [18–20].

The control system consists of a power control loop, a rotor current control loop, a stator current harmonics control loop, PLL, synchronous rotating reference frame transformation, and also a space vector modulation (SVM) module.

In Figure 8.2, the subscription "r" denotes the variables on the rotor side while "s" denotes the stator-side variables. For each variable, $abc \rightarrow \alpha\beta$ and $\alpha\beta \rightarrow dq$ frame transformations are used to get the dq components in the dq rotating frame synchronous to the grid fundamental frequency. The grid frequency is obtained by a PLL, and the rotor frequency is obtained by an optical encoder.

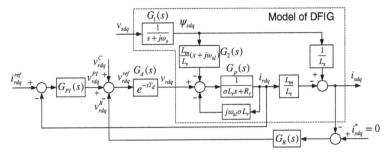

Figure 8.3   Current control block diagram of the DFIG with stator harmonics elimination control.

Under harmonically distorted grid conditions, the fifth- and seventh-order harmonics in the grid voltage are transformed into sixth-order harmonics by the rotating frame transformation synchronous to the grid fundamental frequency. Therefore, the current control loops of the control system shown in Figure 8.2 are required to have the capability of suppressing the impacts of the sixth-order harmonic voltage.

As shown in Figure 8.2, the voltage reference for the rotor-side converter v_{rdq}^{ref} is composed of the output of the PI controller of rotor current control loop v_{rdq}^{PI}, the output of the resonant controller of stator current harmonics control loop v_{rdq}^{R}, and the rotor current decoupling term v_{rdq}^{C}, which can be expressed as

$$v_{rdq}^{ref} = v_{rdq}^{PI} - v_{rdq}^{R} + v_{rdq}^{C} \tag{8.5}$$

In equation (8.5), the minus sign for v_{rdq}^{R} is due to the opposite reference directions for stator and rotor currents. In steady state, v_{rdq}^{PI} is a DC component and v_{rdq}^{R} is an AC component with the frequency six times the grid frequency.

8.3.3   Control Model in *dq* Frame

Combining the model of the DFIG and the stator current harmonics control scheme shown in Figure 8.2, the current control block diagram of the DFIG can be depicted as Figure 8.3.

In a digital control system, the calculation of the control commands may cause a delay of one sampling period [23]. When the sampling frequency is comparable to the harmonic frequency, the digital control delay may cause large phase lag for the harmonics control and thus affect the control performance. Therefore, the control delay must be taken into consideration in the harmonics control system. In Figure 8.3, the control delay is denoted as $G_d(s) = e^{-sT_d}$, where $T_d = T_s$ and T_s is the sampling period.

As shown in Figure 8.3, the stator current $i_{sqd}(s)$ is influenced by the rotor current control loop, the stator current harmonics control loop, and the grid voltage

$v_{sdq}(s)$. The transfer functions of the stator and rotor currents can be expressed as

$$
\begin{cases}
i_{sdq}(s) = \left\{ \left[\left(i_{rdq}^{ref}(s) - i_{rdq}(s) \right) G_{PI}(s) - \left(i_{sdq}^{ref}(s) - i_{sdq}(s) \right) G_R(s) \right] G_d(s) \right. \\
\qquad\qquad \left. - v_{sdq}(s)G_1(s)G_2(s) \right\} G_p(s)\dfrac{L_m}{L_s} - v_{sdq}(s)G_1(s)\dfrac{1}{L_s} \\
i_{rdq}(s) = v_{sdq}(s)G_1(s)\dfrac{1}{L_m} - \dfrac{L_s}{L_m}i_{sdq}(s)
\end{cases}
\tag{8.6}
$$

where $G_1(s) = 1/(s + j\omega_s)$, $G_2(s) = (s + j\omega_{sl}) L_m/L_s$, and $G_p(s) = 1/(\sigma L_r s + R_r)$ are included in the model of the DFIG which is shown in Figure 8.3, $G_{PI}(s)$ is the PI controller of the rotor current control loop, and $G_R(s)$ is the resonant controller of the stator current harmonics control loop. The expressions for $G_{PI}(s)$ and $G_R(s)$ are as follows.

$$
G_{PI}(s) = K_p + \frac{K_i}{s}
\tag{8.7}
$$

$$
G_R(s) = \frac{2K_r\omega_c s}{s^2 + 2\omega_c s + \omega_o^2}
\tag{8.8}
$$

Substituting $i_{rdq}(s)$ in the expression of $i_{sdq}(s)$, the following equation can be obtained for the transfer function between the stator current $i_{sdq}(s)$ and the input variables.

$$
i_{sdq}(s) = v_{sdq}(s)G_{vis}(s) - i_{rdq}^{ref}(s)G_{rs}(s) + i_{sdq}^{ref}(s)G_{ss}(s)
\tag{8.9}
$$

Compared to the traditional control scheme, an extra term of stator current harmonics control $i_{sdq}^{ref}(s)G_{ss}(s)$ is added in equation (8.9), where $G_{ss}(s)$ is the transfer function from stator current reference $i_{sdq}^{ref}(s)$ to stator current $i_{sdq}(s)$. $G_{rs}(s)$ is the transfer function from rotor current reference $i_{rdq}^{ref}(s)$ to stator current $i_{sdq}(s)$, and $G_{vis}(s)$ is the transfer function from the grid voltage $v_{sdq}(s)$ to the stator current $i_{sdq}(s)$. The transfer functions $G_{rs}(s)$, $G_{ss}(s)$, and $G_{vis}(s)$ can be expressed as

$$
G_{rs}(s) = \frac{i_{sdq}(s)}{i_{rdq}^{ref}(s)} = \frac{G_{PI}(s)G_d(s)G_p(s)L_m/L_s}{1 + G_{PI}(s)G_d(s)G_p(s) + G_R(s)G_d(s)G_p(s)L_m/L_s}
\tag{8.10}
$$

$$
G_{ss}(s) = \frac{i_{sdq}(s)}{i_{sdq}^{ref}(s)} = \frac{G_R(s)G_d(s)G_p(s)L_m/L_s}{1 + G_{PI}(s)G_d(s)G_p(s) + G_R(s)G_d(s)G_p(s)L_m/L_s}
\tag{8.11}
$$

$$
G_{vis}(s) = \frac{i_{sdq}(s)}{v_{sdq}(s)}
$$

$$
= \frac{G_1(s)/L_s + G_1(s)G_{PI}(s)G_d(s)G_p(s)/L_s + G_1(s)G_2(s)G_p(s)L_m/L_s}{1 + G_{PI}(s)G_d(s)G_p(s) + G_R(s)G_d(s)G_p(s)L_m/L_s}
\tag{8.12}
$$

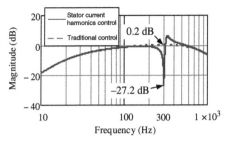

Figure 8.4   Amplitude–frequency characteristics of $G_{vis}(s)$ with the traditional control and with the stator harmonics control.

$G_{vis}(s)$ represents the control system's rejection capability to the grid voltage disturbances. Compared to the traditional control method, in equation (8.12), an extra term $G_R(s)G_d(s)G_p(s)L_m/L_s$ is added to the denominator of the expression for $G_{vis}(s)$. This added term is determined by the resonant controller. If the resonant controller gain K_r is properly designed, $G_{vis}(s)$ may have the required damping at the harmonic frequencies. The effect of stator current harmonics control on the system performance will be discussed in more detail in the next section.

8.3.4   Control Effect

The amplitude–frequency characteristics of $G_{vis}(s)$ with the traditional control method and with the stator harmonics control scheme are compared in Figure 8.4. The calculations are based on a 1.5 MW DFIG wind turbine, the parameters of which can be found in Tables 5.1 and 5.2. The sampling frequency used in the digital controller is 4 kHz. To depict the figure, the parameters of the PI controller are chosen as $K_p = 0.85$ and $K_i = 80$, the parameters of the resonant controller are set as $K_r = 20$ and $\omega_c = 5$ rad/s. It can be seen from Figure 8.4 that by implementing the stator current harmonics control, the DFIG system has obtained a -27.2 dB damping for the fifth- and seventh-order harmonics of the grid voltage.

Assume the amplitudes of the fifth- and seventh-order harmonic voltages in the grid are both 0.05 pu, the amplitude of the stator current harmonics can be calculated according to the following equations:

$$i_{sdq}^{(5)} = i_{sdq}\left(-j6\omega_s\right) = \frac{v_{sdq}\left(-j6\omega_s\right)}{X_{vis}\left(-j6\omega_s\right)} \tag{8.13}$$

$$i_{sdq}^{(7)} = i_{sdq}\left(j6\omega_s\right) = \frac{v_{sdq}\left(j6\omega_s\right)}{X_{vis}\left(j6\omega_s\right)} \tag{8.14}$$

where $X_{vis}\left(j\omega_s\right) = 1/G_{vis}\left(j\omega_s\right)$, $v_{sdq}\left(-j6\omega_s\right)$, and $v_{sdq}\left(j6\omega_s\right)$ are the negative-sequence fifth-order and positive-sequence seventh-order harmonic voltages, respectively.

The amplitudes of the stator current harmonics with different resonant controller gains K_r can be calculated as shown in Figure 8.5. As shown in the figure,

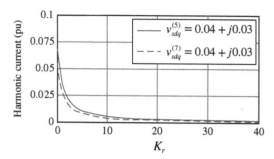

Figure 8.5   Amplitudes of the fifth- and seventh-order stator current harmonics with the stator current harmonics control, the injected voltage $\left| v_{sdq} \left(\pm j6\omega_s \right) \right| = 0.05$ pu for different K_r values.

when $K_r = 0$, the resonant controller is disabled, which means that the traditional control scheme is adopted. When the resonant controller starts working, as K_r increases, the amplitudes of the stator current harmonics decrease rapidly. The stator current harmonics are almost eliminated when K_r goes beyond 20.

As explained in Chapter 6, the harmonically distorted grid voltage may also cause fluctuations in the electromagnetic torque of the DFIG. Figure 8.6 shows the sixth-order fluctuation in the electromagnetic torque with different K_r values of the resonant controller for stator current harmonics.

It is shown that the electromagnetic torque fluctuation drops from 0.06 pu at $K_r = 0$ to below 0.01 pu when K_r is larger than 20. Thus, the electromagnetic torque fluctuation is eliminated by the stator current harmonics control as well.

According to equation (8.5), with the implementation of the stator current harmonics control, the rotor voltage consists of not only the fundamental components but also the harmonic components as well. Therefore, the required DC voltage may become higher.

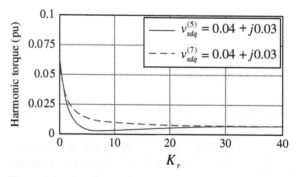

Figure 8.6   Amplitude of the sixth-order fluctuations in electromagnetic torque with stator current harmonics control, the injected voltage $\left| v_{sdq} \left(\pm j6\omega_s \right) \right| = 0.05$ pu for different K_r values.

Referring to the DFIG model described in Chapter 4, the rotor current can be expressed as

$$i_{rdq} = \frac{\Psi_{sdq} - L_s i_{sdq}}{L_m} \tag{8.15}$$

If the fifth- and seventh-order harmonics in the stator current are controlled to be zero, that is $i_{sdq}^{(5)} = i_{sdq}^{(7)} = 0$, then the rotor current harmonics can be deduced as follows:

$$i_{rdq}^{(5)} = \frac{\Psi_{sdq}^{(5)}}{L_m} = -\frac{v_{sdq}^{(5)}}{j5\omega_s L_m} \tag{8.16}$$

$$i_{rdq}^{(7)} = \frac{\Psi_{sdq}^{(7)}}{L_m} = \frac{v_{sdq}^{(7)}}{j7\omega_s L_m} \tag{8.17}$$

Therefore, the complete form of the rotor current expression can be derived as the following equation:

$$i_{rdq} = i_{rdq}^{(1)} + i_{rdq}^{(5)} e^{-j6\omega_s t} + i_{rdq}^{(7)} e^{j6\omega_s t} = i_{rdq}^{(1)} - \frac{v_{sdq}^{(5)}}{j5\omega_s L_m} e^{-j6\omega_s t} + \frac{v_{sdq}^{(7)}}{j7\omega_s L_m} e^{j6\omega_s t} \tag{8.18}$$

Neglecting the resistance of the rotor windings, the steady-state expression of the rotor voltage can be expressed as [7, 17]

$$v_{rdq} = R_r i_{rdq} + \sigma L_r \left(\frac{d}{dt} + j\omega_{sl} \right) i_{rdq} + \frac{L_m}{L_s} \left(\frac{d}{dt} + j\omega_{sl} \right) \Psi_{sdq}$$

$$\approx \sigma L_r \left(\frac{d}{dt} + j\omega_{sl} \right) \left(i_{rdq}^{(1)} - \frac{v_{sdq}^{(5)}}{j5\omega_s L_m} e^{-j6\omega_s t} + \frac{v_{sdq}^{(7)}}{j7\omega_s L_m} e^{j6\omega_s t} \right)$$

$$+ \frac{L_m}{L_s} \left(\frac{d}{dt} + j\omega_{sl} \right) \left(\frac{v_{sdq}^{(1)}}{j\omega_s} - \frac{v_{sdq}^{(5)}}{j5\omega_s} e^{-j6\omega_s t} + \frac{v_{sdq}^{(7)}}{j7\omega_s} e^{j6\omega_s t} \right)$$

$$= \left(j\omega_{sl}\sigma L_r i_{rdq}^{(1)} + \frac{L_m}{L_s} \frac{\omega_{sl}}{\omega_s} v_{sdq}^{(1)} \right)$$

$$+ \left(\frac{\sigma L_r}{L_m} + \frac{L_m}{L_s} \right) \left(\frac{d}{dt} + j\omega_{sl} \right) \left(-\frac{v_{sdq}^{(5)}}{j5\omega_s} e^{-j6\omega_s t} + \frac{v_{sdq}^{(7)}}{j7\omega_s} e^{j6\omega_s t} \right)$$

$$\approx \frac{L_m}{L_s} \frac{\omega_{sl}}{\omega_s} v_{sdq}^{(1)} + \frac{L_r}{L_m} \left(\frac{-6\omega_s + \omega_{sl}}{-5\omega_s} v_{sdq}^{(5)} e^{-j6\omega_s t} + \frac{6\omega_s + \omega_{sl}}{7\omega_s} v_{sdq}^{(7)} e^{j6\omega_s t} \right)$$

$$= S_l^{(1)} \frac{L_m}{L_s} v_{sdq}^{(1)} + \frac{L_r}{L_m} S_l^{(5)} v_{sdq}^{(5)} e^{-j6\omega_s t} + \frac{L_r}{L_m} S_l^{(7)} v_{sdq}^{(7)} e^{j6\omega_s t}$$

$$= v_{rdq}^{(1)} + v_{rdq}^{(5)} e^{-j6\omega_s t} + v_{rdq}^{(7)} e^{j6\omega_s t} \tag{8.19}$$

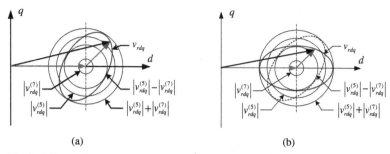

Figure 8.7   Trajectory of the rotor voltage vector when the stator current harmonics control is activated: (a) Arbitrary initial phase angle of the harmonic vector; (b) initial phase angle of the harmonic vector equal to the fundamental.

where $S_l^{(1)}$ is the slip ratio of the positive-sequence fundamental component, $S_l^{(5)}$ is the slip ratio of the negative-sequence fifth-order harmonic component, and $S_l^{(7)}$ is the slip ratio of the positive-sequence seventh-order harmonic component. The definitions of the slip ratios are given below.

$$
\begin{cases}
S_l^{(1)} = \dfrac{\omega_s - \omega_r}{\omega_s} = \dfrac{\omega_{sl}}{\omega_s} = 1 - \dfrac{\omega_r}{\omega_s} \\[2mm]
S_l^{(5)} = \dfrac{-5\omega_s - \omega_r}{-5\omega_s} = \dfrac{-6\omega_s + \omega_{sl}}{-5\omega_s} = 1 + \dfrac{\omega_r}{5\omega_s} \\[2mm]
S_l^{(7)} = \dfrac{7\omega_s - \omega_r}{7\omega_s} = \dfrac{6\omega_s + \omega_{sl}}{7\omega_s} = 1 - \dfrac{\omega_r}{7\omega_s}
\end{cases}
\tag{8.20}
$$

During the normal operation of the DFIG, the rotor speed ω_r varies in the range of 0.7–1.3 times the synchronous speed. Therefore, the slip ratios are in the following ranges: $-0.3 \leq S_l^{(1)} \leq 0.3$, $1.14 \leq S_l^{(5)} \leq 1.26$, and $0.81 \leq S_l^{(7)} \leq 0.9$.

As illustrated in equation (8.19), the rotor voltage vector is composed of two parts, the fundamental vector and the harmonic vector. In steady state, the amplitude and direction of fundamental vector $v_{rdq}^{(1)}$ remain constant in the synchronous rotating dq reference frame. The trajectory of the harmonic vector $v_{rdq}^{(5)} e^{-j6\omega_s t} + v_{rdq}^{(7)} e^{j6\omega_s t}$ is an ellipse of which the semi-major axis is $\left| v_{rdq}^{(5)} \right| + \left| v_{rdq}^{(7)} \right|$ and the semi-minor axis is $\left| v_{rdq}^{(5)} \right| - \left| v_{rdq}^{(7)} \right|$, as shown in Figure 8.7. In Figure 8.7a, the initial phase angle of the harmonic vector is arbitrary. According to the vector composition principle, the following relationship can be derived.

$$
\left| v_{rdq} \right| \leq \left(\left| v_{rdq}^{(1)} \right| + \left| v_{rdq}^{(5)} e^{-j6\omega_s t} + v_{rdq}^{(7)} e^{j6\omega_s t} \right| \right) \leq \left(\left| v_{rdq}^{(1)} \right| + \left| v_{rdq}^{(5)} \right| + \left| v_{rdq}^{(7)} \right| \right)
\tag{8.21}
$$

Only when the major axis of the harmonic vector trajectory coincides with the fundamental vector, as illustrated in Figure 8.7b, $\left| v_{rdq} \right|$ may reach its maximum value:

$$
\left. \left| v_{rdq} \right| \right|_{(\text{max})} = \left| v_{rdq}^{(1)} \right| + \left| v_{rdq}^{(5)} \right| + \left| v_{rdq}^{(7)} \right|
\tag{8.22}
$$

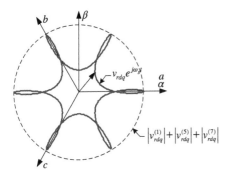

Figure 8.8   Trajectory of rotor voltage vector in the static reference frame when the amplitude of rotor vector reaches its maximum value.

Transforming Figure 8.7b from the synchronous rotating dq reference frame to the static $abc/\alpha\beta$ reference frame, the trajectory is shown in Figure 8.8. It is seen that the maximum amplitude of the rotor voltage is $\left|v_{rdq}^{(1)}\right| + \left|v_{rdq}^{(5)}\right| + \left|v_{rdq}^{(7)}\right|$.

The variables used in the previous analysis are referred to the stator and therefore the actual maximum amplitude of rotor voltage would be the following [7, 17].

$$
\begin{aligned}
V_{rm(max)} &= \frac{1}{n_{sr}}\left|v_{rdq}\right|_{(max)} = V_{rm}^{(1)} + V_{rm}^{(5)} + V_{rm}^{(7)} \\
&= \frac{1}{n_{sr}}\left(\left|S_l^{(1)}\frac{L_m}{L_s}v_{sdq}^{(1)}\right| + \left|\frac{L_r}{L_m}S_l^{(5)}v_{sdq}^{(5)}\right| + \left|\frac{L_r}{L_m}S_l^{(7)}v_{sdq}^{(7)}\right|\right) \\
&= \frac{1}{n_{sr}}\left(\frac{L_m}{L_s}\left|S_l^{(1)}\right|V_{sm}^{(1)} + \frac{L_r}{L_m}S_l^{(5)}V_{sm}^{(5)} + \frac{L_r}{L_m}S_l^{(7)}V_{sm}^{(7)}\right)
\end{aligned} \tag{8.23}
$$

where n_{sr} is the turns ratio between stator and rotor of the DFIG; $V_{rm}^{(1)}$, $V_{rm}^{(5)}$, $V_{rm}^{(7)}$ are the amplitudes of fundamental, fifth- order, and seventh-order harmonic rotor voltages, respectively; $V_{sm}^{(1)}$, $V_{sm}^{(5)}$, $V_{sm}^{(7)}$ are the amplitudes of fundamental, fifth-order, and seventh-order harmonic stator voltages, respectively. Since the slip ratios for the harmonics $S_l^{(5)}$ and $S_l^{(7)}$ are around 1, the actual amplitude of rotor harmonic voltage is approximately $(1/n_{sr})$ of the stator harmonic voltage. For a DFIG, n_{sr} is usually less than 0.4; thus rotor-side converter has to output large voltage harmonics to compensate the impact of grid voltage harmonics.

When $S_l^{(1)}$ is at its lower limit, $S_l^{(1)} = -0.3$, the required rotor voltage reaches the maximum. By setting the amplitude of the grid voltage fundamental component as 1 pu, and assuming that the amplitudes of fifth- and seventh-order harmonics are 0.04 pu and 0.03 pu, respectively, the curve of the rotor voltage amplitude with respect to the grid voltage THD can be depicted as shown in Figure 8.9. As shown in Figure 8.9, a higher rotor voltage amplitude is required to compensate for larger grid voltage THD.

According to the principle of SVM, the DC-link voltage of the RSC must be larger than rotor voltage, $v_{dc} \geq V_{rm(max)}$. With the increase of rotor voltage amplitude, the margin of voltage modulation for the RSC decreases, and the output of the SVM

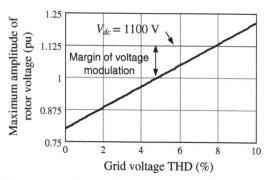

Figure 8.9   Relationship between the maximum amplitude of rotor voltage and grid voltage THD (only considering fifth- and seventh-order harmonics).

approaches saturation. The DC-link voltage of a 690 V_{AC} system is usually set as 1100 V for normal grid conditions, which may be insufficient for the compensation of grid voltage harmonics when the grid voltage THD is relatively high. To meet the requirement of the rotor voltage modulation under harmonically distorted grid conditions, a higher DC-link voltage may be necessary.

However, a high DC-link voltage not only leads to higher losses of the converters but also reduces the operation lifetime of the DC-link capacitors. Since the chance of the converter to operate under heavily harmonically distorted grid conditions is rather scarce, in order to avoid the DC-link voltage from remaining at a high level during normal operation, a flexible controlled DC-link voltage is proposed in [3]. Such a control method only raises the DC-link voltage to meet the requirement of SVM under abnormal grid conditions, while keeping the DC-link voltage at a relatively lower level during normal operation, which mitigates the negative effects of higher DC-link voltage brought by the stator current harmonics control.

To verify the performance of the stator current harmonics control under harmonically distorted grid condition, a simulation is done by using Matlab/Simulink. The schematic description of the simulation model is shown in Figure 8.10. The

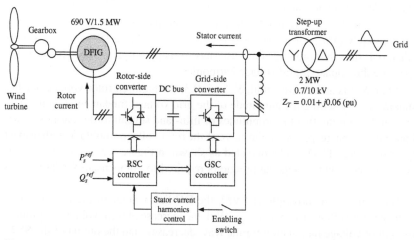

Figure 8.10   Schematic diagram of the simulation model for the DFIG system.

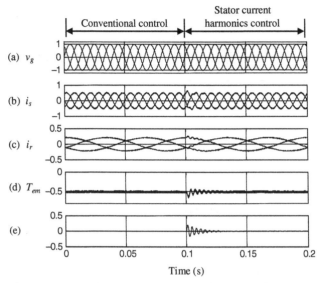

Figure 8.11   Steady-state performances of the conventional control scheme and the stator current harmonics control scheme under normal grid condition, stator current $i_{sdq} = 0.5$ pu. (a) Grid voltage v_g (pu); (b) stator current i_s (pu); (c) rotor current i_r (pu); (d) electromagnetic torque T_{em} (pu); (e) output of resonant controller (pu).

voltage and power ratings of the DFIG are 690 V/1.5 MW, and the grid frequency is 50 Hz. The parameters of the generator can be found in Tables 5.1 and 5.2. The stator of the DFIG is connected to a 10 kV grid via a step-up transformer of which the leakage inductance is $Z_T = 0.01 + j0.06$ (pu). The converters for the DFIG system include the rotor-side and grid-side converters which are connected by a common DC-bus capacitor of 16 mF. The switching frequency of the converters is set to 2 kHz, the sampling frequency is 4 kHz, and the DC-bus voltage is 1150 V. The control system consists of controllers for RSC and GSC. The RSC controller controls the DFIG, and the GSC controller maintains the DC-bus voltage at a constant level. The PI controller for RSC is set to $K_p = 0.85$ and $K_i = 80$, and the parameters for the resonant controller are $K_r = 20$, $\omega_c = 5$ rad/s. In the simulations, to present the rotor currents more explicitly, the rotor speed during steady-state operation is kept at 1.2 pu. It should be noted that in practical situations, the wind power generation system often operates under rated conditions. If not specifically addressed, the harmonic components in the grid voltage used in the simulations are 4% of negative-sequence fifth-order and 3% of positive-sequence seventh-order harmonics, and grid voltage THD is 5%. In Figure 8.10, when the enabling switch is switched on, the stator current harmonics control is enabled, otherwise the stator current harmonics control is disabled and the conventional control scheme is adopted. The simulation system is using discrete models, and the variables are presented in per unit values.

First, the performances of the conventional control scheme and the stator current harmonics control scheme are simulated under normal grid condition. The results are shown in Figure 8.11. The simulation is done under half-load condition with a stator current of 0.5 pu. From 0 to 0.1 s, the conventional control scheme is adopted,

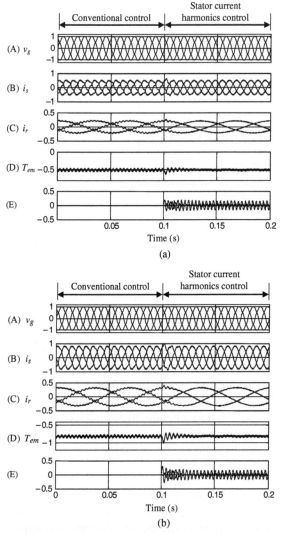

Figure 8.12   Steady-state performances of conventional control scheme and the stator current harmonics control scheme under harmonically distorted grid condition. The test conditions are $v_{sdq}^{(5)} = 0.04$ pu, $v_{sdq}^{(7)} = 0.03$ pu; (a) stator current i_s ($dq = 0.5$ pu), (b) stator current i_s ($dq = 0.8$ pu). (A) Grid voltage v_g (pu); (B) stator current i_s (pu); (C) rotor current i_r (pu); (D) electromagnetic torque T_{em} (pu); (E) output of resonant controller (pu).

and the stator current harmonics control is enabled at 0.1 s. From the waveforms of stator and rotor currents shown in Figure 8.11, it can be seen that both control schemes work well under normal grid condition.

The performances of the conventional control scheme and the stator current harmonics control scheme are tested under harmonically distorted grid condition and they are compared as shown in Figure 8.12. In Figure 8.12, Figure 8.12a is the

TABLE 8.1 Stator current harmonics and electromagnetic torque fluctuations under conventional control and stator current harmonics control

Stator current	0.5 pu		0.8 pu	
	Conventional control	Stator current harmonics control	Conventional control	Stator current harmonics control
THD (%)	11.2	3.1	6.8	2.0
Fifth-order harmonic (%)	9.5	1.1	5.7	0.8
Seventh-order harmonic (%)	4.7	0.4	3.1	0.2
EM torque fluctuation (pu)	± 0.05	± 0.01	± 0.05	± 0.01

simulation result under half-load condition with a stator current of 0.5 pu, and Figure 8.12b shows the results under full-load condition with a stator current of 0.8 pu. From 0 to 0.1 s, the conventional control scheme is adopted for the DFIG system. It is shown that because of the grid voltage harmonics, the corresponding current harmonics occur in the stator current, and the electromagnetic torque suffers from a sixth-order fluctuation, which shows the inability of the conventional control scheme to suppress the impact of the grid voltage harmonics. As soon as the stator current harmonics control is enabled at 0.1 s, the resonant controller starts to provide the sixth-order harmonic voltage command to compensate for the grid voltage harmonics, and the harmonics in the stator current and electromagnetic torque are mitigated.

The harmonics in the stator current and electromagnetic torque under conventional control and stator current harmonics control are listed in Table 8.1. The results verify that the proposed stator current harmonics control method is able to effectively suppress the impacts of grid voltage harmonics on the DFIG and largely enhance the compatibility of the DFIG under harmonically distorted grid conditions.

The performance of the resonant controller on suppressing the stator current harmonics when the grid frequency increases linearly from 49 to 51 Hz is demonstrated in Figure 8.13. It is shown that when the grid frequency deviates from its normal value, the stator current remains to be a high-quality sinusoidal wave, which illustrates that the self-adaptive resonant controller is able to accommodate the changes of the grid frequency.

The susceptibility of the stator current harmonics control to the changes in the generator parameters is also simulated and it is shown in Figure 8.14. The reference value of the magnetizing inductance is 3.84 pu, and the performances are tested with 50% increase and 50% decrease of the magnetizing inductance. From the stator current waveform, the proposed control method works well despite the changes of the generator parameters.

8.3.5 Experimental Results

To verify the effectiveness of the stator current harmonics control under distorted grid conditions, the experimental analysis is conducted on a 30 kW DFIG wind power generation system.

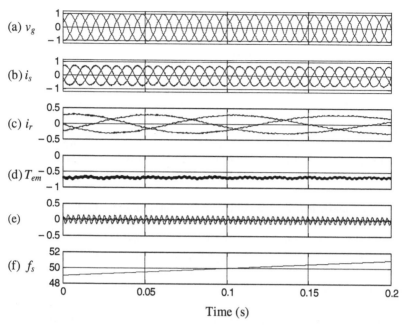

Figure 8.13 Performance of stator current harmonics control when the grid frequency linearly increases from 49 to 51 Hz, test condition $v_{sdq}^{(5)} = 0.04$ pu, $v_{sdq}^{(7)} = 0.03$ pu, stator current $i_{sdq} = 0.7$ pu. (a) Grid voltage v_g (pu); (b) stator current i_s (pu); (c) rotor current i_r (pu); (d) electromagnetic torque T_{em} (pu); (e) output of resonant controller (pu); (f) grid frequency f_s (Hz).

The schematic diagram of the test system is shown in Figure 8.15. As shown in Figure 8.15, the system consists of a DFIG, a driving motor, two back-to-back converters, a motor drive converter, control system, and a computer. The wind turbine is emulated by the driving motor and motor drive converter. The back-to-back converter is composed of an RSC and GSC. The RSC and GSC together with their control circuits adopt modular design, which means the hardware setups of the converters are identical. The RSC is in charge of the vector control of the DFIG and the stator current harmonics control introduced above. The GSC controls the DC-bus voltage. The converters are connected to CAN communication.

In the test system, the rated power of the DFIG is 30 kW, the rated frequency is 50 Hz, and the synchronous speed is 1500 r/min. The detailed parameters of the DFIG can be found in Tables 5.1 and 5.2. The rated power of the back-to-back converter is 10 kW, the switching frequency is 2 kHz, and the sampling frequency is 4 kHz. The DC-bus capacitor is $C = 1600$ μF, and the AC inductance of the GSC is $L_g = 13$ mH. The stator of the DFIG is directly connected to the 380 V power grid. The DC-bus voltage is controlled to 650 V by the GSC.

Figure 8.16 shows the steady-state performance of the system when the rotor speed of the DFIG is 1200 r/min (sub-synchronous) and with 10 kW stator active power and no reactive power. In Figure 8.16a, the conventional current control is

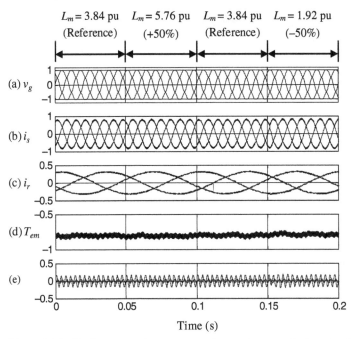

Figure 8.14 Performance of stator current harmonics control when the changes occur in the generator parameters when the test conditions are $v_{sdq}^{(5)} = 0.04$ pu, $v_{sdq}^{(7)} = 0.03$ pu, and the stator current $i_{sdq} = 0.8$ pu. (a) Grid voltage v_g (pu); (b) stator current i_s (pu); (c) rotor current i_r (pu); (d) electromagnetic torque T_{em} (pu); (e) output of resonant controller (pu).

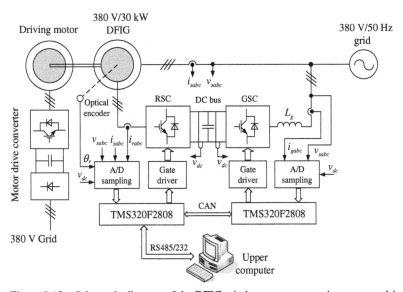

Figure 8.15 Schematic diagram of the DFIG wind power system using a motor drive.

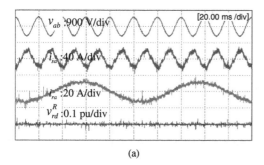

(a)

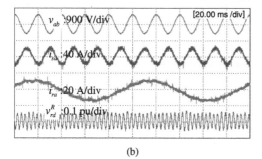

(b)

Figure 8.16   Static performances of the conventional current control and stator current harmonics control under distorted grid voltage. Test conditions: stator output power 10 kW, rotor speed 1200 r/min (0.8 pu). From top to bottom: Grid line voltage v_{ab}, stator current i_{sa}, rotor current i_{ra}, d-axis output of the resonant controller v_{rd}^R: (a) Conventional current control; (b) stator current harmonics control.

adopted, while in Figure 8.16b the stator current harmonics control is introduced. Comparing Figures 8.16a and 8.16b, it can be seen that since the conventional control method is not able to eliminate the impact caused by the grid voltage harmonics, larger harmonics can be observed in the stator current, while after the introduction of stator current harmonics control, the stator current harmonics are largely suppressed, which results in good quality of the stator current. The details of the harmonic analysis of the waveforms in Figure 8.16 are given in Table 8.2.

Figure 8.17 shows the steady-state performance of the system when the rotor speed of the DFIG is 1800 r/min (super-synchronous) and with 20 kW of the stator

TABLE 8.2   Measured harmonics in the grid voltage and stator current when the stator power is 10 kW (corresponding to Figure 8.16)

	Conventional control		Stator current harmonics control	
	Grid voltage	Stator current	Grid voltage	Stator current
THD (%)	1.8	13.1	1.9	4.9
Fifth-harmonic (%)	1.2	12.2	1.2	2.6
Seventh-harmonic (%)	0.5	3.2	0.5	1.0

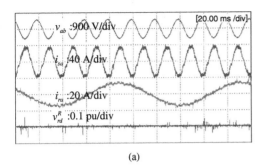

(a)

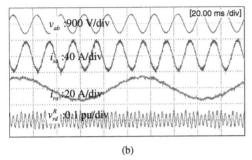

(b)

Figure 8.17 Static performance of the conventional current control and stator current harmonics control under distorted grid voltage. Test conditions: stator output power 20 kW, rotor speed 1800 r/min (1.2 pu). From top to bottom: Grid line voltage v_{ab}, stator current i_{sa}, rotor current i_{ra}, d-axis output of the resonant controller v_{rd}^R: (a) Conventional current control; (b) stator current harmonics control.

active power and no reactive power. In Figure 8.17a, the conventional current control is adopted, while in Figure 8.17b the stator current harmonics control is introduced. Similar to Figure 8.16, the quality of stator current waveform is largely improved with the adoption of stator current harmonics control. The detailed harmonic analysis of the waveforms in Figure 8.17 are given in Table 8.3.

The dynamic response of the DFIG system when the stator current harmonics control is enabled is shown in Figure 8.18, where the waveforms from top to bottom are stator current, rotor current, electromagnetic torque, and d-axis output of the resonant controller. As shown in Figure 8.18, the settling time of the transient process

TABLE 8.3 Measured harmonics in the grid voltage and stator current when the stator power is 20 kW (corresponding to Figure 8.17)

	Conventional control		Stator current harmonics control	
	Grid voltage	Stator current	Grid voltage	Stator current
THD (%)	1.9	6.1	2.0	2.4
Fifth harmonic (%)	0.6	3.8	0.6	0.7
Seventh harmonic (%)	1.1	4.2	1.2	1.0

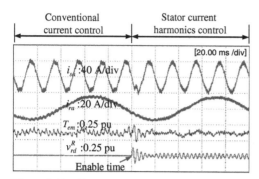

Figure 8.18 Dynamic response of the DFIG system when the stator current harmonics control is enabled. Test conditions: Stator output power 20 kW, rotor speed 1800 r/min (1.2 pu). From top to bottom: Stator current i_{sa}, rotor current i_{ra}, electromagnetic torque T_{em}, the d-axis output of the resonant controller v_{rd}^R.

after the enabling time is about 20 ms. Also, it is shown in the figure that when the stator current harmonics control is enabled, the sixth-order fluctuation in the electromagnetic torque is largely suppressed.

The dynamic performances of the conventional current control method and the stator current harmonics control method are compared in Figure 8.19. In the test, the d-axis component of the rotor current command abruptly changes from 0 to 0.5 pu, and the stator reactive power is controlled to be zero by the q-axis component of rotor current command. Figure 8.19a shows the test result under conventional current control, while in Figure 8.19b the stator current harmonics control is adopted. The rise time of the rotor current in Figure 8.19a is 4 ms, while in Figure 8.19b it is reduced to 3 ms. Besides, with the introduction of stator current harmonics control, the settling time of step response increases. Therefore, the dynamic performance of the rotor current control loop is influenced by the stator current harmonics control. The introduction of the resonant controller reduces the phase margin and cut-off frequency of the rotor current control loop, thus reducing the damping of the system causing a shorter rise time of rotor current step response. Meanwhile, the reduction of the phase margin and cut-off frequency causes the settling time to increase.

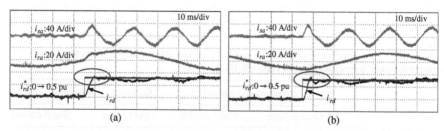

Figure 8.19 Dynamic response of the DFIG to the step change of rotor current command (rotor current command changes from 0 to 0.5 pu). From top to bottom: Stator current i_{sa}, rotor current i_{ra}, d-axis component of rotor current i_{rd}, d-axis command of rotor current i_{rd}^*: (a) conventional current control; (b) stator current harmonics control.

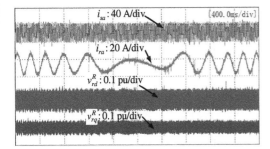

Figure 8.20   Dynamic performance of DFIG system when the rotor speed changes. From top to bottom: Stator current i_{sa}, rotor current i_{ra}, the d-axis output of the resonant controller v_{rd}^R, q-axis output of the resonant controller v_{rq}^R.

The dynamic test result when the speed of DFIG changes from 1350 r/min (0.9 pu) to 1650 r/min (1.1 pu) is shown in Figure 8.20. In the test, the active output power of stator is kept as 10 kW, while the reactive power is 0. As shown in the test result, the waveforms of the stator and rotor currents are quite smooth without oscillation and overshoot in the whole process, thus verifying that the stator current harmonics control scheme is able to operate steadily when the rotor speed goes through the synchronous speed.

8.4   INFLUENCE ON NORMAL CONTROL LOOP

Harmonics control performance of the stator current harmonics control loop is determined by the resonant controller. On the basis of the analysis given in the last section, this section gives more insight into the influence of the resonant controller used in the stator current harmonics control loop on the static and dynamic performance as well as the stability of the control system for DFIG wind power generation. In this section, the DFIG under analysis is rated to be 30 kW. The PI controller for the normal rotor current control loop is set to $K_p = 0.42$ and $K_i = 40$, which makes the open-loop crossover frequency of the PI controller to be 300 Hz.

8.4.1   Static Performance

The frequency response of the closed-loop transfer function of the stator current harmonics control loop $G_{ss}(s)$ is given in Figure 8.21. As illustrated in the figure, at the resonant frequency of 300 Hz, the magnitude and phase of the closed-loop transfer function are 1 and 0, respectively, which indicates that the stator current harmonics control loop is able to track the stator current harmonic commands at the resonant frequency. Also, it is shown that with the increase of K_r, the curves of magnitude and phase become flatter around 300 Hz, which indicates that a larger K_r is able to mitigate the frequency selective characteristic of the stator current harmonics control.

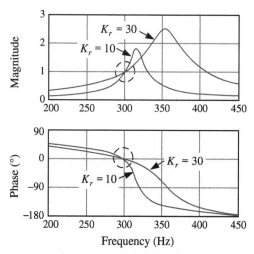

Figure 8.21 Frequency response of the closed-loop transfer function of the stator current harmonics control loop.

When the grid is harmonically distorted, according to equation (8.9), in steady state, the magnitude of stator current at the hth-order harmonic frequency ω_h can be expressed as given by the following equation [7, 17].

$$
\left| i_{sdq}(j\omega_h) \right| = \left| i_{sdq}^{ref}(j\omega_h)G_{ss}(j\omega_h) - i_{rdq}^{ref}(j\omega_h)G_{rs}(j\omega_h) \right.
$$

$$
\left. + v_{sdq}(j\omega_h)G_{uis}(j\omega_h) \right|_{i_{sdq}^{ref}(j\omega_h)=0,\, i_{rdq}^{ref}(j\omega_h)=0}
$$

$$
= \left| G_{vis}(j\omega_h) \right| \left| v_{sdq}(j\omega_h) \right| = \frac{\left| v_{sdq}(j\omega_h) \right|}{\left| X_{vis}(j\omega_h) \right|} \tag{8.24}
$$

Therefore, given the amplitude of harmonic voltage in the grid voltage, the amplitude of stator harmonic current is determined by the impedance $X_{vis}(j\omega_h)$ or admittance $G_{vis}(j\omega_h)$ of the system. From the expression of $G_{vis}(s)$ given in (8.12) and neglecting the influence of the control delay $G_d(s)$, the detailed expression of $|G_{vis}(s)|$ is given as [7, 17]

$$
\left| G_{vis}(s) \right| = \left| \frac{\dfrac{1}{L_s s + j\omega_s L_s} + \dfrac{K_p + K_i/s}{L_s s + j\omega_s L_s}\dfrac{1}{\sigma L_r s + R_r} + \dfrac{L_m}{L_s}\dfrac{s + j\omega_{sl}}{s + j\omega_s}\dfrac{1}{\sigma L_r s + R_r}}{1 + \dfrac{K_p + K_i/s}{\sigma L_r s + R_r} + \dfrac{2K_r\omega_c s}{s^2 + 2\omega_c s + \omega_o^2}\dfrac{1}{\sigma L_r s + R_r}} \right|
$$

$$
= \left| \frac{\dfrac{\sigma L_r s + R_r}{L_s s + j\omega_s L_s} + \dfrac{K_p + K_i/s}{L_s s + j\omega_s L_s} + \dfrac{L_m}{L_s}\dfrac{s + j\omega_{sl}}{s + j\omega_s}}{\sigma L_r s + R_r + K_p + K_i/s + \dfrac{2K_r\omega_c s}{s^2 + 2\omega_c s + \omega_o^2}} \right| \tag{8.25}
$$

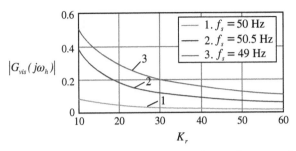

Figure 8.22   Capability of stator current harmonic suppression with the variation of K_r and $\omega_c = 5$ rad/s.

The magnitude at the harmonic frequency ω_h is [7, 17]

$$|G_{vis}(j\omega_h)| = \left| \frac{\frac{j\sigma L_r\omega_h + R_r}{jL_s\omega_h + j\omega_s L_s} + \frac{K_p + K_i/(j\omega_h)}{jL_s\omega_h + j\omega_s L_s} + \frac{L_m}{L_s}\frac{j\omega_h + j\omega_{sl}}{j\omega_h + j\omega_s}}{j\sigma L_r\omega_h + R_r + K_p + K_i/(j\omega_h) + \frac{j2K_r\omega_c\omega_h}{(j\omega_h)^2 + j2\omega_c\omega_h + \omega_o^2}} \right| \qquad (8.26)$$

In a typical DFIG wind power system, $K_r \gg K_p$, $K_r \gg \sigma L_r$, $\omega_h \gg K_i$, and $L_m \approx L_s$; the expression of $|G_{vis}(j\omega_h)|$ can be simplified as [7, 17]

$$|G_{vis}(j\omega_h)| \approx \left| \frac{\frac{j\omega_h + j\omega_{sl}}{j\omega_h + j\omega_s}}{\frac{j2K_r\omega_c\omega_h}{(j\omega_h)^2 + j2\omega_c\omega_h + \omega_o^2}} \right| = S_l^{(h)} \frac{\sqrt{\left[\omega_h\left(1 - (\omega_o/\omega_h)^2\right)\right]^2 + (2\omega_c)^2}}{2K_r\omega_c} \qquad (8.27)$$

where $S_l^{(h)}$ is the slip ratio of the hth-order harmonic in the synchronous rotating dq reference frame:

$$S_l^{(h)} = \frac{\omega_h + \omega_{sl}}{\omega_h + \omega_s} = \frac{(\omega_h + \omega_s) - \omega_r}{\omega_h + \omega_s} \qquad (8.28)$$

When the resonant frequency ω_o of the resonant controller is equal to the grid harmonic frequency ω_h, equation (8.27) can be further simplified as

$$|G_{vis}(j\omega_h)| \approx \frac{S_l^{(h)}}{K_r} \qquad (8.29)$$

Equations (8.27) and (8.29) determine the capability of suppressing the grid harmonics by the control system. When the grid harmonic frequency coincides with the resonant frequency, the harmonic suppression capability increases with a larger gain of the resonant controller K_r. When the grid harmonic frequency deviates from the resonant frequency, the harmonic suppression capability is inversely proportional to $K_r\omega_c$. Therefore, larger K_r enhances the harmonic suppression capability.

Currently, most grid codes require wind power systems to operate properly when the grid frequency fluctuates in a certain range, for example, 49.5–50.5 Hz in Denmark, 49–50.5 Hz in Germany, 48–50.5 Hz in Spain, and 49.5–50.5 Hz in China (for detailed information about grid codes, please refer to Chapter 3). Figure 8.22

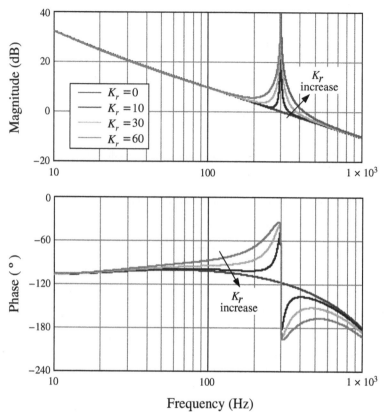

Figure 8.23   Frequency response of the open-loop transfer function $D(s)$.

shows the curves of $\left|G_{vis}(j\omega_h)\right|$ with the variation of K_r under different grid frequencies. It can be seen that with the increase of K_r, a better suppression capability is obtained, and the influence of the grid frequency deviation decreases. Therefore, from the perspectives of harmonic suppression and reducing the susceptibility to grid frequency variation, K_r should be as large as possible. However, large K_r may affect the dynamic performance of the system or even cause instability of the system.

8.4.2   Stability of the System

Referring to the current control block diagram of the DFIG shown in Figure 8.3. The transfer function shown in equations (8.10) –(8.12) have the same characteristic polynomial $\Delta = 1 + G_{PI}(s)G_d(s)G_p(s) + G_R(s)G_d(s)G_p(s)L_m/L_s$. To evaluate the dynamic response and stability of the DFIG control system, the following open-loop transfer function is introduced:

$$D(s) = G_{PI}(s)G_d(s)G_p(s) + G_R(s)G_d(s)G_p(s)L_m/L_s \qquad (8.30)$$

The frequency response of the open-loop transfer function $D(s)$ is given in Figure 8.23. It is shown that with the introduction of resonant control ($K_r \neq 0$), an abrupt change occurs in the phase–frequency curve. The abrupt change of the phase may cause a reduction of the phase margin. When K_r equals 0, 10, 30, 60, the open-loop crossover frequencies are 300 Hz, 340 Hz, 380 Hz, and 425 Hz, respectively, while the phase margins are 60°, 34°, 20°, and 8°, respectively. Thus, with the increase of K_r, the open-loop crossover frequency increases while the phase margin decreases, which may cause oscillations in the control system.

To obtain the analytical expression of the open-loop crossover frequency and phase margin with regard to the resonant controller gain, substituting the expressions of $G_{PI}(s)$, $G_R(s)$, $G_d(s)$, and $G_p(s)$ in equation (8.30), the expression of $D(s)$ can be rearranged as given below [7, 17, 18]:

$$D(s) = G_{PI}(s)G_d(s)G_p(s) + G_R(s)G_d(s)G_p(s)L_m/L_s$$

$$= \left(\frac{K_p + K_i/s}{\sigma L_r s + R_r} + \frac{2K_r\omega_c s}{s^2 + 2\omega_c s + \omega_o^2} \frac{L_m/L_s}{\sigma L_r s + R_r} \right) e^{-sT_d}$$

$$= \frac{K_i}{R_r} \left[\frac{\frac{K_p}{K_i} + \frac{1}{s}}{(\sigma L_r/R_r)s + 1} + \frac{2\omega_c \frac{K_r L_m/L_s}{K_i}s}{(s/\omega_o)^2 + 2\omega_c/\omega_o^2 s + 1} \frac{1}{(\sigma L_r/R_r)s + 1} \right] e^{-sT_d}$$

$$= \frac{K_i}{R_r} \left[\frac{\left(\frac{K_p}{K_i}s + 1\right)\left[(s/\omega_o)^2 + 2\omega_c/\omega_o^2 s + 1\right] + 2\omega_c \frac{K_r L_m/L_s}{K_i}s^2}{s(\tau_r s + 1)\left[(s/\omega_o)^2 + 2\omega_c/\omega_o^2 s + 1\right]} \right] e^{-sT_d}$$

$$= \frac{K_i}{R_r} \left[\frac{\frac{K_p}{K_i\omega_o^2}s^3 + \left(\frac{1}{\omega_o^2} + \frac{2\omega_c(K_r L_m/L_s + K_p)}{K_i\omega_o^2}\right)s^2 + \left(\frac{K_p}{K_i} + \frac{2\omega_c}{\omega_o^2}\right)s + 1}{s(\tau_r s + 1)\left[(s/\omega_o)^2 + 2\omega_c/\omega_o^2 s + 1\right]} \right] e^{-sT_d} \quad (8.31)$$

where $\tau_r = \sigma L_r/R_r$ is the time constant of the rotor of the DFIG. In a typical control system, usually $\omega_o^2 \gg \omega_c$ and $\omega_o^2 \gg 1$, therefore equation (8.31) can be further simplified as [7, 17, 18]

$$D(s) \approx \frac{K_i}{R_r} \left[\frac{\frac{K_p}{K_i\omega_o^2}s^3 + \frac{2\omega_c(K_r + K_p)}{K_i\omega_o^2}s^2 + \frac{K_p}{K_i}s + 1}{s(\tau_r s + 1)\left[(s/\omega_o)^2 + 2\omega_c/\omega_o^2 s + 1\right]} \right] e^{-sT_d} \quad (8.32)$$

In the denominator of equation (8.32), $2\omega_c/\omega_o^2 s = (2\omega_c/\omega_o)(s/\omega_o)$, since $(2\omega_c/\omega_o) \ll 0.1$, and in the range of the switching frequency, $(\omega/\omega_o) < 10$, the $2\omega_c/\omega_o^2 s$ term in the denominator can be neglected, and equation (8.32) can be further simplified to [7, 17, 18]

$$D(s) \approx \frac{K_i}{R_r} \left[\frac{\frac{K_p}{K_i\omega_o^2}s^3 + \frac{2\omega_c(K_r + K_p)}{K_i\omega_o^2}s^2 + \frac{K_p}{K_i}s + 1}{s(\tau_r s + 1)\left[(s/\omega_o)^2 + 1\right]} \right] e^{-sT_d} \quad (8.33)$$

Taking $s = j\omega_{cr}$, the frequency characteristic of the open-loop transfer function $D(s)$ at the crossover frequency ω_{cr} can be expressed as [7, 17, 18]

$$D(j\omega_{cr}) \approx \frac{K_i - 2\left(K_r + K_p\right)\omega_c(\omega_{cr}/\omega_o)^2 + jK_p\omega_{cr}\left[1 - (\omega_{cr}/\omega_o)^2\right]}{jR_r\omega_{cr}(j\omega_{cr}\tau_r + 1)\left[1 - (\omega_{cr}/\omega_o)^2\right]}e^{-j\omega_{cr}T_d} \quad (8.34)$$

As shown in Figure 8.23, the open-loop crossover frequency ω_{cr} after introducing the resonant controller is larger than the resonant frequency ω_o of the resonant controller, thus $2K_r\omega_c(\omega_{cr}/\omega_o)^2 > K_i$ and $\omega_{cr}\tau_r \gg 1$, and for the sake of simplicity, the influence of K_i is neglected in the analysis. Equation (8.34) is simplified as [7, 17, 18]

$$D(j\omega_{cr}) \approx \frac{2\omega_c(K_r + K_p)(\omega_{cr}/\omega_o)^2 + jK_p\omega_{cr}\left[(\omega_{cr}/\omega_o)^2 - 1\right]}{\omega_{cr}^2\sigma L_r\left[1 - (\omega_{cr}/\omega_o)^2\right]}e^{-j\omega_{cr}T_d} \quad (8.35)$$

The phase angle of the open-loop transfer function at the crossover frequency can be obtained as [7, 17, 18]

$$\angle D(j\omega_{cr}) \approx -\pi + \arctan\left(\frac{K_p\omega_{cr}}{2\omega_c(K_r + K_p)}\left[1 - (\omega_o/\omega_{cr})^2\right]\right) - \omega_{cr}T_d \quad (8.36)$$

From equation (8.36), the phase margin can be expressed as [7, 17, 18]

$$\phi_m = \pi + \angle D(j\omega_{cr}) \approx \arctan\left(\frac{K_p\omega_{cr}}{2\omega_c(K_r + K_p)}\left[1 - (\omega_o/\omega_{cr})^2\right]\right) - \omega_{cr}T_d \quad (8.37)$$

Therefore, the phase margin is not only determined by the PI controller but influenced by the resonant controller as well. If the resonant frequency ω_o approaches the crossover frequency ω_{cr}, the value of polynomial $\left[1 - (\omega_o/\omega_{cr})^2\right]$ decreases significantly, which causes a reduction of the phase margin. The following conclusions can be drawn from equation (8.37) :

1. The phase margin is inversely proportional to K_r and ω_c of the resonant controller;

2. When the resonant frequency ω_o approaches the crossover frequency ω_{cr}, the stability of the system may be affected.

From conclusion (1), in order to avoid an inadequate phase margin, the product $K_r\omega_c$ must be kept small. Since K_r determines the gain of the resonant controller, ω_c is required to be as small as possible. However, for a fixed-point digital signal processor (DSP), considering the quantization errors and rounding errors in the digital system, it is difficult to realize a very small value for ω_c. Taking both the stability and digital realization into consideration, for a 32-bit fixed-point DSP, experimental experiences suggest that ω_c is better to be in the range of 3–10 rad/s.

From conclusion (2), when the resonant frequency ω_o approaches crossover frequency ω_{cr}, the influence of resonant controller on the stability of the system must be considered. On the contrary, if $\omega_o/\omega_{cr} < 1/3$, the resonant controller has little impact on the stability.

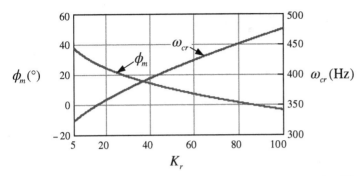

Figure 8.24   Phase margin and open-loop crossover frequency under different resonant controller gains.

The gain of the open-loop transfer function at the crossover frequency equals 1, that is [7, 17, 18]:

$$|D(j\omega_{cr})| \approx \left| \frac{2\omega_c(K_r + K_p)(\omega_{cr}/\omega_o)^2 + jK_p\omega_{cr}\left[(\omega_{cr}/\omega_o)^2 - 1\right]}{\omega_{cr}^2\sigma L_r\left[1 - (\omega_{cr}/\omega_o)^2\right]} e^{-j\omega_{cr}T_d} \right| = 1$$

(8.38)

Equation (8.38) can be deduced to the following equation [7, 17, 18]:

$$(K_r + K_p)\omega_c \approx 0.5\omega_{cr}\left[1 - (\omega_o/\omega_{cr})^2\right]\sqrt{(\sigma L_r\omega_{cr})^2 - K_p^2}$$

(8.39)

Combining (8.37) and (8.39), the following equation is derived [7, 17, 18]:

$$\omega_{cr} \approx K_p\sqrt{1 + 1/\tan^2(\phi_m + \omega_{cr}T_d)}\bigg/(\sigma L_r)$$

(8.40)

The phase margin and open-loop crossover frequency under different resonant controller gains can be calculated as shown in Figure 8.24, where $K_p = 0.45$, $K_i = 40$, $\omega_c = 5$ rad/s. It can be seen that the phase margin decreases with increasing K_r, while the crossover frequency increases with increasing K_r. When $K_r > 80$, the phase margin becomes negative which indicates instability of the system. Therefore, both the static performance and stability have to be taken into consideration when choosing the appropriate value for K_r.

8.4.3   Dynamic Performance

To suppress the stator current harmonics, the stator harmonic current command $i_{sdq}^{ref}(s)$ is set to zero, and thus equation (8.9) can be rearranged as

$$i_{sdq}(s) = -i_{rdq}^{ref}(s)G_{rs}(s) + v_{sdq}(s)G_{vis}(s)$$

(8.41)

It can be seen that the static output of stator current is related to the rotor current command and grid voltage, while the dynamic performance is determined by the

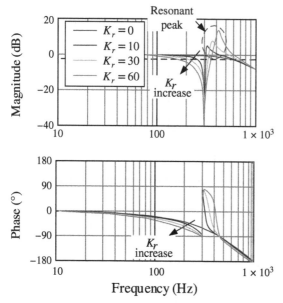

Figure 8.25   Frequency characteristic of the closed-loop transfer function of the rotor current $G_{rs}(s)$.

closed-loop transfer function $G_{rs}(s)$ of the rotor current control loop, the frequency characteristic of which is shown in Figure 8.25. With the increase of K_r, the cut-off frequency of the rotor current control loop decreases. For example, if K_r increases from 0 to 30, the cut-off frequency will decrease from 600 to 215 Hz. Since the dynamic settling time of the system is usually inversely proportional to the cut-off frequency, the decrease of cut-off frequency may cause longer settling time. This is a negative effect brought by the stator current harmonics control.

Also, as shown in Figure 8.25 a high resonant peak occurs at the resonant frequency, and the peak value increases with the increase of K_r. Because of the resonant peak, oscillations may be induced in the rotor currents in a frequency range around the resonant peak, and a very large resonant peak may even cause instability. This is another negative effect on the performance of rotor current control brought by the stator current harmonics control loop.

8.5 DESIGN AND OPTIMIZATION OF CURRENT CONTROLLER

8.5.1 Systematic Design Procedure

According to the analysis given in the previous section, when the resonant frequency of the resonant controller approaches the open-loop crossover frequency of the system, the dynamic performance of the system is significantly influenced by the resonant controller. The phase margin is related to K_p and $K_r\omega_c$. For the sake of static performance, higher K_r helps to reduce static errors and the susceptibility of the resonant controller to the variations in the grid frequency. Therefore, taking both the dynamic and static performances into consideration, the systematic design procedure for the resonant controller is given as follows [21, 22]:

1. Choose the appropriate phase margin and open-loop crossover frequency. The recommended phase margin is in the range of [30°, 60°], and the crossover frequency is better to be less than 1/4 of the switching frequency. The recommended crossover frequency is in the range of $[f_{sw}/10, f_{sw}/5]$.

2. Substitute the required phase margin ϕ_m and open-loop crossover frequency ω_{cr} into equation (8.40), obtaining K_p.

3. Substitute ϕ_m, ω_{cr}, and K_p into equation (8.39), obtaining $(K_r + K_p)\omega_c$.

4. Choose the appropriate value for ω_c according to the accuracy of the digital system. For a 32-bit fixed-point DSP, ω_c is recommended to be 3–10 rad/s.

5. From the above-obtained K_p, $(K_r + K_p)\omega_c$, and ω_c, K_r can be calculated according to $K_r = \left[(K_r + K_p)\omega_c\right]/\omega_c - K_p$.

If the obtained K_r is not able to meet the requirement for static performance, then new values for phase margin and crossover frequency should be chosen, and a new K_r can be obtained by performing the design procedure above. The design cycle continues until the obtained K_r is able to achieve the required balance between the dynamic and static performances.

Example: Controller Design for a 30 kW System
The switching frequency is 2 kHz, and the sampling frequency is 4 kHz. According to the design procedure given above, the controller parameters can be obtained as

1. The phase margin is chosen to be 30°, and the crossover frequency is set to 1/6 of switching frequency (333 Hz).

2. From equation (8.40), K_p is 0.42.

3. From equation (8.39), $(K_r + K_p)\omega_c$ is 54 rad/s.

4. Set $\omega_c = 5$ rad/s.

5. From steps (2), (3), and (4), $K_r = 10$ can be derived.

Therefore, the parameters for the control system are $K_p = 0.42$, $K_r = 10$, and $\omega_c = 5$ rad/s, and $K_i = 40$ is chosen. The frequency characteristics of the open-loop transfer function shown in equation (8.30) adopting these parameters can be depicted as shown in Figure 8.26.

8.5.2   Phase Compensation Methods for the Resonant Controller

In the example given in the last section, when the phase margin is 30°, the gain of resonant controller reaches only 10. According to Figure 8.22, if the resonant controller gain is too small, the static performance of the system is affected. However,

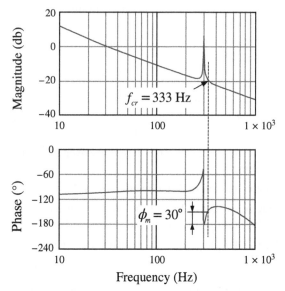

Figure 8.26 Frequency characteristics of the open-loop transfer function shown in equation (8.30) when $K_p = 0.42$, $K_i = 40$, $K_r = 10$, and $\omega_c = 5$ rad/s.

according to Figure 8.24, a very large resonant controller gain may cause insufficient phase margin, which may affect stability and dynamic performance of the system. To achieve the best balance between static and dynamic performances, methods of increasing the phase margin while keeping a large resonant controller gain will be introduced in this section.

The controller parameters used in this section resemble those adopted in the example given in Section 16, $K_p = 0.42$, $K_i = 40$, and $\omega_c = 5$ rad/s, but K_r is increased from 10 to 30. The frequency characteristics of the open-loop transfer function are depicted in Figure 8.27. It is shown that the crossover frequency is 373 Hz, while the phase margin is only 17°. Therefore, methods of the phase compensation are needed to increase the phase margin of the system.

8.5.2.1 Phase Lead Compensation Method

The harmonics control loop with lead compensation based on the dq reference frame is shown in Figure 8.28, in which the angle of the inverse dq transformation becomes $6\theta_s + \phi_k$ by adding ϕ_k to the original angle. The block diagram shown in Figure 8.28 can be used to substitute the resonant controller $G_R(s)$ shown in Figure 8.2 to improve the stability of the system. The added angle ϕ_k makes the phase angle of the resonant controller at the resonant frequency $\omega_o = 6\omega_s$ to increase by ϕ_k. In Figure 8.28, e_{sdq} is the input error of stator current harmonics control loop as shown in Figure 8.2, $e_{sdq} = i_{sdq}^{ref} - i_{sdq}$. Multiplying e_{sdq} with the rotating factor $e^{-j6\theta_s}$ transfers it into the negative-sequence fifth-order dq frame while multiplying e_{sdq}

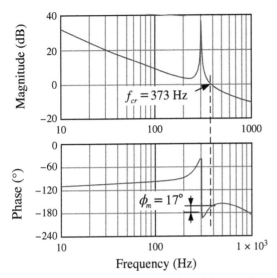

Figure 8.27 Frequency characteristics of the open-loop transfer function when $K_p = 0.42$, $K_i = 40$, $K_r = 30$, and $\omega_c = 5$ rad/s.

with the rotating factor $e^{j6\theta_s}$ transfers it into the positive-sequence seventh-order dq frame.

The transfer function of the control loop can be derived from Figure 8.28 as [9, 10]:

$$
\begin{aligned}
G_{R\_\phi}(s) = \frac{v_{rdq}^R(s)}{e_{sdq}(s)} &= G_{dq}^{(5)}(s + j\omega_o)e^{-j\phi_k} + G_{dq}^{(7)}(s - j\omega_o)e^{j\phi_k} \\
&= \cos\phi_k \left[G_{dq}^{(7)}(s - j\omega_o) + G_{dq}^{(5)}(s + j\omega_o) \right] \\
&\quad + j\sin\phi_k \left[G_{dq}^{(7)}(s - j\omega_o) - G_{dq}^{(5)}(s + j\omega_o) \right]
\end{aligned}
\tag{8.42}
$$

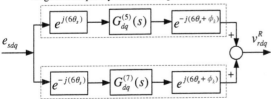

Figure 8.28 Harmonics control loop with lead compensation in dq reference frame.

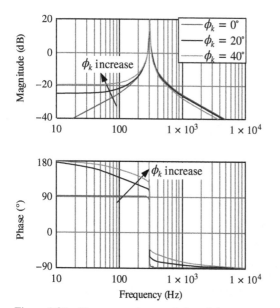

Figure 8.29 Frequency characteristics of the resonant controllers with phase lead compensation.

Taking $G_{dq}^{(5)}(s) = G_{dq}^{(7)}(s) = K_r/(1 + s/\omega_c)$, the resonant controller with phase lead compensation can be expressed as the equation below [9, 10, 22, 23].

$$G_{R\_\phi}(s) = \cos\phi_k \left[\frac{K_r}{1 + (s - j\omega_o)/\omega_c} + \frac{K_r}{1 + (s + j\omega_o)/\omega_c} \right]$$

$$+ j\sin\phi_k \left[\frac{K_r}{1 + (s - j\omega_o)/\omega_c} - \frac{K_r}{1 + (s + j\omega_o)/\omega_c} \right]$$

$$= \cos\phi_k \frac{2K_r\omega_c(s + \omega_c)}{s^2 + 2\omega_c s + \omega_c^2 + \omega_o^2} + j\sin\phi_k \frac{j2K_r\omega_c\omega_o}{s^2 + 2\omega_c s + \omega_c^2 + \omega_o^2}$$

$$= 2K_r\omega_c \frac{(s + \omega_c)\cos\phi_k - \omega_o\sin\phi_k}{s^2 + 2\omega_c s + \omega_c^2 + \omega_o^2} \tag{8.43}$$

Since $\omega_c \ll \omega_o$, equation (8.43) can be simplified as

$$G_{R\_\phi}(s) = 2K_r\omega_c \frac{s\cos\phi_k - \omega_o\sin\phi_k}{s^2 + 2\omega_c s + \omega_o^2} \tag{8.44}$$

Equation (8.44) is the expression of the resonant controller with phase lead compensation. When $\phi_k = 0$, equation (8.44) gets the original form of the resonant controller as shown in equation (8.1). Figure 8.29 shows the frequency characteristics of the resonant controllers with 0°, 20°, and 40° of the phase lead compensation. It can be seen that the phases at the resonant frequency of 300 Hz are increased by

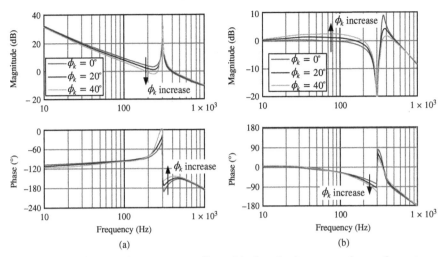

Figure 8.30   Influence of resonant controllers with phase lead compensation on the system dynamic performance: (a) Open-loop transfer function; (b) rotor current closed-loop transfer function.

$0°$, $20°$, and $40°$. Therefore, a resonant controller with phase lead compensation can enhance the phase margin of the control system.

However, the phase lead compensation causes the magnitude and phase of the resonant controller to increase in the frequency range below 300 Hz. Since such compensation method adds a constant angle ϕ_k to the signals in the whole frequency range, and since the delay angle decreases with lower frequency, overcompensation is caused for the signals with frequencies below 300 Hz, which might influence the dynamic performance of the control system.

Figure 8.30 illustrates the influence of the resonant controllers with phase lead compensation on system dynamic performance. In Figure 8.30a, when ϕ_k is $0°$, $20°$, and $40°$, the phase margin is $17°$, $26°$, and $34°$, respectively. The phase margin increases with increasing ϕ_k. From this perspective, a suitable phase lead compensation is able to enhance the stability of the control system. However, from Figure 8.30b it is seen that by introducing phase lead compensation, the rotor closed-loop gain in the low frequency range (10–200 Hz) goes beyond 0 dB, which means overshoot may occur in the rotor current response in such frequency range, affecting the dynamic tracking performance of the rotor current control loop.

8.5.2.2 Lead–Lag Compensation Method
The lead–lag compensator is only effective in the required frequency range, while the characteristics in other frequency ranges are not influenced, thus eliminating the overcompensation problem with the phase lead compensator.

The expression of the lead–lag compensator is shown as equation (8.45), where $\left(1 + \alpha T_1 s\right)/\left(1 + T_1 s\right)$ is a leading compensator, $\left(1 + \frac{T_2}{\alpha}s\right)/\left(1 + T_2 s\right)$ is a lagging compensator, and K_c is the gain of the compensator. The lead compensator enhances

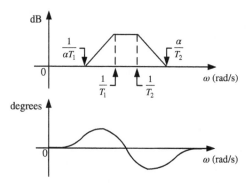

Figure 8.31 Frequency characteristics of the lead–lag compensator.

the phase margin of the system, and the lag compensator is used to counteract the effects of the lead compensator on the characteristics in high frequency range. The schematic diagram of the frequency characteristics of the lead–lag compensator is shown in Figure 8.31.

$$G_c(s) = K_c \left(\frac{1 + \alpha T_1 s}{1 + T_1 s} \right) \left(\frac{1 + \frac{T_2}{\alpha} s}{1 + T_2 s} \right) \qquad \text{where } \alpha > 1, T_1 > T_2 \qquad (8.45)$$

The transfer function of the resonant controller with lead–lag compensation is expressed as

$$G_{R\_c}(s) = G_R(s)G_c(s) \qquad (8.46)$$

The DFIG wind power control loop with phase compensation is shown in Figure 8.32. In the stator current harmonics control loop, when the selection switch is put on 1, phase lead compensation is adopted in the system, and when the selection switch is put on 2, lead–lag compensation is adopted.

To compare the control performances without phase compensation, with phase lead compensation and with lead–lag compensation, the following conditions are specified: the open-loop crossover frequency before and after the compensation should be kept the same, and the phase margin gained by the two compensation methods should be identical.

The dynamic performances of the systems without phase compensation, with phase lead compensation and with lead–lag compensation are compared in Figure 8.33, in which the adopted controller parameters are given as below:

1. Resonant controller without phase compensation $G_R(s)$: $K_R = 30$, $\omega_c = 5$ rad/s;
2. Resonant controller with phase lead compensation $G_{R\_c}(s)$:

$$K_R = 30, \ \omega_c = 5 \, \text{rad/s}, \ K_c = 0.4, \ \alpha = 3, \ T_1 = 232 \mu\text{s}, \ T_2 = 58 \mu\text{s};$$

3. Resonant controller with lead–lag compensation $G_{R\_\phi}(s)$: $K_R = 20$, $\omega_c = 5$ rad/s, $\phi_k = 30°$.

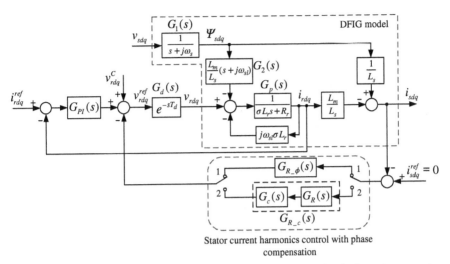

Stator current harmonics control with phase
compensation

Figure 8.32   DFIG wind power control loop with phase compensation in the stator current harmonics control.

Applying the above parameters, the open-loop crossover frequency of the system is 373 Hz, and the phase margin without phase compensation is 17°, while the phase margin with phase compensation is 35°. Figure 8.33a shows the open-loop frequency characteristics of the three systems. In Figure 8.33, curves 1, 2, and 3 are corresponding to the systems without phase compensation, with phase lead compensation, and with lead–lag compensation, respectively. Compared to that without

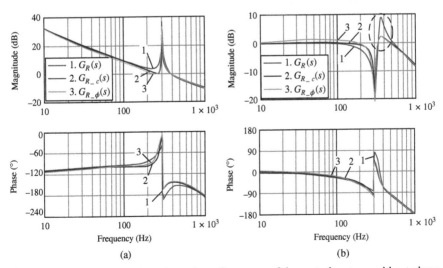

Figure 8.33   Comparison of the dynamic performances of the control systems without phase compensation, with phase lead compensation, and with lead–lag compensation: (a) Open-loop transfer function; (b) rotor current closed-loop transfer function.

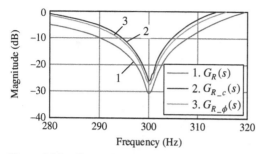

Figure 8.34 Comparison of the static performances of the control systems without phase compensation, with phase lead compensation, and with lead–lag compensation.

phase compensation, the phase margin is enhanced with both the compensation methods. Therefore, the stability of the system is improved with the phase compensation methods.

It can be seen from the frequency characteristics of the rotor current closed-loop transfer function shown in Figure 8.33b that by applying the phase compensation methods, the resonant peak near the resonant frequency is reduced to around 0 dB and the resonant peak is significantly suppressed. However, comparing curves 2 and 3 in the low frequency range, it is observed that in the frequency range of 10–200 Hz, curve 3 is above the 0 dB line, while curve 2 almost overlaps the 0 dB line. The lead–lag compensation method overcomes the tracking errors in the low frequency range, which are introduced by the phase lead compensation method and thereby achieving better dynamic tracking performance.

As shown in equation (8.24), the amplitude of current harmonic at steady state is determined by the magnitude of the transfer function from grid voltage to stator current at the harmonic frequency $|G_{vis}(j\omega_h)|$. Figure 8.34 shows the $|G_{vis}(j\omega_h)|$ curves with different phase compensation methods. In the figure, curves 1, 2, and 3 are corresponding to the systems without phase compensation, with phase lead compensation, and with lead–lag compensation, respectively, in which the values of $|G_{vis}(j\omega_h)|$ at 300 Hz are -31 dB, -26 dB, and -27 dB, respectively. The damping ratios provided by $|G_{vis}(j\omega_h)|$ with the two compensation methods are slightly reduced compared to that without compensation but still below -20 dB. By selecting the appropriate controller parameters, resonant controllers with the phase compensation methods are also able to provide sufficient damping for the current harmonics.

It has to be noted that when achieving the same phase margin, the damping ratio provided by $|G_{vis}(j\omega_h)|$ without phase compensation is much lower than those achieved with the phase compensation methods. For example, when the phase margin is achieved as 35°, $|G_{vis}(j\omega_h)|$ with phase compensation is -26 dB, while without phase compensation, it is only -15 dB. Therefore, from the perspective of keeping the phase margin at a certain value, the phase compensation methods are able to enhance the static performance of the system.

In general, the phase compensation methods can overcome the contradiction between static and dynamic performances of the resonant controller, giving the control system good static and dynamic performances at the same time.

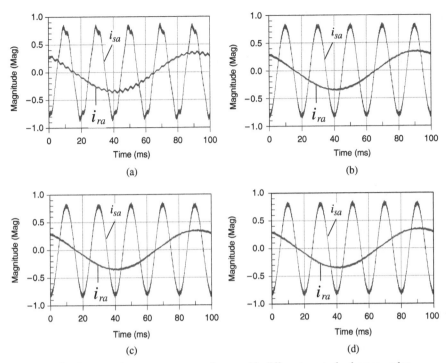

Figure 8.35 Stator and rotor current waveforms with different control schemes under harmonically distorted grid conditions and load conditions of 0.8 pu stator current: (a) Without harmonics control; (b) resonant control without phase compensation; (c) resonant control with phase lead compensation; (d) resonant control with lead–lag compensation.

8.5.3 Simulation Results of Phase Compensation

The phase compensation methods introduced in the previous section are verified with simulations. The control structure used in the simulation is identical to that shown in Figure 8.32. The power rating of the DFIG used in the simulation is 30 kW, the grid voltage is 380 V/50 Hz, the DC-bus voltage is 650 V, the switching frequency is 2 kHz, and the sampling frequency is 4 kHz. The controller parameters are identical to those given in the previous section. In the simulation, the grid voltage contains 1.6% of the negative-sequence fifth-order harmonic voltage and 1.2% of positive-sequence seventh-order harmonic voltage, and the total harmonic distortion (THD) is 2%.

Figure 8.35 shows the static simulation results of the DFIG system under harmonically distorted grid condition when the output stator current is 0.8 pu. Figure 8.35a shows the steady-state simulation results of the stator and rotor currents without applying harmonics control, in which the harmonic distortion can be observed in the stator current waveforms. The waveforms shown in Figures 8.35b–8.35d correspond to the results with resonant controllers without phase compensation, with phase lead compensation, and with lead–lag compensation methods, in which good harmonic suppression performances can be observed.

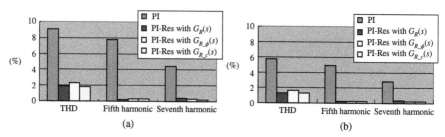

Figure 8.36   Stator current harmonics under conventional control, resonant control without phase compensation, resonant control with phase lead compensation, and resonant control with lead–lag compensation at different load conditions: (a) Stator current 0.5 pu; (b) stator current 0.8 pu.

The detailed values of stator current harmonics under these four conditions are given in Figure 8.36. It is shown that without resonant control, the fifth- and seventh-order harmonics in the stator current are quite large causing the THD to be larger than 5%, while by adopting resonant control, either with or without phase compensation, the fifth- and seventh-order harmonics are significantly mitigated, thus achieving low THD at both half-load and full-load conditions.

The dynamic responses of the DFIG system with different control schemes under harmonically distorted grid condition are shown in Figure 8.37. In the simulation, the d component of stator current is kept at 0.5 pu, and the q component is set to 0 pu, and the stator current harmonics control loop starts to operate at 40 ms. As shown in Figure 8.37, without phase compensation, the enabling of stator current harmonics control causes oscillation in the currents, and the settling time is around 20 ms. With the application of phase compensation, the settling time of oscillation is reduced.

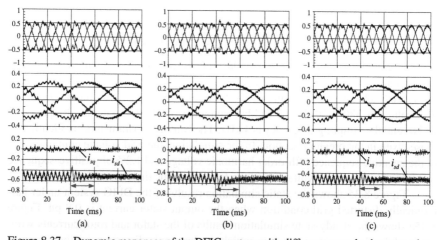

Figure 8.37   Dynamic responses of the DFIG system with different control schemes under harmonically distorted grid condition and stator current 0.5 pu. From top to bottom: three-phase stator current, three-phase rotor current, d and q components of stator current: (a) Without phase compensation; (b) phase lead compensation; (c) lead–lag compensation.

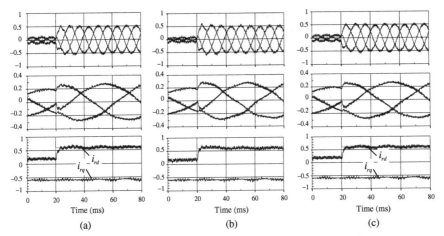

Figure 8.38   Dynamic responses of the DFIG system for the step change of rotor current command with different control schemes under harmonically distorted grid conditions; rotor current command changes from 0.2 to 0.6 pu. From top to bottom: three-phase stator current, three-phase rotor current, d and q components of rotor current: (a) Without phase compensation; (b) phase lead compensation; (c) lead–lag compensation.

However, a significant overshoot is also caused by the phase lead compensation. The settling time is reduced to 15 ms with the application of lead–lag compensation, and the amplitude of the oscillation is also mitigated.

Figure 8.38 shows the dynamic responses of the DFIG system to a step change in rotor current command with different control schemes under harmonically distorted grid conditions, in which the rotor current command i_{rd}^* abruptly changes from 0.2 to 0.6 pu at 40 ms. As shown in Figure 8.38, oscillations occur without phase compensation, and an overshoot occurs with phase lead compensation, while neither oscillation nor overshoot occurs when the lead–lag compensation is applied. Therefore, it is verified that by applying the lead–lag compensation method, the control system has achieved better dynamic performance.

8.6   SUMMARY

To eliminate the negative effects brought to the DFIG system by the harmonic distortion in the grid voltage, the stator current harmonics control based on the resonant controller is introduced in this chapter. The resonant control scheme has the advantages of simple structure and high robustness. Analytical, simulation, and experimental results are presented to verify that the stator current harmonics controller is able to effectively suppress the stator current harmonics and reduce the electromagnetic torque fluctuations caused by the grid voltage harmonic distortion. Thus the operational reliability of the DFIG system under distorted grid conditions can be improved. A systematic design procedure for the controller parameters is given taking both the static and dynamic characteristics into consideration. To further improve the performance of resonant control, phase lead compensation and lead–lag compensation methods are used to enhance the phase margin of the system thus improving the dynamic performance of the control system by adopting the proposed resonant controller.

REFERENCES

[1] R. Teodorescu, F. Blaabjerg, M. Liserre, and P. Loh, "Proportional-resonant controllers and filters for grid-connected voltage-source converters," *IEE Proc. Electr. Power Appl.*, vol. 153, no. 5, pp. 750–762, 2006.

[2] A. Timbus, M. Ciobotaru, R. Teodorescu, and F. Blaabjerg, "Adaptive resonant controller for grid-connected converters in distributed power generation systems," in *Proc. Appl. Power Electron. Conf. Exposition*, Dallas, TX, 2006, pp. 1601–1606.

[3] C. Liu, X. Huang, M. Chen, and D. Xu, "Flexible control of DC-link voltage for doubly fed induction generator during grid voltage swell, " in *Proc. Energy Convers. Congr. Exposition*, Atlanta, GA, 2010, pp. 3091–3095.

[4] C. Liu, F. Blaabjerg, W. Chen, and D. Xu, "Stator current harmonic control with resonant controller for doubly fed induction generator," *IEEE Trans. Power Electron.*, vol. 27, no. 7, pp. 3207–3220, 2012.

[5] C. Liu, W. Chen, F. Blaabjerg, and D. Xu, "Optimized design of resonant controller for stator current harmonic compensation in DFIG wind turbine systems," in *Proc. IEEE Appl. Power Electron. Conf. Exposition*, Orlando, FL, 2012, pp. 2038–2044.

[6] X. Yuan, W. Merk, H. Stemmler, and J. Allmeling, "Stationary-frame generalized integrators for current control of active power filters with zero steady-state error for current harmonics of concern under unbalanced and distorted operating conditions," *IEEE Trans. Ind. Appl.*, vol. 38, no. 2, pp. 523–532, 2002.

[7] R. Teodorescu, M. Liserre, and P. Rodriguez, *Grid Converters for Photovoltaic and Wind Power Systems*. Piscataway, NJ: John Wiley & Sons, 2011.

[8] P. Mattavelli, "A closed-loop selective harmonic compensation for active filters," *IEEE Trans. Ind. Appl.*, vol. 37, no. 1, pp. 81–89, 2001.

[9] R. Venturini, P. Mattavelli, P. Zanchetta, and M. Sumner, "Variable frequency adaptive selective compensation for active power filters," in *Proc. IET Int. Conf. Power Electron., Mach. Drives*, York, UK, 2008, pp. 16–21.

[10] R. Bojoi, G. Griva, V. Bostan, M. Guerriero, F. Farina, and F. Profumo, "Current control strategy for power conditioners using sinusoidal signal integrators in synchronous reference frame," *IEEE Trans. Power Electron.*, vol. 20, no. 6, pp. 1402–1412, 2005.

[11] R. Bojoi, L. R. Limongi, F. Profumo, D. Roiu, and A. Tenconi, "Analysis of current controllers for active power filters using selective harmonic compensation schemes," *IEEJ Trans. Electr Electron. Eng.*, vol. 4, no. 2, pp. 139–157, 2009.

[12] C. Lascu, L. Asiminoaei, I. Boldea, and F. Blaabjerg, "High performance current controller for selective harmonic compensation in active power filters," *IEEE Trans. Power Electron.*, vol. 22, no. 5, pp. 1826–1835, Sep. 2007.

[13] M. Castilla, J. Miret, J. Matas, L. Garcia de Vicuna, and J. M. Guerrero, "Control design guidelines for single-phase grid-connected photovoltaic inverters with damped resonant harmonic compensators," *IEEE Trans. Ind. Electron.*, vol. 56, no. 11, pp. 4492–4501, Nov. 2009.

[14] D. N. Zmood and D. G. Holmes, "Stationary frame current regulation of PWM inverters with zero steady-state error," *IEEE Trans. Power Electron.*, vol. 18, no. 3, pp. 814–822, May 2003.

[15] D. N. Zmood, D. G. Holmes, and G. H. Bode, "Frequency-domain analysis of three-phase linear current regulators," *IEEE Trans. Ind. Appl.*, vol. 37, no. 2, pp. 601–610, Mar./Apr. 2001.

[16] J. Hu, Y. He, L. Xu, and B. W. Williams, "Improved control of DFIG systems during network unbalance using PI–R current regulators," *IEEE Trans. Ind. Electron.*, vol. 56, no. 2, pp. 439–451, Feb. 2009.

[17] J. Hu, H. Nian, H. Xu, and Y. He, "Dynamic modeling and improved control of DFIG under distorted grid voltage conditions," *IEEE Trans. Energy Convers.*, vol. 26, no. 1, pp. 163–175, Mar. 2011.

[18] V. T. Phan and H. H. Lee, "Performance enhancement of stand-alone DFIG systems with control of rotor and load side converters using resonant controllers," *IEEE Trans. Ind. Appl.*, vol. 48, no. 1, pp. 199–210, Jan.–Feb. 2012.

[19] A. Timbus, M. Liserre, R. Teodorescu, P. Rodriguez, and F. Blaabjerg, "Evaluation of current controllers for distributed power generation systems," *IEEE Trans. Power Electron.*, vol. 24, no. 3, pp. 654–664, Mar. 2009.

[20] M. Liserre, R. Teodorescu, and F. Blaabjerg, "Multiple harmonics control for three-phase grid converter systems with the use of PI-RES current controller in a rotating frame," *IEEE Trans. Power Electron.*, vol. 21, no. 3, pp. 836–841, May 2006.

[21] D. G. Holmes, T. A. Lipo, B. P. McGrath, and W. Y. Kong, "Optimized design of stationary frame three phase AC current regulators," *IEEE Trans. Power Electron.*, vol. 24, no. 11, pp. 2417–2426, Nov. 2009.

[22] C. Lascu, L. Asiminoaei, I. Boldea, and F. Blaabjerg, "High performance current controller for selective harmonic compensation in active power filters," *IEEE Trans. Power Electron.*, vol. 22, no. 5, pp. 1826–1835, Sep. 2007.

[23] A. G. Yepes, F. D. Freijedo, J. Doval-Gandoy, Ó. López, J. Malvar, and P. Fernandez-Comesaña, "Effects of discretization methods on the performance of resonant controllers," *IEEE Trans. Power Electron.*, vol. 25, no. 7, pp. 1692–1712, Jul. 2010.

[24] G. Shen, X. Zhu, J. Zhang, and D. Xu, "A new feedback method for PR current control of LCL-filter-based grid-connected inverter," *IEEE Trans. Ind. Electron.*, vol. 57, no. 6, pp. 2033–2041, Jun. 2010.

[25] R. Teodorescu, F. Blaabjerg, U. Borup, and M. Liserre, "A new control structure for grid-connected LCL PV inverters with zero steady-state error and selective harmonic compensation," in *Proc. IEEE Appl. Power Electron. Conf. Exposition*, vol. 1, Anaheim, CA, 2004, pp. 580–586.

CHAPTER 9

DFIG UNDER UNBALANCED GRID VOLTAGE

9.1 INTRODUCTION

Under unbalanced grid conditions, a small negative-sequence voltage may cause a large negative-sequence current since the negative-sequence impedance of the DFIG is small, and cause fluctuations in the electromagnetic torque and DC-bus voltage in the back-to-back converter, thereby affecting the safe operation of the DFIG wind power system. In this chapter, the influence of unbalanced grid voltage on the performance of the DFIG system will be discussed, including the impacts on the GSC, RSC, and the DFIG. The limitations of the conventional control methods under unbalanced grid voltage will also be analyzed.

Under unbalanced grid, negative-sequence components are present in the grid voltage, so the stator current of the DFIG and the AC current of the GSC also contain negative-sequence components, which will cause imbalance in three-phase currents and fluctuations in the power and electromagnetic torque of the system, as well as the DC-bus voltage. This chapter will also give an analysis of these effects. All the calculations given in this chapter are based on the 1.5 MW DFIG system shown in Tables 5.1 and 5.2.

9.2 RSC AND DFIG UNDER UNBALANCED GRID VOLTAGE

To give the readers an overall perspective of the system, a schematic diagram of the DFIG wind power system with the back-to-back converter is shown in Figure 9.1.

An unbalanced three-phase grid voltage contains the positive-sequence, negative-sequence, and zero-sequence components. In a three-phase three-wire system, zero-sequence voltage does not cause zero-sequence current and therefore, the influence of zero-sequence component is neglected. Thus, the unbalanced grid

Advanced Control of Doubly Fed Induction Generator for Wind Power Systems, First Edition.
Dehong Xu, Frede Blaabjerg, Wenjie Chen, and Nan Zhu.
© 2018 by The Institute of Electrical and Electronics Engineers, Inc. Published 2018 by John Wiley & Sons, Inc.

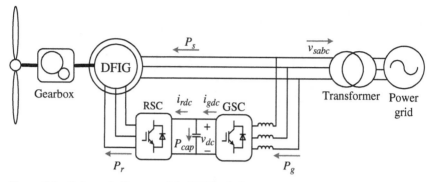

Figure 9.1   Schematic diagram of the DFIG wind power system with the back-to-back converter.

voltage can be expressed in the positive-sequence synchronous rotating dq reference frame by the following equation [4]:

$$v_{sdq} = v_{sdq}^{(+1)} + v_{sdq}^{(-1)} e^{-j2\omega_s t} \tag{9.1}$$

where $v_{sdq}^{(+1)}$ is the vector of the grid voltage positive-sequence component in the positive-sequence dq frame, $v_{sdq}^{(-1)}$ is the vector of grid voltage negative-sequence component in the negative-sequence dq frame, the rotating factor $e^{-j2\omega_s t}$ represents the negative-sequence voltage vector $v_{sdq}^{(-1)}$ is rotating in the negative direction at an angular speed of $2\omega_s$ in the positive-sequence dq frame. In this chapter, the magnetic flux and current vectors of the stator and rotor are expressed similarly as the grid voltage. The superscript $(+1)$ represents the positive fundamental component, and the superscript (-1) represents the negative fundamental component. The severity of voltage unbalance is expressed by the voltage unbalance factor τ_v.

$$\tau_v = \frac{V_{sm}^{(-1)}}{V_{sm}^{(+1)}} \times 100\% = \frac{\left| v_{sdq}^{(-1)} \right|}{\left| v_{sdq}^{(+1)} \right|} \times 100\% \tag{9.2}$$

The magnitudes of the positive-sequence voltage $V_{sm}^{(+1)}$ and the negative-sequence voltage $V_{sm}^{(-1)}$ under the unbalanced grid can be calculated using the following equation [2], where κ is the proportion of voltage dip, γ represents the type of voltage unbalance, for single-phase voltage dip $\gamma = 3$, and for two-phase voltage dip $\gamma = 2$.

$$\begin{cases} V_{sm}^{(+1)} = \left(1 - \dfrac{\kappa}{\gamma} \right) V_{sm} \\[4mm] V_{sm}^{(-1)} = \dfrac{\kappa}{\gamma} V_{sm} \end{cases} \tag{9.3}$$

9.2.1 Rotor and Stator Currents

According to the DFIG model described in Chapter 5, if the stator resistance is neglected, the steady-state expression of the stator flux ψ_{sdq} can be written as [1, 2]

$$\psi_{sdq} = \psi_{sdq}^{(+1)} + \psi_{sdq}^{(-1)} e^{-j2\omega_s t} = \frac{v_{sdq}^{(+1)}}{j\omega_s} - \frac{v_{sdq}^{(-1)}}{j\omega_s} e^{-j2\omega_s t} \tag{9.4}$$

where $\psi_{sdq}^{(+1)}$ and $\psi_{sdq}^{(-1)}$ are the positive- and negative-sequence components of the stator flux.

Assume that the steady-state rotor current can be expressed as

$$i_{rdq} = i_{rdq}^{(+1)} + i_{rdq}^{(-1)} e^{-j2\omega_s t} \tag{9.5}$$

where $i_{rdq}^{(+1)}$ and $i_{rdq}^{(-1)}$ are the positive-sequence and negative-sequence components of the rotor current, respectively.

According to the DFIG model described in Chapter 5 and neglecting the rotor resistance, the positive-sequence component of the rotor voltage $v_{rdq}^{(+1)}$ and the negative-sequence component of the rotor current $i_{rdq}^{(-1)}$ can be expressed as [1, 2, 4, 8]

$$
\begin{aligned}
v_{rdq} &= \left(R_r + \sigma L_r \frac{d}{dt} + j\omega_{sl}\sigma L_r \right) i_{rdq} + \frac{L_m}{L_s}\left(\frac{d}{dt} + j\omega_{sl} \right)\psi_{sdq} \\[2mm]
&\cong \left(\sigma L_r \frac{d}{dt} + j\omega_{sl}\sigma L_r \right)\left(i_{rdq}^{(+1)} + i_{rdq}^{(-1)} e^{-j2\omega_s t} \right) \\[2mm]
&\quad + \frac{L_m}{L_s}\left(\frac{d}{dt} + j\omega_{sl} \right)\left(\frac{v_{sdq}^{(+1)}}{j\omega_s} + \frac{v_{sdq}^{(-1)}}{-j\omega_s} e^{-j2\omega_s t} \right) \\[2mm]
&= j\omega_{sl}\sigma L_r i_{rdq}^{(+1)} + j\left(-2\omega_s + \omega_{sl} \right)\sigma L_r i_{rdq}^{(-1)} e^{-j2\omega_s t} \\[2mm]
&\quad + \frac{L_m}{L_s}\left(\frac{\omega_{sl}}{\omega_s} v_{sdq}^{(+1)} + \frac{-2\omega_s + \omega_{sl}}{-\omega_s} v_{sdq}^{(-1)} e^{-j2\omega_s t} \right) \\[2mm]
&= \left(j\omega_{sl}\sigma L_r i_{rdq}^{(+1)} + \frac{L_m}{L_s} S_l v_{sdq}^{(+1)} \right) \\[2mm]
&\quad + \left(j\left(-2\omega_s + \omega_{sl} \right)\sigma L_r i_{rdq}^{(-1)} + \frac{-2\omega_s + \omega_{sl}}{-\omega_s}\frac{L_m}{L_s} v_{sdq}^{(-1)} \right) e^{-j2\omega_s t} \tag{9.6}
\end{aligned}
$$

where L_m is the equivalent mutual inductance between the stator and rotor windings, L_r and L_s are the equivalent self-inductances of the rotor and stator windings, respectively, $\sigma = 1 - L_m^2/(L_s L_r)$, R_r is the resistance of the rotor winding, ω_{sl} is the slip angular velocity of the DFIG, ω_s is the angular velocity of the stator voltage, and S_l is the slip ratio.

The rotor voltage is controlled by the RSC, and if the negative-sequence voltage is not controlled by the RSC, the rotor voltage is symmetrical and the rotor voltage can be expressed as

$$v_{rdq} = v_{rdq}^{(+1)} + v_{rdq}^{(-1)} e^{-j2\omega_s t} = v_{rdq}^{(+1)} \tag{9.7}$$

Combining equations (9.6) and (9.7), the following equation can be derived [1, 2, 4, 8]:

$$
\begin{cases}
v_{rdq}^{(+1)} = S_l \dfrac{L_m}{L_s} v_{sdq}^{(+1)} + j\omega_{sl} \sigma L_r i_{rdq}^{(+1)} \approx S_l v_{sdq}^{(+1)} \\
j\left(-2\omega_s + \omega_{sl}\right) \sigma L_r i_{rdq}^{(-1)} + \dfrac{-2\omega_s + \omega_{sl}}{-\omega_s} \dfrac{L_m}{L_s} v_{sdq}^{(-1)} = 0 \\
\Rightarrow i_{rdq}^{(-1)} = \dfrac{L_m}{L_s} \dfrac{v_{sdq}^{(-1)}}{j\omega_s \sigma L_r} = -\dfrac{L_m}{L_s} \dfrac{v_{sdq}^{(-1)}}{X_{vir}(-j\omega_s)}
\end{cases}
\tag{9.8}
$$

where $X_{vir}(-j\omega_s) = -j\omega_s \sigma L_r$ is the negative-sequence impedance from the grid voltage to the rotor current. Therefore, under the unbalanced grid, the negative-sequence components may occur in the rotor current. The complete expression of the rotor current is given as [1, 2, 4, 8]

$$i_{rdq} = i_{rdq}^{(+1)} + \frac{L_m}{L_s} \frac{v_{sdq}^{(-1)}}{j\omega_s \sigma L_r} = i_{rdq}^{(+1)} - \frac{L_m}{L_s} \frac{v_{sdq}^{(-1)}}{X_{vir}(-j\omega_s)} e^{-j2\omega_s t} \tag{9.9}$$

Substituting equations (9.4) and (9.9) in the DFIG model described in Chapter 5, the steady-state expression of stator current can be derived as the following equation, where $X_{vis}(-j\omega_s) = -j\omega_s \sigma L_s$ is the negative-sequence impedance from the grid voltage to the stator current. It is usually in the range of 0.1–0.2 pu. Therefore, a small negative-sequence voltage may cause large negative-sequence current at the AC side of the GSC. [1, 2, 4, 8]

$$
\begin{aligned}
i_{sdq} &= \frac{\Psi_{sdq}}{L_s} - \frac{L_m}{L_s} i_{rdq} \\
&= \frac{v_{sdq}^{(+1)}}{j\omega_s L_s} - \frac{v_{sdq}^{(-1)}}{j\omega_s L_s} e^{-j2\omega_s t} - \left(\frac{L_m}{L_s}\right)^2 \frac{v_{sdq}^{(-1)}}{j\omega_s \sigma L_r} e^{-j2\omega_s t} \\
&= \frac{v_{sdq}^{(+1)}}{j\omega_s L_s} - \frac{L_m}{L_s} i_{rdq}^{(+1)} - \left(1 + \frac{L_m^2}{L_s L_r \sigma}\right) \frac{v_{sdq}^{(-1)}}{j\omega_s L_s} e^{-j2\omega_s t} \\
&= i_{sdq}^{(+1)} - \frac{v_{sdq}^{(-1)}}{j\omega_s \sigma L_s} e^{-j2\omega_s t} \\
&= i_{sdq}^{(+1)} + \frac{v_{sdq}^{(-1)}}{X_{vis}(-j\omega_s)} e^{-j2\omega_s t}
\end{aligned}
\tag{9.10}
$$

where $X_{vis}(-j\omega_s) = -j\omega_s\sigma L_s$ is the negative-sequence impedance from the grid voltage to stator current and it is usually in the range of 0.1–0.2 pu. Therefore, a small negative-sequence voltage may cause a large negative-sequence current in the stator of the DFIG as well.

9.2.2 Active and Reactive Powers

According to the expressions of the grid voltage and stator current, the stator active power under unbalanced grid can be written as [6, 7]

$$P_s = 1.5\,\mathrm{Re}\left(v_{sdq}i^*_{sdq}\right) = 1.5\,\mathrm{Re}\left(v^{(+1)}_{sdq}i^{(+1)*}_{sdq} + v^{(-1)}_{sdq}i^{(-1)*}_{sdq}\right)$$

$$+ 1.5\,\mathrm{Re}\left(v^{(+1)}_{sdq}i^{(-1)*}_{sdq}e^{j2\omega_s t} + v^{(-1)}_{sdq}i^{(+1)*}_{sdq}e^{-j2\omega_s t}\right)$$

$$= P^{(0)}_s + P^{(2)}_s \tag{9.11}$$

where

$$\begin{cases} P^{(0)}_s = 1.5\,\mathrm{Re}\left(v^{(+1)}_{sdq}i^{(+1)*}_{sdq} + v^{(-1)}_{sdq}i^{(-1)*}_{sdq}\right) \\[4mm] \quad = 1.5\,\mathrm{Re}\left(v^{(+1)}_{sdq}i^{(+1)*}_{sdq} + \dfrac{v^{(-1)}_{sdq}v^{(-1)*}_{sdq}}{X^*_{vis}(-j\omega_s)}\right) \\[6mm] P^{(2)}_s = 1.5\,\mathrm{Re}\left(v^{(+1)}_{sdq}i^{(-1)*}_{sdq}e^{j2\omega_s t} + v^{(-1)}_{sdq}i^{(+1)*}_{sdq}e^{-j2\omega_s t}\right) \\[4mm] \quad = 1.5\,\mathrm{Re}\left(\dfrac{v^{(+1)}_{sdq}v^{(-1)*}_{sdq}}{X^*_{vis}(-j\omega_s)}e^{j2\omega_s t} + v^{(-1)}_{sdq}i^{(+1)*}_{sdq}e^{-j2\omega_s t}\right) \end{cases} \tag{9.12}$$

$P^{(0)}_s$ is the fundamental component of the stator active power, which is composed of the positive and negative-sequence components of the fundamental active power. $P^{(2)}_s$ is the second-order fluctuation component of the stator active power, which is caused by the interaction between positive- and negative-sequence voltages and currents. As shown in the equation, under unbalanced grid conditions, a second-order fluctuation component is superimposed on the fundamental component of the stator active power.

Similar to the active power, the stator reactive power can be expressed as

$$Q_s = 1.5\,\mathrm{Im}\left(v_{sdq}i^*_{sdq}\right) = Q^{(0)}_s + Q^{(2)}_s \tag{9.13}$$

where

$$
\begin{cases}
Q_s^{(0)} = 1.5 \operatorname{Im} \left(v_{sdq}^{(+1)} i_{sdq}^{(+1)*} + v_{sdq}^{(-1)} i_{sdq}^{(-1)*} \right) = 1.5 \operatorname{Im} \left(v_{sdq}^{(+1)} i_{sdq}^{(+1)*} + \dfrac{v_{sdq}^{(-1)} v_{sdq}^{(-1)*}}{X_{vis}^*(-j\omega_s)} \right) \\[4mm]
Q_s^{(2)} = 1.5 \operatorname{Im} \left(v_{sdq}^{(+1)} i_{sdq}^{(-1)*} e^{j2\omega_s t} + v_{sdq}^{(-1)} i_{sdq}^{(+1)*} e^{-j2\omega_s t} \right) \\[4mm]
\qquad = 1.5 \operatorname{Im} \left(\dfrac{v_{sdq}^{(+1)} v_{sdq}^{(-1)*}}{X_{vis}^*(-j\omega_s)} e^{j2\omega_s t} + v_{sdq}^{(-1)} i_{sdq}^{(+1)*} e^{-j2\omega_s t} \right)
\end{cases}
\tag{9.14}
$$

$Q_s^{(0)}$ is the fundamental component of stator reactive power, which is composed of the positive- and negative-sequence components of the fundamental reactive power. $Q_s^{(2)}$ is the second-order fluctuation component of stator reactive power, which is caused by the interaction between positive- and negative-sequence voltages and currents.

As shown in the expression of the second-order fluctuation components in the stator active and reactive powers given in equations (9.12) and (9.14), it can be seen that due to the small value of the negative-sequence impedance $X_{vis}(-j\omega_s)$, a small negative-sequence voltage may cause large second-order fluctuations in the stator active and reactive powers.

According to the expressions of rotor voltage and current given in equations (9.6) and (9.9), the rotor active power can be derived as

$$
P_r = 1.5 \operatorname{Re} \left(v_{rdq} i_{rdq}^* \right)
$$

$$
= 1.5 \operatorname{Re} \left[v_{rdq}^{(+1)} \left(i_{rdq}^{(+1)*} - \frac{L_m}{L_s} \frac{v_{sdq}^{(-1)*}}{X_{vir}^*(-j\omega_s)} e^{j2\omega_s t} \right) \right]
$$

$$
= 1.5 \operatorname{Re} \left(v_{rdq}^{(+1)} i_{rdq}^{(+1)*} \right) - 1.5 \frac{L_m}{L_s} \operatorname{Re} \left(\frac{v_{rdq}^{(+1)} v_{sdq}^{(-1)*}}{X_{vir}^*(-j\omega_s)} e^{j2\omega_s t} \right)
\tag{9.15}
$$

Since $v_{rdq}^{(+1)} \approx S_l v_{sdq}^{(+1)}$ and $\frac{L_m}{L_s} \approx 1$, the rotor active power can be written as a function of the grid voltage:

$$
P_r = P_r^{(0)} + P_r^{(2)} \approx 1.5 S_l \operatorname{Re} \left(v_{sdq}^{(+1)} i_{rdq}^{(+1)*} \right) - 1.5 S_l \operatorname{Re} \left(\frac{v_{sdq}^{(+1)} v_{sdq}^{(-1)*}}{X_{vir}^*(-j\omega_s)} e^{j2\omega_s t} \right)
\tag{9.16}
$$

where

$$
\begin{cases}
P_r^{(0)} \approx S_l P_s^{(0)} \\[4mm]
P_r^{(2)} \approx -1.5 S_l \, \mathrm{Re} \left(\dfrac{v_{sdq}^{(+1)} v_{sdq}^{(-1)*}}{X_{vir}^*(-j\omega_s)} e^{j2\omega_s t} \right)
\end{cases}
\tag{9.17}
$$

Therefore, the rotor active power consists also of the fundamental component $P_r^{(0)}$ and the second-order fluctuation component $P_r^{(2)}$. The second-order fluctuation in the rotor active power will flow into the DC bus through the three-phase bridge for the RSC and causes fluctuations in the DC-bus voltage.

9.2.3 Electromagnetic Torque

Combining the expression of stator flux shown in equation (9.4) and the expression of stator current shown in equation (9.10), the electromagnetic torque can be derived as [1, 2, 4, 8]

$$
\begin{aligned}
T_{em} &= 1.5 n_p \, \mathrm{Re} \left(j \psi_{sdq} i_{sdq}^{*} \right) \\[3mm]
&= 1.5 n_p \, \mathrm{Re} \left[j \left(\frac{v_{sdq}^{(+1)}}{j\omega_s} - \frac{v_{sdq}^{(-1)}}{j\omega_s} e^{-j2\omega_s t} \right) \left(i_{sdq}^{(+1)*} + \frac{v_{sdq}^{(-1)*}}{X_{vis}^*(-j\omega_s)} e^{j2\omega_s t} \right) \right] \\[3mm]
&= 1.5 \frac{n_p}{\omega_s} \, \mathrm{Re} \left(v_{sdq}^{(+1)} i_{sdq}^{(+1)*} - \frac{v_{sdq}^{(-1)} v_{sdq}^{(-1)*}}{X_{vis}^*(-j\omega_s)} \right) \\[3mm]
&\quad + 1.5 \frac{n_p}{\omega_s} \, \mathrm{Re} \left(\frac{v_{sdq}^{(+1)} v_{sdq}^{(-1)*}}{X_{vis}^*(-j\omega_s)} e^{j2\omega_s t} - v_{sdq}^{(-1)} i_{sdq}^{(+1)*} e^{-j2\omega_s t} \right) \\[3mm]
&= T_{em}^{(0)} + T_{em}^{(2)}
\end{aligned}
\tag{9.18}
$$

where $T_{em}^{(0)}$ and $T_{em}^{(2)}$ are the fundamental and second-order fluctuation components of the electromagnetic torque, respectively, which can be expressed as

$$
\begin{cases}
T_{em}^{(0)} = 1.5 \dfrac{n_p}{\omega_s} \, \mathrm{Re} \left(v_{sdq}^{(+1)} i_{sdq}^{(+1)*} - v_{sdq}^{(-1)} i_{sdq}^{(-1)*} \right) = 1.5 \dfrac{n_p}{\omega_s} \, \mathrm{Re} \left(v_{sdq}^{(+1)} i_{sdq}^{(+1)*} - \dfrac{v_{sdq}^{(-1)} v_{sdq}^{(-1)*}}{X_{vis}^*(-j\omega_s)} \right) \\[5mm]
T_{em}^{(2)} = 1.5 \dfrac{n_p}{\omega_s} \, \mathrm{Re} \left(v_{sdq}^{(+1)} i_{sdq}^{(-1)*} e^{j2\omega_s t} - v_{sdq}^{(-1)} i_{sdq}^{(+1)*} e^{-j2\omega_s t} \right) \\[5mm]
\qquad = 1.5 \dfrac{n_p}{\omega_s} \, \mathrm{Re} \left(\dfrac{v_{sdq}^{(+1)} v_{sdq}^{(-1)*}}{X_{vis}^*(-j\omega_s)} e^{j2\omega_s t} - v_{sdq}^{(-1)} i_{sdq}^{(+1)*} e^{-j2\omega_s t} \right)
\end{cases}
\tag{9.19}
$$

Note that the expression of $T_{em}^{(2)}$ is composed of two terms: (1) the product of the fundamental positive-sequence voltage and the fundamental negative-sequence current, (2) the product of the fundamental negative-sequence voltage and the fundamental positive-sequence current. Since the value of $X_{vis}(s)$ is usually in the range of 0.1–0.2 pu, under rated stator current, the first term is more than five times the second term. Thus, under unbalanced grid voltage, the main factor inducing the second-order fluctuation in the electromagnetic torque is the negative-sequence stator current. The second-order fluctuation in the electromagnetic torque will result in vibration of the motor shaft, causing fatigue in mechanic components such as the shaft and the gearbox, affecting the safe operation of the DFIG system. If the negative-sequence stator current can be suppressed, the second-order fluctuation in electromagnetic torque will be largely reduced, thus mitigating the impact of grid unbalance on the DFIG system.

9.2.4 Simulation on the Influence of Grid Voltage Unbalance

A transient simulation is done on a 1.5 MW DFIG system to observe the influence of grid voltage unbalance on the stator current, the rotor current, and the electromagnetic torque. The transient response of the DFIG system when a 30% single-phase voltage dip happens at 0.1 s is shown in Figure 9.2.

As shown in Figure 9.2, a large negative-sequence stator current and electromagnetic torque fluctuations are induced after the voltage dip, which largely affects the power quality and even safe operation of the wind power system.

9.3 GSC UNDER UNBALANCED GRID VOLTAGE

9.3.1 Grid Current

According to the model of the GSC presented in Chapter 5, the voltage vector equation of the GSC can be written as [1, 2]

$$v_{sdq} = R_g i_{gdq} + L_g \frac{d}{dt} i_{gdq} + j\omega_s L_g i_{gdq} + v_{gdq} \tag{9.20}$$

If the negative-sequence components are not controlled by the GSC, then the voltage at the AC side is symmetrical:

$$v_{gdq} = v_{gdq}^{(+1)} \tag{9.21}$$

Assume that the AC current of the GSC can be expressed as

$$i_{gdq} = i_{gdq}^{(+1)} + i_{gdq}^{(-1)} e^{-j2\omega_s t} \tag{9.22}$$

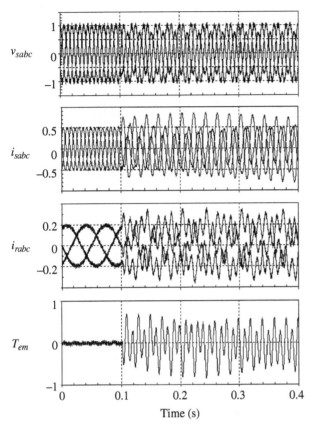

Figure 9.2   Simulation results for a 1.5 MW DFIG system when 30% single-phase voltage dip happens at 0.1 s. From top to bottom: three-phase grid voltage, three-phase stator current, three-phase rotor current, and the electromagnetic torque.

Under unbalanced grid, substituting equations (9.21) and (9.22) in equation (9.20), the following equation can be derived [1, 2, 4, 8]:

$$v_{sdq} \approx L_g \frac{d}{dt} i_{gdq} + j\omega_s L_g i_{gdq} + v_{gdq}$$

$$= L_g \left(\frac{d}{dt} + j\omega_s \right) \left(i_{gdq}^{(+1)} + i_{gdq}^{(-1)} e^{-j2\omega_s t} \right) + v_{gdq}^{(+1)}$$

$$= L_g \left(j\omega_s i_{gdq}^{(+1)} - j\omega_s i_{gdq}^{(-1)} e^{-j2\omega_s t} \right) + v_{gdq}^{(+1)}$$

$$= j\omega_s L_g i_{gdq}^{(+1)} + v_{gdq}^{(+1)} - j\omega_s i_{gdq}^{(-1)} e^{-j2\omega_s t} \qquad (9.23)$$

Combining the expression of the grid voltage shown in equations (9.1) and (9.23), the relations of the GSC voltage $v_{gdq}^{(+1)}$ and GSC negative-sequence current

$i_{gdq}^{(-1)}$ with the grid voltage can be expressed as

$$\begin{cases} v_{sdq}^{(+1)} = j\omega_s L_g i_{gdq}^{(+1)} + v_{gdq}^{(+1)} \Rightarrow v_{gdq}^{(+1)} \approx v_{sdq}^{(+1)} \\ v_{sdq}^{(-1)} = -j\omega_s i_{gdq}^{(-1)} \Rightarrow i_{gdq}^{(-1)} = -\dfrac{v_{sdq}^{(-1)}}{j\omega_s L_g} = \dfrac{v_{sdq}^{(-1)}}{X_{vig}(-j\omega_s)} \end{cases} \tag{9.24}$$

where $X_{vig}(-j\omega_s)$ is the negative-sequence impedance from the grid voltage to the grid-side current of the GSC, $X_{vig}(-j\omega_s) = -j\omega_s L_g$, which is typically about 0.1–0.2 pu. Therefore, a small negative-sequence voltage may cause large negative-sequence current at the AC side of the GSC.

9.3.2 Active Power of the Generator

The active power flowing into the GSC is [6, 7]

$$P_g = 1.5\,\mathrm{Re}\big(v_{gdq} i_{gdq}^*\big) = 1.5\,\mathrm{Re}\left(v_{gdq}^{(+1)} i_{gdq}^{(+1)*} + v_{gdq}^{(+1)} i_{gdq}^{(-1)*} e^{j2\omega_s t}\right)$$
$$= P_g^{(0)} + P_g^{(2)} \tag{9.25}$$

where $P_g^{(0)}$ is the grid-side fundamental active power, $P_g^{(2)}$ is the grid-side second-order fluctuation active power, of which the expressions are given by the following [6, 7]:

$$\begin{cases} P_g^{(0)} = 1.5\,\mathrm{Re}\left(v_{gdq}^{(+1)} i_{gdq}^{(+1)*}\right) \approx 1.5\,\mathrm{Re}\left(v_{sdq}^{(+1)} i_{sdq}^{(+1)*}\right) \\ P_g^{(2)} = 1.5\,\mathrm{Re}\left(v_{gdq}^{(+1)} i_{gdq}^{(-1)*} e^{j2\omega_s t}\right) \approx 1.5\,\mathrm{Re}\left(\dfrac{v_{sdq}^{(+1)} v_{sdq}^{(-1)*}}{X_{vig}^*(-j\omega_s)} e^{j2\omega_s t}\right) \end{cases} \tag{9.26}$$

From the expression of $P_g^{(2)}$, it can be seen that a small negative-sequence voltage may cause a large second-order power fluctuation. The fluctuation power will flow into the DC-bus capacitor through the GSC causing fluctuations in the DC-bus voltage as well.

9.3.3 DC-Bus Current and Voltage

In a back-to-back converter system connected by a common DC bus, the active power flowing into the DC-bus capacitor consists of active powers from the rotor side and the grid side. A schematic diagram of the power flow definitions in a DFIG system is shown in Figure 9.3.

As shown in the figure, on neglecting the losses of power converters, the active power flowing into the DC-bus capacitor can be written as

$$P_{cap} = P_g - P_r = \left(P_g^{(0)} - P_r^{(0)}\right) + \left(P_g^{(2)} - P_r^{(2)}\right) \tag{9.27}$$

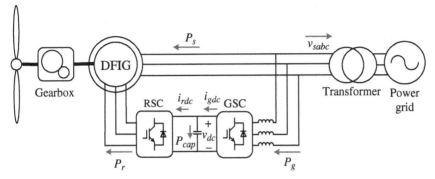

Figure 9.3   Schematic diagram of the power flow definitions in a DFIG system.

By controlling the DC voltage, the GSC is usually able to keep good track of the fundamental power of the RSC, that is, $P_g^{(0)} = P_r^{(0)}$, and the DC-bus voltage is controlled at a constant value. Given the above-mentioned conditions and substituting equations (9.17) and (9.26) in (9.27), the active power flowing into the DC-bus capacitor can be derived as [6, 7]

$$P_{cap} = P_g^{(2)} - P_r^{(2)} \approx 1.5\,\text{Re}\left(\frac{v_{sdq}^{(+1)}v_{sdq}^{(-1)*}}{X_{vig}^*(-j\omega_s)}e^{j2\omega_s t}\right) + 1.5 S_l\,\text{Re}\left(\frac{v_{sdq}^{(+1)}v_{sdq}^{(-1)*}}{X_{vis}^*(-j\omega_s)}e^{j2\omega_s t}\right)$$

$$= 1.5\,\text{Re}\left[v_{sdq}^{(+1)}v_{sdq}^{(-1)*}\left(\frac{1}{X_{vig}^*(-j\omega_s)} + \frac{S_l}{X_{vis}^*(-j\omega_s)}\right)e^{j2\omega_s t}\right] \tag{9.28}$$

As shown in equation (9.28), under unbalanced grid conditions, a second-order fluctuation is present in the active power flowing into the DC-bus capacitor, of which the magnitude is [6, 7]

$$P_{capm} \approx 1.5 V_{sm}^{(+1)} V_{sm}^{(-1)}\left|\frac{1}{X_{vig}^*(-j\omega_s)} + \frac{S_l}{X_{vis}^*(-j\omega_s)}\right| \tag{9.29}$$

Thus, the magnitude of the second-order harmonic current in the DC-bus capacitor is

$$I_{capm}^{(2)} = \frac{P_{capm}}{V_{dc}} \approx 1.5\frac{V_{sm}^{(+1)} V_{sm}^{(-1)}}{V_{dc}}\left|\frac{1}{X_{vig}^*(-j\omega_s)} + \frac{S_l}{X_{vis}^*(-j\omega_s)}\right| \tag{9.30}$$

Under unbalanced grid condition, the current of the DC-bus capacitor is proportional to the negative-sequence voltage $V_{sm}^{(-1)}$ and decreases with larger negative-sequence impedances. Since the negative-sequence impedances of the DFIG and GSC are small, large second-order harmonic current will flow into the DC-bus capacitor causing temperature rise in the capacitor and since the lifetime of the electrolytic capacitor may be influenced by the temperature and a large second-order harmonic current may cause significant reduction of the capacitor lifetime.

The second-order harmonic current in the DC-bus capacitor causes also second-order fluctuations in the DC-bus voltage. Assume that the DC-bus voltage can be expressed as

$$v_{dc} = V_{dc} + v_{dc}^{(2)} = V_{dc} + V_{dcm}^{(2)} \sin(2\omega_s t + \varphi_0)$$ (9.31)

Combining with the equation of the DC-bus capacitor $i_{cap} = C\frac{dv_{dc}}{dt}$, the magnitude of second-order fluctuation in DC-bus voltage is derived as

$$V_{dcm}^{(2)} = \frac{I_{capm}^{(2)}}{2\omega_s C} \approx 1.5 \frac{V_{sm}^{(+1)} V_{sm}^{(-1)}}{2\omega_s C V_{dc}} \left| \frac{1}{X_{vig}^*(-j\omega_s)} + \frac{S_l}{X_{vis}^*(-j\omega_s)} \right|$$ (9.32)

Since the AC-side voltage of the converter is the product of switching function and DC-bus voltage, the second-order harmonic component in the DC-bus voltage will cause a third-order harmonic voltage in the AC side of the GSC, which will induce a third harmonic current on the AC side of the GSC. Meanwhile, low frequency harmonic currents and voltages will be induced at the output of the RSC as well, and the fluctuation of the DC-bus voltage will also affect the lifetime of the DC-bus capacitor.

9.4 CONTROL LIMITATIONS UNDER UNBALANCED GRID VOLTAGE

According to the analysis given in the previous section, the negative effects caused by the grid voltage unbalances are:

1. Second-order fluctuation in the electromagnetic torque of the DFIG, causing fatigue in mechanical components such as gearbox and shaft;
2. Large negative-sequence current causes increase in the winding loss of the DFIG, thereby reducing the efficiency;
3. Large second-order harmonic current flowing through the DC-bus capacitor, reducing the capacitor lifetime;
4. Second-order harmonic component in the DC-bus voltage, causing third-order harmonics in the AC-side output voltages of the back-to-back converter and affecting the quality of the output current.

The electromagnetic torque fluctuation, second-order harmonic current in the DC-bus capacitor, and second-order fluctuation in the DC-bus voltage are given in equations (9.19), (9.30), and (9.32), respectively. It can be seen from these expressions that in order to reduce the impact of the grid voltage unbalance on the DFIG system, the impedance for negative-sequence voltage of the system must be increased by improving the overall control scheme.

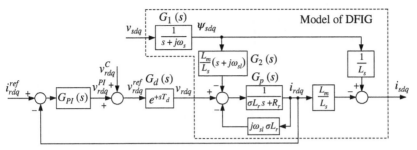

Figure 9.4   Current control block diagram of the DFIG using a PI controller.

9.4.1   Control Limitations of RSC

If the conventional vector control scheme is adopted for the RSC as discussed in Chapter 5, the current control block diagram of the DFIG can be depicted as shown in Figure 9.4.

The impedance for the fundamental negative-sequence voltage of the system can be written as

$$
G_{vis}(-j2\omega_s) = \frac{i_{sdq}(-j2\omega_s)}{v_{sdq}(-j2\omega_s)} = \frac{1}{X_{vis}(-j2\omega_s)}
$$

$$
= \frac{G_1(s)/L_s + G_1(s)G_{PI}(s)G_d(s)G_p(s)/L_s + G_1(s)G_2(s)G_p(s)L_m/L_s}{1 + G_{PI}(s)G_d(s)G_p(s)} \Bigg|_{s=-j2\omega_s}
$$

$$(9.33)$$

The second-order fluctuation in the electromagnetic torque is

$$
T_{em}^{(2)} = 1.5\frac{n_p}{\omega_s} \mathrm{Re}\left(v_{sdq}^{(+1)} i_{sdq}^{(-1)*} e^{j2\omega_s t} - v_{sdq}^{(-1)} i_{sdq}^{(+1)*} e^{-j2\omega_s t}\right)
$$

$$
= 1.5\frac{n_p}{\omega_s} \mathrm{Re}\left(G_{vis}^*(-j2\omega_s)v_{sdq}^{(+1)} v_{sdq}^{(-1)*} e^{j2\omega_s t} - v_{sdq}^{(-1)} i_{sdq}^{(+1)*} e^{-j2\omega_s t}\right) \quad (9.34)
$$

Take the 1.5 MW DFIG system shown in Tables 5.1 and 5.2 as an example; if the controller gain K_p of the PI controller ranges from 0.3 to 1.5, the control bandwidth of the current loop changes from 100 to 500 Hz, and the control performance of the DFIG system under unbalanced grid voltage is as shown in Figure 9.5.

If the slip ratio is 0.2, when the control bandwidth varies, the suppression of the DFIG system on the fundamental negative-sequence voltage is shown in Figure 9.5a. As shown in Figure 9.5a, the suppression of the negative-sequence voltage increases with an increasing gain of the PI controller, which indicates that the PI-controller-based vector control is able to increase the negative-sequence impedance to some extent. However, all the values of $\left|G_{vis}(-j2\omega_s)\right|$ shown in Figure 9.5a are above 0 dB, which means that the conventional control method based on the PI controller is not able to effectively suppress the impact of the grid voltage unbalance

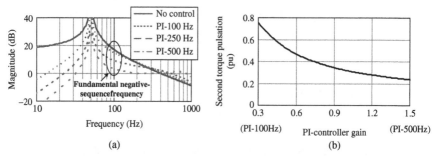

Figure 9.5   Control performance of the DFIG system under unbalanced grid voltage when the crossover frequency ranges from 100 to 500 Hz: (a) The suppression of voltage disturbance by the control system $G_{vis}(s)$; (b) second-order fluctuation in the electromagnetic torque $T_{em}^{(2)}$.

on the DFIG system. If the voltage unbalance factor is 20% (positive-sequence voltage $\left|v_{sdq}^{(+1)}\right| = 0.83$pu, the negative-sequence voltage $\left|v_{sdq}^{(-1)}\right| = 0.17$pu), and the fundamental positive-sequence current is set to $i_{sdq}^{(+1)} = 0.5$pu, Figure 9.5b shows the second-order fluctuation in the electromagnetic torque under different bandwidths of PI current control loop. It can be seen that with the increase of control bandwidth, the second-order fluctuation of the electromagnetic torque decreases. However, when the bandwidth is as large as 500 Hz, the magnitude of the electromagnetic torque fluctuation is still quite as large as 0.25 pu. Therefore, adopting the conventional PI control method, the negative effects of the grid voltage unbalance are not significantly mitigated.

9.4.2   Control Limitations of GSC

According to the model and control system of the GSC introduced in Chapter 5 and taking the control delay into consideration, the block diagram of the GSC control system can be depicted as shown in Figure 9.6.

In Figure 9.6, $G_{gp}(s) = 1/(L_g s + R_g)$, $K_u = 1.5 V_{sd}/V_{dc}$.

As shown in Figure 9.6, the grid voltage v_{sdq} and rotor-side DC current i_{rdc} are the disturbances of the GSC control system. Under unbalanced grid, v_{sdq} and i_{rdc} contain the second-order harmonic components, which may cause second-order harmonic components in the capacitor current i_{cap}. According to Figure 9.6, neglecting

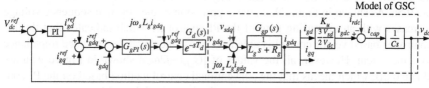

Figure 9.6   Block diagram of GSC control system using a PI controller.

the influence of control delay on the current decoupling terms, the equation of the current control loop can be written as

$$i_{cap}(s) = \left[\left(i_{gd}^{ref}(s) - i_{gd}(s)\right) G_{PI}(s)G_d(s) + v_{sd}(s)\right] G_{gp}(s)K_u - i_{rdc}(s) \quad (9.35)$$

Also, as shown in Figure 9.6,

$$i_{gd}(s) = \left(i_{cap}(s) + i_{rdc}(s)\right)/K_u \quad (9.36)$$

Substituting (9.36) in (9.35), the capacitor current is written as

$$i_{cap}(s) = G_{gc}(s)i_{gd}^{ref}(s) + G_{vc}(s)v_{sd}(s) - G_{rc}(s)i_{rdc}(s) \quad (9.37)$$

where $G_{gc}(s)$ is the transfer function from the d-axis reference of the grid current $i_{gd}^{ref}(s)$ to the capacitor current $i_{cap}(s)$, which represents the tracking performance of current control loop. $G_{vc}(s)$ is the transfer function from the grid voltage $v_{sd}(s)$ to the capacitor current $i_{cap}(s)$, which represents the suppression capability of the current control loop on the grid voltage disturbance. $G_{rc}(s)$ is the transfer function from rotor-side DC current $i_{rdc}(s)$ to capacitor current $i_{cap}(s)$, which represents the suppression capability of the current control loop on the rotor-side DC current disturbance. Based on the block diagram shown in Figure 9.6, a detailed expression of these transfer functions are shown below:

$$G_{gc}(s) = \left.\frac{i_{cap}(s)}{i_{gd}^{ref}(s)}\right|_{v_{sd}=0,i_{rdc}=0} = \frac{K_u G_{gp}(s)G_{PI}(s)G_d(s)}{1 + G_{gp}(s)G_{PI}(s)G_d(s)} \quad (9.38)$$

$$G_{vc}(s) = \left.\frac{i_{cap}(s)}{v_{sd}(s)}\right|_{i_{gdq}^{ref}=0,i_{rdc}=0} = \frac{K_u G_{gp}(s)}{1 + G_{gp}(s)G_{PI}(s)G_d(s)} \quad (9.39)$$

$$G_{rc}(s) = \left.\frac{i_{cap}(s)}{i_{rdc}(s)}\right|_{i_{gdq}^{ref}=0,v_{sd}=0} = 1 \quad (9.40)$$

Given the above expressions, the closed-loop transfer function of the grid current $G_{icl}(s)$ and the transfer function $G_{vig}(s)$ from the grid voltage $v_{sd}(s)$ to the grid current $i_{gd}(s)$ can be derived as

$$G_{icl}(s) = \left.\frac{i_{gd}(s)}{i_{gd}^{ref}(s)}\right|_{v_{sd}=0,i_{rdc}=0} = \frac{G_{gc}(s)}{K_u} = \frac{G_{gp}(s)G_{PI}(s)G_d(s)}{1 + G_{gp}(s)G_{PI}(s)G_d(s)} \quad (9.41)$$

$$G_{uig}(s) = \left.\frac{i_{gd}(s)}{v_{sd}(s)}\right|_{i_{gdq}^{ref}=0,i_{rdc}=0} = \frac{G_{vc}(s)}{K_u} = \frac{G_{gp}(s)}{1 + G_{gp}(s)G_{PI}(s)G_d(s)} \quad (9.42)$$

Figure 9.7 shows the suppression capability of the GSC on the grid voltage and rotor-side DC current disturbances. Similar to the RSC, with an increase of current

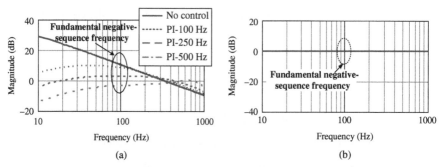

Figure 9.7   Suppression capability of the GSC on grid voltage and rotor-side DC current disturbances: (a) Suppression of grid voltage disturbance $G_{vig}(s)$; (b) suppression of rotor-side DC current $G_{rc}(s)$.

loop bandwidth, the suppression capability of negative-sequence grid voltage disturbance is improved to some extent. However, the suppression is still relatively low when the bandwidth is increased to values as large as 500 Hz. As shown in Figure 9.7b and equation (9.40), the current control loop cannot suppress the disturbance of rotor-side DC current, and the magnitude–frequency characteristic is the 0 dB line. Therefore, the conventional current control scheme is not able to mitigate the impact of the negative-sequence grid voltage on the AC current of the GSC and the DC-bus capacitor current.

Moreover, the DC voltage outer loop shown in Figure 9.6 also contains the disturbance terms v_{sdq} and i_{rdc}, thus the suppression capability of the DC voltage outer loop on the rotor-side DC current disturbance i_{rdc} must be considered. Substituting the inner current loop with the expression given in equation 9.35, the block diagram of DC voltage outer loop can be depicted, as shown in Figure 9.8.

In Figure 9.8, $G_c(s) = 1/Cs$. The transfer functions from the disturbance terms i_{rdc} and v_{sdq} to the capacitor current can be derived according to Figure 9.8:

$$G_{rc\_o}(s) = \frac{i_{cap}(s)}{i_{rdc}(s)} = \frac{G_{rc}}{1 + G_{PI}(s)G_{gc}(s)G_c(s)} \tag{9.43}$$

$$G_{vc\_o}(s) = \frac{i_{cap}(s)}{v_{sd}(s)} = \frac{G_{vc}}{1 + G_{PI}(s)G_{gc}(s)G_c(s)} \tag{9.44}$$

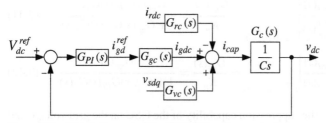

Figure 9.8   Block diagram of the GSC DC voltage outer loop.

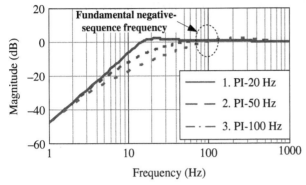

Figure 9.9   Suppression capability of voltage outer loop shown in Figure 9.6 on the disturbances.

With the introduction of the DC voltage outer loop, compared to the original current loop, the damping of disturbance is increased by a factor of $\left(1 + G_{PI}(s)G_{gc}(s)G_c(s)\right)$. Therefore, the suppression capability of the voltage loop on the disturbances is determined by the following equation:

$$G_{vcl}(s) = \frac{1}{1 + G_{PI}(s)G_{gc}(s)G_c(s)} \tag{9.45}$$

Usually, the crossover frequency of the outer loop is lower than 1/5 of the inner loop. If the crossover frequency of the inner current loop is set to 100 Hz, 250 Hz, and 500 Hz, the appropriate outer loop crossover frequency should be 20 Hz, 50 Hz, and 100 Hz, respectively. Figure 9.9 shows the magnitude–frequency characteristics of $G_{vcl}(s)$ under these three crossover frequencies. As shown in Figure 9.8, the outer loop has good suppression for low frequency disturbances below 10 Hz. However, at the fundamental negative-sequence frequency of 100 Hz, $|G_{vcl}(s)|$ is near 0 dB, which means the voltage outer loop is not able to suppress the disturbances with the frequency of 100 Hz. In other words, the voltage outer loop utilizing the PI controller does not have any suppression capability on the disturbances caused by unbalanced grid voltage.

9.4.3   DC-Bus Capacitor Current and Voltage

Considering the effects of control, the actual output voltages of the GSC and the RSC contain negative-sequence components. According to the definitions of power flow shown in Figure 9.3, the actual active powers flowing into the GSC and RSC are [6, 7]:

$$P_g = 1.5 \, \text{Re} \left(v_{gdq} i_{gdq}^* \right) = 1.5 \, \text{Re} \left(v_{gdq}^{(+1)} i_{gdq}^{(+1)*} + v_{gdq}^{(-1)} i_{gdq}^{(-1)*} \right)$$

$$+ 1.5 \, \text{Re} \left(v_{gdq}^{(+1)} i_{gdq}^{(-1)*} e^{j2\omega_s t} + v_{gdq}^{(-1)} i_{gdq}^{(+1)*} e^{-j2\omega_s t} \right)$$

$$= P_g^{(0)} + P_g^{(2)} \tag{9.46}$$

$$P_r = 1.5 \, \mathrm{Re} \left(v_{rdq} i_{rdq}^* \right) = 1.5 \, \mathrm{Re} \left(v_{rdq}^{(+1)} i_{rdq}^{(+1)*} + v_{rdq}^{(-1)} i_{rdq}^{(-1)*} \right)$$

$$+ 1.5 \, \mathrm{Re} \left(v_{rdq}^{(+1)} i_{rdq}^{(-1)*} e^{j2\omega_s t} + v_{rdq}^{(-1)} i_{rdq}^{(+1)*} e^{-j2\omega_s t} \right)$$

$$= P_r^{(0)} + P_r^{(2)} \tag{9.47}$$

Then the active power flowing into the DC-bus capacitor can be written as

$$P_{cap} = P_g^{(2)} - P_r^{(2)}$$

$$= 1.5 \, \mathrm{Re} \left(v_{gdq}^{(+1)} i_{gdq}^{(-1)*} e^{j2\omega_s t} + v_{gdq}^{(-1)} i_{gdq}^{(+1)*} e^{-j2\omega_s t} \right)$$

$$- 1.5 \, G_{rc}(-j2\omega_s) \, \mathrm{Re} \left(S_l v_{sdq}^{(+1)} i_{rdq}^{(-1)*} e^{j2\omega_s t} + v_{rdq}^{(-1)} i_{rdq}^{(+1)*} e^{-j2\omega_s t} \right) \tag{9.48}$$

It should be stated that since the value of $G_{rc}(-j2\omega_s)$ is mostly a real number, it can be directly multiplied with the active power that is flowing into the DC capacitor through the RSC, as shown in equation (9.48). In equation (9.48), the terms $v_{gdq}^{(+1)}$, $i_{gdq}^{(+1)}$, and $i_{rdq}^{(+1)}$ are determined by the conditions of the grid and wind turbine which are independent of the controllability of the voltage unbalance of the converters. However, the negative-sequence output voltages of the converters $v_{gdq}^{(-1)}$, $v_{rdq}^{(-1)}$ and the negative-sequence currents $i_{gdq}^{(-1)}$, $i_{rdq}^{(-1)}$ are closely related to the converters' controllability of the negative-sequence components.

According to the RSC control scheme shown in Figure 9.4, the transfer function from the grid voltage to the rotor current can be written as

$$G_{vir}(s) = \frac{i_{rdq}(s)}{v_{sdq}(s)} = -\frac{G_1(s)G_2(s)G_p(s)}{1 + G_{PI}(s)G_d(s)G_p(s)} \tag{9.49}$$

With the introduction of closed-loop control, the negative-sequence currents $i_{gdq}^{(-1)}$, $i_{rdq}^{(-1)}$ can be obtained with the following equations:

$$\begin{cases} i_{gdq}^{(-1)} = v_{sdq}^{(-1)} G_{vig}(-j2\omega_s) \\ i_{rdq}^{(-1)} = v_{sdq}^{(-1)} G_{vir}(-j2\omega_s) \end{cases} \tag{9.50}$$

According to the block diagrams of RSC and GSC control systems shown in Figures 9.4 and 9.6, and considering that there is no negative-sequence component in the current reference in the conventional control scheme, the negative-sequence output voltage of the GSC and the RSC can be obtained by the following equations:

$$\begin{cases} v_{gdq}^{(-1)} = \left\{ -G_{PI}^g(s)G_d(s) \left(i_{gdq}^{ref}(s) - i_{gdq}(s) \right) \right\}_{s=-j2\omega_s} \\ \qquad = G_{PI}^g(-j2\omega_s)G_d(-j2\omega_s)i_{gdq}^{(-1)} \\ v_{rdq}^{(-1)} = \left\{ G_{PI}^r(s)G_d(s) \left(i_{rdq}^{ref}(s) - i_{rdq}(s) \right) \right\}_{s=-j2\omega_s} \\ \qquad = -G_{PI}^r(-j2\omega_s)G_d(-j2\omega_s)i_{rdq}^{(-1)} \end{cases} \tag{9.51}$$

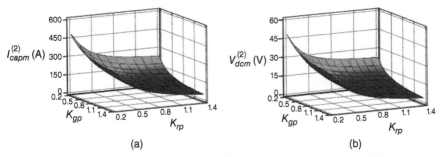

Figure 9.10   Magnitudes of capacitor current (a) and second-order harmonic DC voltage (b) when the voltage unbalance factor is 20% and slip ratio is 0.2.

In equation (9.51), to distinguish the PI controllers of GSC and RSC, the superscript "g" stands for the GSC and the superscript "r" stands for the RSC. Substituting (9.50) in (9.51), the following equations can be derived:

$$
\begin{cases}
v_{gdq}^{(-1)} = G_{PI}^{g}(-j2\omega_s)G_d(-j2\omega_s)G_{vig}(-j2\omega_s)v_{sdq}^{(-1)} \\
v_{rdq}^{(-1)} = -G_{PI}^{r}(-j2\omega_s)G_d(-j2\omega_s)G_{vir}(-j2\omega_s)v_{sdq}^{(-1)}
\end{cases}
\tag{9.52}
$$

Substituting (9.50) and (9.52) in (9.48):

$$
\begin{aligned}
P_{cap} = 1.5\,\mathrm{Re}\Big\{& v_{sdq}^{(+1)}v_{sdq}^{(-1)*}G_{vig}^{*}(-j2\omega_s)e^{j2\omega_s t} \\
&+ \Big(i_{gdq}^{(+1)*}v_{sdq}^{(-1)}G_{PI}^{g}(-j2\omega_s)G_d(-j2\omega_s) \Big)\,G_{vig}(-j2\omega_s)e^{-j2\omega_s t} \Big\} \\
-1.5G_{rc}(-j2\omega_s)\,\mathrm{Re}\Big\{& v_{sdq}^{(+1)}v_{sdq}^{(-1)*}S_l G_{vir}^{*}(-j2\omega_s)e^{j2\omega_s t} \\
&- \Big(i_{rdq}^{(+1)*}v_{sdq}^{(-1)}G_{PI}^{r}(-j2\omega_s)G_d(-j2\omega_s) \Big) \\
&\times G_{vir}(-j2\omega_s)e^{-j2\omega_s t} \Big\}
\end{aligned}
\tag{9.53}
$$

In equation (9.53), the suppression capability of the GSC on the negative-sequence disturbances $G_{vig}(-j2\omega_s)$ and $G_{rc}(-j2\omega_s)$ are determined by the control parameters of GSC, and the suppression capability of the RSC on the negative-sequence disturbances $G_{vir}(-j2\omega_s)$ is determined by the control parameters of RSC. Obviously, if $G_{vig}(-j2\omega_s)$ and $G_{rc}(-j2\omega_s)$ can provide good damping characteristics, the active power flowing into the capacitor P_{cap} can be largely reduced.

The relationship between the second-order harmonic current of the DC capacitor and P_{cap} can be expressed as

$$
I_{capm}^{(2)} = \frac{\left| P_{cap} \right|}{V_{dc}}
\tag{9.54}
$$

When the voltage unbalance factor is 20%, Figures 9.10–9.12 give the relationship between the magnitudes of capacitor current and second-order harmonic DC

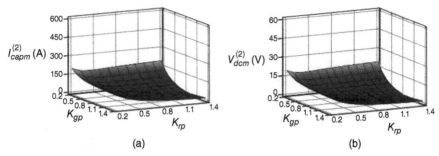

Figure 9.11 Magnitudes of capacitor current (a) and second-order harmonic DC voltage (b) when voltage unbalance factor is 20% and slip ratio is 0.

voltage with the control parameters when the slip ratios are 0.2, 0, and − 0.2, respectively. In the figures, K_{gp} and K_{rp} are the proportional gains in the current PI controllers of the GSC and RSC. The values of K_{gp} and K_{rp} are ranging from 0.2 to 1.4, meaning that the open-loop bandwidth of the current loop ranges from 70 to 500 Hz. As shown in Figures 9.10–9.12, when the DFIG is operating under a sub-synchronous condition with a slip ratio of 0.2 and under a super-synchronous condition with a slip ratio of − 0.2, the magnitudes of the capacitor current and DC voltage fluctuation are quite large. At synchronous speed, since $S_l = 0$, the positive-sequence rotor voltage, the RSC, and GSC active power are all zero. The second-order harmonic current in the capacitor is only introduced by the product of negative-sequence current and a positive voltage of the GSC and the product of positive-sequence current and negative-sequence voltage of the RSC; therefore the magnitudes of the capacitor current and DC voltage fluctuations are relatively small at synchronous speed. As it can be seen from Figures 9.10–9.12, under unbalanced grid conditions, the DFIG system with conventional vector control experiences large second-order harmonic current in the DC-bus capacitor which might increase the losses in the capacitor, increase the core temperature of the capacitor, and significantly shorten the lifetime.

9.5 SUMMARY

According to the analysis given in this chapter, the DFIG system under unbalanced grid condition suffers from the second-order fluctuation in the electromagnetic

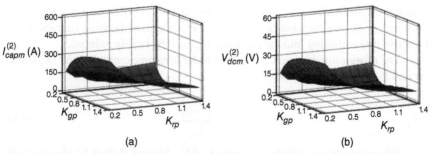

Figure 9.12 Magnitudes of capacitor current (a) and second-order harmonic DC voltage (b) when voltage unbalance factor is 20% and slip ratio is − 0.2.

torque, second-order harmonic current in DC-bus capacitor, second-order fluctuation in DC-bus voltage, and third-order harmonics in the grid current. These negative effects may cause significant lifetime reduction of mechanical components and the DC-bus capacitor, as well as deterioration of the power quality. As shown in the analysis given in this chapter, if the conventional current control scheme is adopted for the DFIG system, the system impedance in the fundamental negative-sequence voltage is not largely increased, the second-order fluctuations in the electromagnetic torque, DC voltage, and DC capacitor current are not significantly mitigated. Therefore, the compatibility of the DFIG system to the unbalanced grid condition does not meet the requirements if the conventional current control scheme is applied. Further improvements on the control algorithm have to be made.

REFERENCES

[1] L. Xu and Y. Wang, "Dynamic modeling and control of DFIG-based wind turbines under unbalanced network conditions," *IEEE Trans. Power Syst.*, vol. 22, no. 1, pp. 314–323, 2007.

[2] J. López, E. Gubía, P. Sanchis, X. Roboam, and L. Marroyo, "Wind turbines based on doubly fed induction generator under asymmetrical voltage dips," *IEEE Trans. Energy Convers.*, vol. 23, no. 1, pp. 321–330, 2008.

[3] L. Moran, P. Ziogas, and G. Joos, "Design aspects of synchronous PWM rectifier-inverter systems under unbalanced input voltage conditions," *IEEE Trans. Ind. Appl.*, vol. 28, no. 6, pp. 1286–1293, 1992.

[4] A. Luna, K. Lima, F. Corcoles, E. Watanabe, P. Rodriguez, and R. Teodorescu, "Control of DFIG-WT under unbalanced grid voltage conditions," in *Proc. IEEE Energy Convers.* Congr. Exposition, San Jose, CA, 2009, pp. 370–377.

[5] T. K. A. Brekken and N. Mohan, "Control of a doubly fed induction wind generator under unbalanced grid voltage conditions," *IEEE Trans. Energy Convers.*, vol. 22, no. 1, pp. 129–135, 2007.

[6] M. I. Martinez, G. Tapia, A. Susperregui, and H. Camblong, "DFIG power generation capability and feasibility regions under unbalanced grid voltage conditions," *IEEE Trans. Energy Convers.*, vol. 26, no. 4, pp. 1051–1062, 2011.

[7] Y. Zhou, P. Bauer, J. Ferreira, and J. Pierik, "Operation of grid-connected DFIG under unbalanced grid voltage condition," *IEEE Trans. Energy Convers.*, vol. 24, no. 1, pp. 240–246, 2009.

[8] J. Hu and Y. He, "Modeling and control of grid-connected voltage-sourced converters under generalized unbalanced operation conditions," *IEEE Trans. Energy Convers.*, vol. 23, no. 3, pp. 903–913, 2008.

[9] A. G. Abo-Khalil, D.-C. Lee, and J.-I. Jang, "Control of back-to-back PWM converters for DFIG wind turbine systems under unbalanced grid voltage," in *Proc. IEEE Int.* Symp. Ind. Electron., Vigo, Spain, 2007, pp. 2637–2642.

[10] A. Yazdani and R. Iravani, "A unified dynamic model and control for the voltage-sourced converter under unbalanced grid conditions," *IEEE Trans. Power Deliv.*, vol. 21, no. 3, pp. 1620–1629, 2006.

[11] L. Xu, "Coordinated control of DFIG's rotor and grid side converters during network unbalance," *IEEE Trans. Power Electron.*, vol. 23, no. 3, pp. 1041–1049, 2008.

[12] J. Hu, Y. He, L. Xu, and B. W. Williams, "Improved control of DFIG systems during network unbalance using PI–R current regulators," *IEEE Trans. Ind. Electron.*, vol. 56, no. 2, pp. 439–451, 2009.

[13] D. Santos-Martin, J. L. Rodriguez-Amenedo, and S. Arnalte, "Direct power control applied to doubly fed induction generator under unbalanced grid voltage conditions," *IEEE Trans. Power Electron.*, vol. 23, no. 5, pp. 2328–2336, 2008.

[14] E. Muljadi, D. Yildirim, T. Batan, and C. P. Butterfield, *"Understanding the unbalanced-voltage problem in wind turbine generation,"* in *Conf. Record IEEE Ind. Appl. Conf.*, vol. 2, Phoenix, AZ, 1999, pp. 1359–1365.

[15] S. Wangsathitwong, S. Sirisumrannukul, S. Chatratana, and W. Deleroi, "Symmetrical components-based control technique of doubly fed induction generators under unbalanced voltages for reduction of torque and reactive power pulsations," in *Proc. Int. Conf. Power Electron. Drive Syst.*, Bangkok, Thailand, 2007, pp. 1325–1330.

[16] P. Xiao, K. A. Corzine, and G. K. Venayagamoorthy, "Multiple reference frame-based control of three-phase PWM boost rectifiers under unbalanced and distorted input conditions," *IEEE Trans. Power Electron.*, vol. 23, no. 4, pp. 2006–2017, 2008.

[17] A. H. Ghorashi, S. S. Murthy, B. P. Singh, and B. Singh, "Analysis of wind driven grid connected induction generators under unbalanced grid conditions," *IEEE Trans. Energy Convers.*, vol. 9, no. 2, pp. 217–223, 1994.

[18] Y. Suh, "Analysis and control of three phase AC/DC PWM converter under unbalanced operating conditions," Ph.D. thesis, University of Wisconsin, Madison, WI, 2004.

[19] R. Cardenas, R. Pena, S. Alepuz, and G. Asher, "Overview of control systems for the operation of DFIGs in wind energy applications," *IEEE Trans. Ind. Electron.*, vol. 60, no. 7, pp. 2776–2798, Jul. 2013.

[20] G. Abad, M. Á. Rodríguez, G. Iwanski, and J. Poza, "Direct power control of doubly-fed-induction-generator-based wind turbines under unbalanced grid voltage," *IEEE Trans. Power Electron.*, vol. 25, no. 2, pp. 442–452, Feb. 2010.

[21] O. Gomis-Bellmunt, A. Junyent-Ferré, A. Sumper, and J. Bergas-Jané, "Ride-through control of a doubly fed induction generator under unbalanced voltage sags," *IEEE Trans. Energy Convers.*, vol. 23, no. 4, pp. 1036–1045, Dec. 2008.

[22] P. S. Flannery and G. Venkataramanan, "Unbalanced voltage sag ride-through of a doubly fed induction generator wind turbine with series grid-side converter," *IEEE Trans. Ind. Appl.*, vol. 45, no. 5, pp. 1879–1887, 2009.

[23] R. Pena, R. Cardenas, E. Escobar, J. Clare, and P. Wheeler, "Control system for unbalanced operation of stand-alone doubly fed induction generators," *IEEE Trans. Energy Convers.*, vol. 22, no. 2, pp. 544–545, Jun. 2007.

[24] J. Hu and Y. He, "Reinforced control and operation of DFIG-based wind-power-generation system under unbalanced grid voltage conditions," *IEEE Trans. Energy Convers.*, vol. 24, no. 4, pp. 905–915, Dec. 2009.

CHAPTER 10

CONTROL OF DFIG WIND POWER SYSTEM UNDER UNBALANCED GRID VOLTAGE

10.1 INTRODUCTION

Many power system operators have issued grid codes that require the grid-connected wind turbine to withstand certain voltage disturbances, such as transient voltage unbalance, without tripping [1–4]. Some grid codes even require the wind turbine system to operate properly with higher than 2% of steady-state grid voltage unbalance. In China, standards issued in 2012 require that large-scale wind turbines should withstand a steady-state voltage unbalance of 2% and a short-time voltage unbalance of 4% without tripping [5], which can also be found in some other international standards, for example, N-50160 [6]. In order to meet these requirements, the wind turbine systems must continuously develop and improve their performances.

In this chapter, the control schemes for DFIG wind power systems under unbalanced grid voltage are discussed. Resonant controllers will be used to resolve the problems caused by unbalanced grid voltage: stator current unbalance and DC voltage fluctuation of the back-to-back converter.

10.2 CONTROL TARGETS

As demonstrated in the previous chapter, grid voltage unbalance causes numerous troubles in the DFIG system. Since the stator of a DFIG is directly connected to the grid, a negative-sequence component is added to the stator flux under unbalanced grid voltage conditions. As a consequence, large negative-sequence currents flow through the stator and the rotor of the DFIG, which causes a significant second-order harmonic fluctuation in the electromagnetic torque and active and reactive powers are also fluctuating in the stator and the rotor. The torque fluctuation causes fatigue in the mechanical parts such as the gearbox and the shaft. In addition, the active power fluctuations, which flow through the DC link capacitors from both GSC and RSC cause

Advanced Control of Doubly Fed Induction Generator for Wind Power Systems, First Edition.
Dehong Xu, Frede Blaabjerg, Wenjie Chen, and Nan Zhu.

a large second-order harmonic current in the DC capacitors as well as voltage ripples in the DC link. It results in higher power losses in the DC link capacitors and higher operation temperature, which will shorten the lifetime of the electrolytic capacitors.

Therefore, to enhance the performance of DFIG system under unbalanced grid condition, the following control targets can be adopted [17–19, 23]:

1. Constant output active power: eliminate the second-order fluctuations in active power;

2. Balanced rotor current: eliminate the negative-sequence component in the rotor current;

3. Balanced stator current: eliminate the negative-sequence component in the stator current;

4. Constant electromagnetic torque: eliminate the second-order fluctuations in the electromagnetic torque.

During the operation of the DFIG system, the four control targets introduced above can be selected according to the specific requirements.

10.3 STATOR CURRENT CONTROL WITH RESONANT CONTROLLER

Many grid codes for grid-connected DFIG system have required that under unbalanced grid voltage dip, the DFIG system has to remain connected to the grid and provide reactive current [7]. For example, the Spanish grid operator Red Eléctrica requires the DFIG system to provide reactive current to help to recover the grid voltage under unbalanced grid voltage dip [8]. To achieve this, the negative-sequence reactive current should be controlled to zero. Under such conditions, the control target of eliminating the negative-sequence component in the stator current can be adopted [20–22].

10.3.1 Control Scheme

Figure 10.1 shows the stator current balancing control scheme based on the resonant controller. On the basis of the conventional stator current vector control scheme, a stator current balancing control loop is added to control the negative-sequence stator current with the second-order resonant controller. The control scheme introduced in Figure 10.1 is a closed-loop control scheme which is independent of the motor parameters and negative-sequence component of the grid voltage.

10.3.2 Analysis of the Controller

Combining the stator current harmonics control introduced in Chapter 8 with the stator current balancing control, the resonant controller consists of the second-order

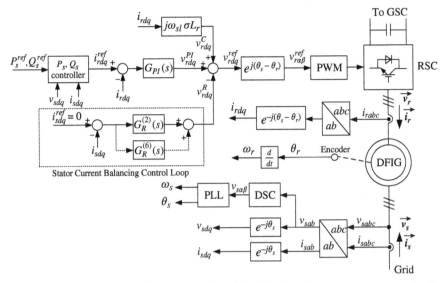

Figure 10.1   Stator current balancing control scheme based on the resonant controller for the rotor-side converter of a DFIG wind power system.

resonant controller and the sixth-order resonant controller (shown in the dashed box in Figure 10.1), of which the transfer function is given by the following equation [14]:

$$G_R(s) = G_R^{(2)}(s) + G_R^{(6)}(s) = \frac{2K_r^{(2)}\omega_c s}{s^2 + 2\omega_c s + (2\omega_s)^2} + \frac{2K_r^{(6)}\omega_c s}{s^2 + 2\omega_c s + (6\omega_s)^2} \quad (10.1)$$

According to the open-loop transfer function given in equation (8.30), the open-loop frequency characteristics of the RSC current loop can be depicted as shown in Figure 10.2. To depict the different figures, the parameters of the second-order resonant controller are chosen as $K_r^{(2)} = 20$ and $\omega = 5 \text{rad/s}$, and the parameters of the sixth-order resonant controller are $K_r^{(6)} = 20$ and $\omega = 5 \text{rad/s}$. Figure 10.2a shows the open-loop frequency characteristics with only the stator current balancing control, that is, $G_R(s) = G_R^{(2)}(s)$. It can be seen that since the resonant frequency is far away from the crossover frequency, the second-order resonant controller does not have much impact on the system phase margin, thus its influence on the system stability can be neglected. Figure 10.2b shows the open-loop frequency characteristics for both the stator current balancing control and stator current harmonics control. Under such conditions, very large controller gains of the current loop appear at frequencies of 100 Hz (the frequency of the negative-sequence stator current in dq frame) and 300 Hz (the frequency of fifth- and seventh-order stator current harmonics in dq frame). Such a feature ensures good control performance of negative-sequence stator current and stator current harmonics. As shown in Figure 10.3, with the introduction of resonant controllers, the damping of the negative-sequence grid voltage increases from 8 to $- 20$ dB, and the damping of the fifth- and seventh-order harmonics increases from 3 to $- 24$ dB.

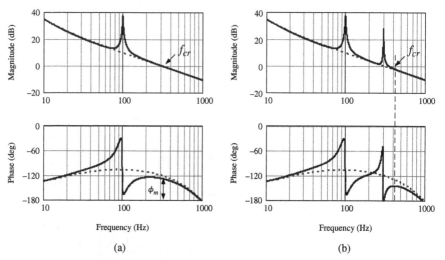

(a)   (b)

Figure 10.2   Open-loop frequency characteristics of the RSC current loop: (a) With only the balanced stator current control; (b) with the balanced stator current control and stator current harmonics control.

By adopting the stator current balancing control, the impedance of the DFIG system on fundamental negative-sequence voltage becomes

$$X_{vis}(-j2\omega_s) = \frac{1}{G_{vis}(-j2\omega_s)}$$

$$= \frac{1 + G_{PI}(s)G_d(s)G_p(s) + G_R(s)G_d(s)G_p(s)L_m/L_s}{G_1(s)/L_s + G_1(s)G_{PI}(s)G_d(s)G_p(s)/L_s + G_1(s)G_2(s)G_p(s)L_m/L_s}\Bigg|_{s=-j2\omega_s}$$

$$(10.2)$$

Compared to the expression without stator current balancing control, the numerator of equation (10.2) has an extra term $G_R(s)G_d(s)G_p(s)L_m/L_s$ which is

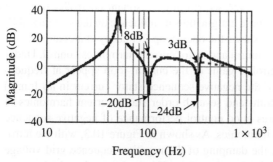

Figure 10.3   Suppression capability of the stator current resonant control on grid voltage disturbances.

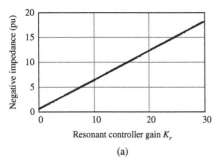 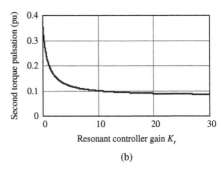

Figure 10.4   Relation between the control performance of DFIG system and the resonant controller gain K_r: (a) System impedance in fundamental negative-sequence voltage; (b) second-order fluctuation in the electromagnetic torque.

determined by the resonant controller. Thus, as long as the gain of $G_R(s)$ at 100 Hz is large enough, the system will have sufficient impedance in the negative-sequence voltage. The negative-sequence impedance and the electromagnetic torque under various gain K_r of the resonant controller can be calculated as shown in Figure 10.4. As shown in Figure 10.4, with the increase of K_r, the system impedance in the fundamental negative-sequence voltage increases linearly, and the second-order fluctuation in the electromagnetic torque decreases. For example, when K_r increases from 0 to 20, $|X_{vis}(-j2\omega_s)|$ increases from 0.5 to 12 pu and the amplitude of the electromagnetic torque second-order fluctuation decreases from 0.25 to 0.05 pu.

Therefore, the adoption of stator current balancing control significantly increases the system impedance in the fundamental negative-sequence voltage and largely suppresses the electromagnetic torque fluctuation and enhances the compatibility of the DFIG system under unbalanced grid condition.

10.3.3   Experiment and Simulation Results

To verify the effectiveness of resonant-controller-based stator current balancing control method under unbalanced grid conditions, the experimental analysis is conducted on a 30 kW DFIG system testbed of which the detailed explanation about the system will be given in "Part V: DFIG Test Bench" of this book. The PI controller parameters of the RSC current control loop are $K_p = 0.36$ and $K_i = 30$. The second-order resonant controller of the stator current balancing control is set as $K_r = 20$ and $\omega_c = 5$ rad/s. In the experiments, phases A and B of the DFIG system are directly connected to the 230 V grid, and phase C is connected to the grid via a transformer which produces an 11% voltage dip for phase C causing a 4% voltage unbalance. The three-phase line-to-line voltages are shown in Figure 10.5.

The output active power for the steady-state experiment is 15 kW, while the reactive power is set to zero, and the rotor speed is kept at 1200 r/min (0.8 pu). The steady-state test results are shown in Figure 10.6. In the test results of the traditional current vector control shown in Figure 10.6a, it can be seen that unbalanced grid voltage not only causes unbalance in stator currents, which may lead to uneven thermal

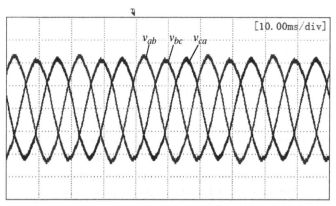

Figure 10.5   Three-phase line-to-line voltage when the voltage unbalance factor is 4% on a 30 kW DFIG test system.

distribution among the windings but also causes large second-order fluctuation in the electromagnetic torque, which affects the lifetime of mechanical components of the DFIG system. When the stator current balancing control is applied, the impedance of the control system in the negative-sequence grid voltage increases significantly and eliminates the negative-sequence component of stator current and the fluctuations in

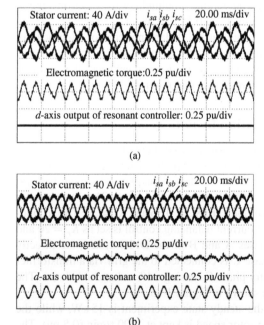

Figure 10.6   Steady-state test results of traditional vector control and stator current balancing control under 4% of grid voltage unbalance: (a) With traditional vector control; (b) with stator current balancing control.

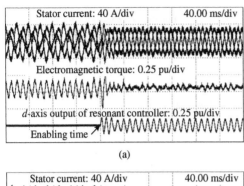

(a)

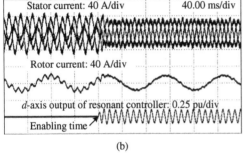

(b)

Figure 10.7   Transient response of the DFIG when the stator current balancing control is enabled: (a) Stator current and electromagnetic torque; (b) Stator current and rotor current.

the electromagnetic torque, as shown in Figure 10.6b. The stator current balancing control provides a good solution for the negative effects caused by unbalanced grid voltage on the lifetime of the DFIG and enhancing the compatibility of the DFIG system in the unbalanced grid condition.

Figure 10.7 shows the transient response of the DFIG when the stator current balancing control is enabled, where Figure 10.7a shows the waveforms of three-phase stator current and the electromagnetic torque, and Figure 10.7b shows the waveforms of three-phase stator and rotor currents.

As shown in Figure 10.7, before the stator current balancing control is enabled, severe unbalance is present in the three-phase stator current, and the electromagnetic torque experiences large fluctuations, and large harmonics can be observed in the rotor current. After switching to the stator current balancing control, the system impedance for the negative-sequence voltage increases significantly and thus the negative-sequence currents are suppressed, leading to balanced stator currents, reduced electromagnetic torque fluctuation, and reduced rotor current harmonics. During the transition, the output of the resonant controller does not experience large oscillations and thus a smooth transition of the control scheme is achieved. The results of transient experiments further verify that the stator current balancing control is able to improve the performance of the DFIG system under unbalanced grid condition.

To verify the effectiveness of stator current balancing control under unbalanced grid voltage dip, simulations are done on a 1.5 MW DFIG system. Figure 10.8 shows

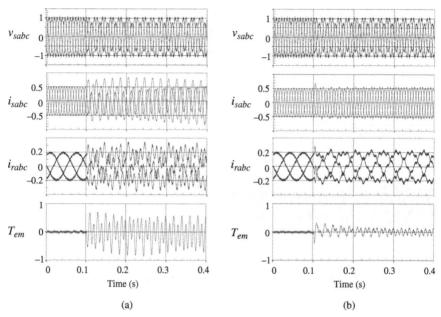

(a)   (b)

Figure 10.8 Simulation results for a 1.5 MW DFIG system using traditional vector control and stator current balancing control when 30% single-phase voltage dip happens at 0.1 s. From top to bottom: three-phase grid voltage, three-phase stator current, three-phase rotor current and electromagnetic torque: (a) With traditional vector control; (b) with stator current balancing control.

the transient responses of the DFIG system when a 30% single-phase voltage dip happens at 0.1 s.

In Figure 10.8a, when the traditional vector control is adopted, large negative-sequence stator current and electromagnetic torque fluctuation are induced after the voltage dip. While in Figure 10.8b, the stator current balancing control is adopted and the impact of unbalanced voltage dip on the stator current and the electromagnetic torque is reduced significantly, and the transition is smooth, which is beneficial for the DFIG system when reacting to the unbalanced grid dip. Moreover, the simulation also verifies that the introduction of stator current balancing control does not have much influence on the stability of the overall DFIG system.

10.4 DC VOLTAGE FLUCTUATION CONTROL BY GSC

According to the analysis given in Chapter 9, under unbalanced grid voltage, a large second-order harmonic current is flowing through the DC bus capacitor causing fluctuations in the DC bus voltage [15, 16, 24]. The fluctuations in the DC bus voltage cause the RSC and GSC to interact through the common DC bus and may destroy the decoupled control of the RSC and GSC.

In this section, the control scheme of the DC bus voltage and capacitor current under unbalanced grid will be discussed and considering the requirements of stability and modular design.

10.4.1 Challenges in the Control of GSC

DC capacitors and switching power devices are the critical components of a power converter. The performance of these devices strongly influences the reliability of the DFIG system and 20 years lifetime is required for the DFIG wind power systems, which demands the improvement of the reliability of DC capacitor to meet such requirement. The major cause of the reduction of lifetime and decline of performance of DC capacitors is the vaporization of electrolyte, which can be accelerated by high temperatures. Take the GXR3 series electrolytic capacitor [9], from Hitachi and widely used in wind power converters, as an example, the relationship between lifetime and operating temperature is given as [9]

$$L = L_0 \times 2^{\frac{T_0 - T}{10}} \times \left(\frac{V_R}{V} \right)^{2.5} \tag{10.3}$$

where L is the estimated lifetime, L_0 is the lifetime when operating at the maximum core temperature and rated ripple current, T_0 is the maximum operating core temperature, T is the actual operating core temperature, V_R is the reference operating voltage, and V is the actual operating voltage. As shown in equation (10.3), the lifetime of the electrolytic capacitor decreases exponentially with the increase of operating temperature.

The operating core temperature of the electrolytic capacitor can be estimated by the following equation [9]:

$$T_{cap} = T_a + (T_0 - T_a) \times \left(\frac{I_{cap}}{I_R} \right)^2 \tag{10.4}$$

where T_a is the ambient temperature, I_R is the rated current under the maximum temperature and I_{cap} is the actual current flowing through the capacitor.

Under unbalanced grid condition, the current flowing through the DC capacitor consists of high frequency switching ripple current and second-order harmonic current. The equivalent RMS value of the switching ripple current on the DC side of a single converter can be calculated using the following equation [10]:

$$I_{pwm} = I_{rms} \sqrt{2M \left[\frac{\sqrt{3}}{4\pi} + \left(\frac{\sqrt{3}}{\pi} - \frac{9}{16}M \right) \cos^2 \varphi_{ui} \right]} \tag{10.5}$$

where I_{rms} is the RMS value of converter output current, M is the voltage modulation ratio of the power converter, which is defined as $M = \frac{V_m}{V_{dc}/2}$, and φ_{ui} is the power

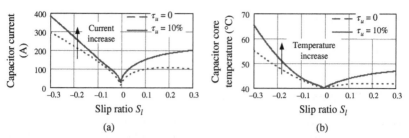

Figure 10.9   Capacitor current and capacitor core temperature when the unbalance factor of the grid voltage is 20%: (a) Equivalent capacitor current, converted to 100 Hz; (b) Capacitor core temperature, ambient temperature 40 °C.

factor angle of the converter. Then the equivalent RMS value of capacitor current caused by both the GSC and RSC can be expressed as [10]

$$I_{cpwm} = \sqrt{I^2_{gpwm} + I^2_{rpwm}} \qquad (10.6)$$

The currents of different frequencies have different influences on the wear and tear of the capacitor. With the same RMS values, low frequency components have larger effects than high frequency components. When evaluating the capacitor temperature, the usually adopted reference frequency of the currents is 120 Hz or 100 Hz by most capacitor manufacturers. In this example, the frequency of the switching ripple current is 2 kHz, and provided by the guide book of GXR3 capacitor, the equivalent RMS value of the switching component when converted to 120 Hz can be calculated as [9]

$$I'_{cpwm} = \frac{I_{cpwm}}{1.3} \qquad (10.7)$$

Under unbalanced grid, the magnitude of the equivalent second-order harmonic current flowing through the capacitor is [10]

$$I_{cap} = I^{(2)}_{cap} + I'_{cpwm} \qquad (10.8)$$

According to equations (10.7) and (10.8) and combining them with the steady-state relations of the DFIG introduced in the previous chapters, the relations of the capacitor current and core temperature to the slip ratio S_l can be depicted as Figure 10.9, where the voltage unbalance factor τ_u was defined in Chapter 9 by equation (9.2), the dashed lines represent the situation under balanced grid ($\tau_u = 0$) while the solid lines represent the results under unbalanced grid with the unbalance factor of 20%. Under balanced grid, the capacitor current only consists of the switching components. It is worth to notice that under balanced grid, when the slip ratio is zero, that is, when the DFIG is rotating at synchronous speed, only zero vectors are effective for RSC and the active current of GSC is zero, and since the RMS capacitor current is not relevant with zero vectors, the RMS value of the capacitor current is also zero.

Compared to the situation under the balanced grid, under unbalanced grid voltage, the capacitor current and core temperature are both increased. Since the lifetime

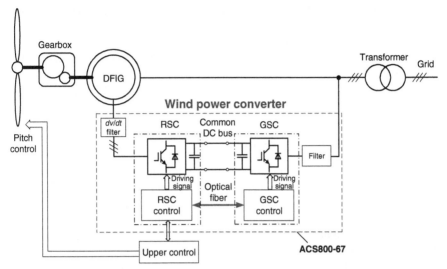

Figure 10.10   Architecture of a back-to-back converter based on a modular design from ABB ACS800-67 series [11].

of the electrolytic capacitor decreases exponentially with the increase of core temperature, unbalanced grid conditions may accelerate the wear-out of the DC capacitors. Thus, from the perspective of enhancing the reliability of DC capacitors, reducing second-order harmonic current in the DC capacitor is required. By over-rating the DC bus capacitors, the fluctuation in the DC voltage can be mitigated, however at the cost of higher material expense and higher weight. It would be desirable to develop an improved control scheme to suppress the DC voltage fluctuations without the necessity of increasing the DC bus capacitance.

10.4.1.1   Requirement of Modular Design

In wind power systems of the MW scale, based on the requirements of reliability, power-scaling extensibility and ease of maintenance, wind power converters usually adopt modular design structure, for example, ACS800-67 series from ABB [11], Prowind series from Converteam [12], and PM3000 series from AMSC [13]. Figure 10.10 shows the hardware architecture of the ACS800-67 converter designed by ABB for MW-scale DFIG wind power generation. The typical modular structure is adopted by the ACS800-67 converter, which is composed of two independent converter modules, the GSC and the RSC, and each of the converter modules consists of its own power circuit and control unit. The DC terminals of the power circuits of the two converter modules are connected by a common DC bus. The control units of the RSC and GSC are connected by a double strand optical fiber.

In the modular design represented by the ACS800-67, the control units of the GSC and RSC are independent of each other and the communication between the two controllers consist only of the transfer of on/off commands and operation status word to coordinate the controllers. Such feature requires the control units of GSC and RSC to use not only independent sets of hardware but also independent and decoupled control algorithms.

According to the analysis given in Chapter 9, under unbalanced grid voltage, the fluctuation of the DC voltage is caused by the power fluctuations of both the GSC and the RSC. Thus, from the perspective of suppressing DC voltage fluctuation, the coupling between the power fluctuations of the GSC and the RSC challenges the modular structure of a back-to-back converter.

10.4.2 DC Current Calculation

To control the DC voltage fluctuation, the current flowing through the DC capacitor should be detected. However, in high power converters, to reduce the stray inductances, the DC bus usually adopts a laminated bus bar structure which makes it difficult for the direct measurement of the capacitor current. Besides, in a back-to-back converter with a modular structure, the DC capacitor current is divided into two parts: the GSC side and the RSC side as shown in Figure 10.10. If the capacitor currents of the GSC and the RSC are measured separately, the modular structure of the back-to-back converter is lost.

To meet the requirement of a modular design, the capacitor current can be measured indirectly by detecting the DC bus voltage. Under unbalanced voltage condition, the capacitor current mainly contains the second-order component, thus only the second-order harmonic capacitor current should be detected. The DC voltage under unbalanced grid can be written as

$$v_{dc} = V_{dc} + v_{dc}^{(2)} = V_{dc} + V_{dcm}^{(2)} \sin(2\omega_s t + \varphi_0) \tag{10.9}$$

The corresponding DC capacitor current is

$$i_{cap}(t) = C\frac{d}{dt}v_{dc}(t) = C\frac{d}{dt}\left(V_{dc} + V_{dcm}^{(2)} \sin(2\omega_s t + \varphi_0)\right) = 2\omega_s C V_{dcm}^{(2)} \cos(2\omega_s t + \varphi_0)$$

$$= -2\omega_s C v_{dc}^{(2)}(t - T_s/8) \tag{10.10}$$

where φ_0 is the initial phase of second-order harmonic capacitor current and T_s is the grid frequency. As shown in equation (10.10), the magnitude of capacitor current i_{cap} is the multiplication of the magnitude of second-order harmonic DC voltage $v_{dc}^{(2)}$ and the constant $2\omega_s C$, and the phase of the capacitor current is that of the second-order harmonic DC voltage $v_{dc}^{(2)}$ with a delay of $T_s/8$. For a grid frequency of 50 Hz, the delay is only 2.5 ms and thus the time required for capacitor current detection is able to meet the requirement of dynamic performance of capacitor current control. It is worth noting that compared to the capacitance of the DC capacitor, the ESR is relatively small with little influence on the accuracy of capacitor current detection as shown in equation (10.10). Combining equations (10.9) and (10.10), the schematic diagram of the capacitor current detection algorithm can be depicted as shown in Figure 10.11. In the detection algorithm, the differential operation is replaced by a time delay, which avoids the introduction of noise which might be brought by the implementation of a differential operation.

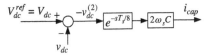

Figure 10.11 Schematic diagram of capacitor current detection algorithm based voltage measurement.

10.4.3 Control Scheme

To enhance the reliability of the DC capacitor, the control scheme should be modified to increase the system's impedance in the negative-sequence voltage so that the impact of the unbalanced grid on DC bus capacitor can be mitigated. Besides, the control scheme must be compatible with the requirements of the modular structure.

As shown in Figure 9.3, two disturbances are included in the current control loop of GSC: grid voltage v_{sdq} and rotor-side DC current i_{rdc}. Under unbalanced grid, both v_{sdq} and i_{rdc} contain second-order fluctuating components. According to the analysis given in the last chapter, conventional GSC control scheme cannot effectively suppress the influence of the second-order fluctuations in the v_{sdq} and i_{rdc} on the capacitor current.

According to control theory, as long as the disturbance is included in the closed loop with sufficient loop gain, a little influence will be caused by the disturbance on the output. In other words, the system has good suppression on the disturbance introduced. Following such principle, as long as v_{sdq} and i_{rdc} are included in the closed loop and the control loop has sufficiently high control gain, the influence of v_{sdq} and i_{rdc} on the i_{cap} can be mitigated. The requirement can be satisfied by adopting PI + Resonant controller for the DC voltage outer loop. However, such a method may affect the system stability, which is not desirable.

The improved control scheme for the suppression of second-order disturbances of v_{sdq} and i_{rdc} is shown in Figure 10.12. On the basis of the GSC current control loop shown in Figure 9.3, a capacitor current control loop is added so that the disturbances v_{sdq} and i_{rdc} are included in the loop. To suppress the second-order components in the disturbances, the controller of the capacitor current control loop is designed to be a second-order resonant controller $G_R^{(2)}(s)$. With a sufficiently large controller gain of $G_R^{(2)}(s)$, the impact of the disturbances on the capacitor current and DC voltage can be

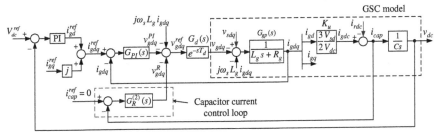

Figure 10.12 Block diagram of the capacitor current control scheme based on the resonant controller.

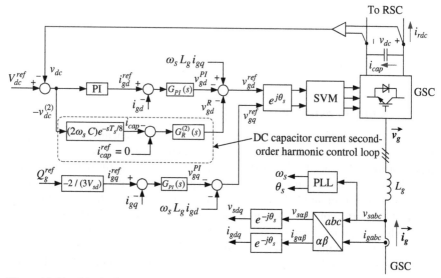

Figure 10.13 Block diagram of DC capacitor current second-order harmonic elimination control based on the resonant controller.

mitigated. Neglecting the feedforward decoupling term, the GSC voltage command v_{gdq}^{ref} is

$$v_{gdq}^{ref} = -\left(v_{gdq}^{PI} + v_{gdq}^{R}\right) \tag{10.11}$$

where $v_{gdq}^{PI} = v_{gd}^{PI} + jv_{gq}^{PI}$ is the output of the PI controller for fundamental current loop, at steady state v_{gdq}^{PI} is a DC component; $v_{gdq}^{R} = v_{gd}^{R} + jv_{gq}^{R}$ is the output of the resonant controller for capacitor current control loop which is used for the control of negative-sequence current, under the unbalanced grid, v_{gdq}^{R} is an AC component with two times the grid frequency.

Adopting the capacitor current calculation method introduced in the Section 10.4.2, combining Figures 10.11 and 10.12, the complete block diagram of the DC capacitor current second-order harmonic elimination control based on the resonant controller can be depicted as Figure 10.13. A capacitor current second-order harmonic control loop is added to the conventional GSC control scheme. Since the capacitor second-order harmonic current is obtained indirectly by measuring the DC voltage, no additional hardware detection circuits are needed. Moreover, the control scheme does not depend on the control variables of the RSC, which is suitable for wind power converters adopting a modular design.

10.4.4 Control Model

As illustrated in Chapter 9, the DC voltage outer loop is only effective in suppressing the disturbances below 10 Hz, and it is not able to suppress the disturbance of 100 Hz.

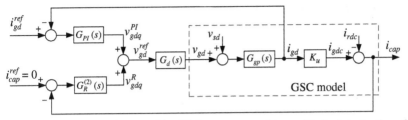

Figure 10.14   Block diagram of the GSC current control loop with the introduction of capacitor current control.

Therefore, when analyzing the suppression capability of the GSC on the disturbance of unbalanced grid voltage, only the control performance of the current loop should be taken into consideration, while the outer voltage loop can be neglected. In Figure 10.13, the capacitor current loop only includes the d-axis loop, that is, $v_{gdq}^R = v_{gd}^R$. Combining with Figure 10.12, and considering that the d- and q-axis are decoupled, the d-axis current control loop of the GSC can be further simplified as shown in Figure 10.14. The d-axis current control loop of the GSC consists of two loops, that is, active current loop and capacitor current loop. The active current control loop controls the fundamental active power of the GSC and maintains a steady DC bus voltage in coordination with the voltage outer loop. The capacitor current control loop is in charge of suppressing the second-order harmonic current flowing through the DC capacitor so that the DC bus voltage can be kept constant.

The loop equation can be derived according to Figure 10.14

$$\left[\left(i_{gd}^{ref}(s) - i_{gd}(s)\right) G_{PI}(s)G_d(s) + \left(i_{cap}^{ref}(s) - i_{cap}(s)\right) G_R^{(2)}(s)G_d(s) + u_{sd}(s)\right]$$
$$\times G_{gp}(s)K_u - i_{rdc}(s) = i_{cap}(s) \tag{10.12}$$

Since

$$i_{gd}(s) = \left(i_{cap}(s) + i_{rdc}(s)\right)/K_u \tag{10.13}$$

substituting (10.13) in (10.12), the following equation can be derived:

$$i_{cap}(s) = G_{gc}(s)i_{gd}^{ref}(s) + G_{cc}(s)i_{cap}^{ref}(s) + G_{vc}(s)v_{sd}(s) - G_{rc}(s)i_{rdc}(s) \tag{10.14}$$

where $G_{gc}(s)$ is the transfer function from grid current d-axis reference $i_{gd}^{ref}(s)$ to the capacitor current $i_{cap}(s)$, which represents the tracking capability of the current loop; $G_{cc}(s)$ is the transfer function from capacitor current reference $i_{cap}^{ref}(s)$ to the capacitor current $i_{cap}(s)$, which represents the tracking capability of the capacitor current loop; $G_{vc}(s)$ is the transfer function from grid voltage $u_{sd}(s)$ to the capacitor current $i_{cap}(s)$, which represents the suppression capability of the current loop on grid voltage disturbance; $G_{rc}(s)$ is the transfer function from rotor-side DC current $i_{rdc}(s)$ to the capacitor current $i_{cap}(s)$, which represents the suppression capability of the current loop on rotor-side DC current disturbance. The detailed expressions of these transfer functions are listed as follows:

$$
G_{gc}(s) = \left. \frac{i_{cap}(s)}{i_{gd}^{ref}(s)} \right|_{i_{cap}^{ref}(s)=0, v_{sd}(s)=0, i_{rdc}(s)=0}
$$

$$
= \frac{K_u G_{gp}(s) G_{PI}(s) G_d(s)}{1 + G_{gp}(s) G_{PI}(s) G_d(s) + K_u G_{gp}(s) G_R^{(2)}(s) G_d(s)} \tag{10.15}
$$

$$
G_{cc}(s) = \left. \frac{i_{cap}(s)}{i_{cap}^{ref}(s)} \right|_{i_{gd}^{ref}(s)=0, v_{sd}(s)=0, i_{rdc}(s)=0}
$$

$$
= \frac{K_u G_{gp}(s) G_R^{(2)}(s) G_d(s)}{1 + G_{gp}(s) G_{PI}(s) G_d(s) + K_u G_{gp}(s) G_R^{(2)}(s) G_d(s)} \tag{10.16}
$$

$$
G_{vc}(s) = \left. \frac{i_{cap}(s)}{v_{sd}(s)} \right|_{i_{gd}^{ref}(s)=0, i_{cap}^{ref}(s)=0, i_{rdc}(s)=0}
$$

$$
= \frac{K_u G_{gp}(s)}{1 + G_{gp}(s) G_{PI}(s) G_d(s) + K_u G_{gp}(s) G_R^{(2)}(s) G_d(s)} \tag{10.17}
$$

$$
G_{rc}(s) = \left. \frac{i_{cap}(s)}{i_{rdc}(s)} \right|_{i_{gd}^{ref}(s)=0, i_{cap}^{ref}(s)=0, v_{sd}(s)=0}
$$

$$
= \frac{1 + G_{gp}(s) G_{PI}(s) G_d(s)}{1 + G_{gp}(s) G_{PI}(s) G_d(s) + K_u G_{gp}(s) G_R^{(2)}(s) G_d(s)} \tag{10.18}
$$

As shown in equation (10.18), when $G_R^{(2)}(s) = 0$, $G_{rc}(s) = 1$, the suppression capability of the GSC on the i_{rdc} disturbance deteriorates to the situation with the conventional control scheme.

According to Figure 10.14, the closed-loop transfer function of GSC d-axis current loop can be written as

$$
G_{icl}(s) = \frac{i_{gd}(s)}{i_{gd}^{ref}(s)} = \frac{1}{K_u} G_{gc}(s) \tag{10.19}
$$

The transfer function from grid voltage $v_{sd}(s)$ to the GSC AC current $i_{gd}(s)$ is

$$
G_{gvi}(s) = \frac{i_{gd}(s)}{v_{sd}(s)} = \frac{1}{K_u} G_{vc}(s) \tag{10.20}
$$

According to control theory, the open-loop transfer function of the GSC current loop can be derived from the denominators of equations (10.15)–(10.18):

$$
G_{go}(s) = G_{gp}(s) G_{PI}(s) G_d(s) + K_u G_{gp}(s) G_R^{(2)}(s) G_d(s) \tag{10.21}
$$

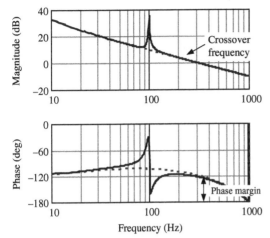

Figure 10.15   Open-loop frequency characteristics of the GSC current loop; solid lines represent the situation with the introduction of capacitor current control, while dashed lines represent the conventional control.

Take the 1.5 MW DFIG as an example, and the PI controller of the GSC current loop is set to $K_{gp} = 1$ and $K_{gi} = 80$, and the second-order harmonic resonant controller is designed as $K_r = 20$ and $\omega_c = 3$ rad/s.

Figure 10.15 shows the open-loop frequency characteristics of the GSC current loop with the introduction of DC capacitor current suppression control. In the figure, solid lines represent the situation with the introduction of capacitor current control, while the dashed lines represent the conventional control. As shown in the figure, with the introduction of capacitor current control, the controller gain of the current loop at the frequency of 100 Hz increases significantly. Besides, from the phase–frequency characteristics, the stability of the current loop is not largely affected by the introduction of the capacitor current control.

Figure 10.16 shows the closed-loop control performance of the current loop, where Figure 10.16a gives the frequency characteristic of the closed-loop transfer function $G_{icl}(s)$ of the d-axis current loop, and Figure 10.16b represents the frequency characteristics of the closed-loop transfer function $G_{cc}(s)$ of capacitor current control loop. As it can be seen from Figure 10.16b, at the frequency of 100 Hz, the magnitude of $G_{cc}(s)$ is 1 (0 dB), and the phase is 0, which illustrates a good capacitor current tracking performance at 100 Hz. When the capacitor current reference $i_{cap}^{ref}(s) = 0$, the second-order current harmonic of the DC capacitor can be controlled to zero.

Figure 10.17 shows the suppression capability of GSC current loop on the disturbances of the grid voltage and rotor-side DC current, where Figure 10.17a shows the frequency characteristics of the transfer function $G_{vc}(s)$ from the d-axis grid voltage to capacitor current, Figure 10.17b shows the frequency characteristic of the transfer function $G_{rc}(s)$ from the rotor-side DC current to the capacitor current. In Figure 10.17, solid lines represent the situation with the introduction of a capacitor current control, while dashed lines represent the conventional control. As shown in

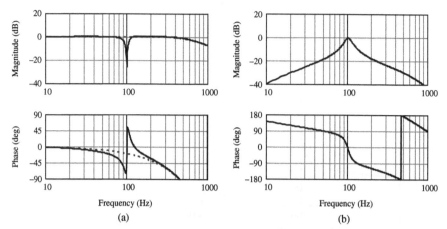

Figure 10.16   Closed-loop control performance of the GSC current loop; solid lines represent the situation with the introduction of capacitor current control, while dashed lines represent the conventional control: (a) Frequency characteristic of d-axis current control closed-loop transfer function; (b) frequency characteristic of capacitor current control closed-loop transfer function.

Figure 10.17, with the introduction of capacitor current loop, the damping of the disturbance at 100 Hz by the current loop increases significantly, which indicates good suppression capability of the GSC current loop on the second-order disturbances in grid voltage and rotor-side DC current.

According to the expression of the magnitude of the second-order harmonic capacitor current shown in Chapter 9, it can be seen from Figure 10.17 that since the magnitudes of $G_{gvi}(-j2\omega_s)$ and $G_{rc}(-j2\omega_s)$ are significantly reduced by the capacitor

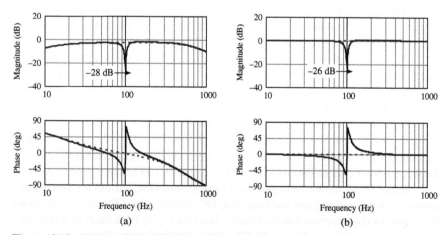

Figure 10.17   Suppression capability of GSC current loop on the disturbances of grid voltage and rotor-side DC current: (a) Suppression on grid voltage disturbance; (b) suppression on rotor-side DC current disturbance.

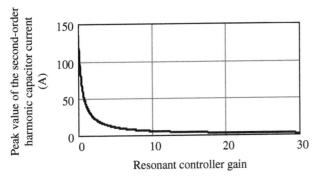

Figure 10.18   Relationship between capacitor current second-order harmonic and resonant controller gain K_r.

current control, the magnitude of capacitor current second-order harmonic would thus be largely reduced. When the slip ratio $S_l = 0.2$ and the grid voltage unbalance factor is 20%, Figures 10.18 and 10.19 show the curves of the capacitor current second-order harmonic and DC voltage second-order fluctuation with respect to controller gain K_r of the resonant controller in a 1.5 MW DFIG system. When the resonant controller gain is zero, the capacitor current control is disabled, that is, the GSC adopts the conventional control scheme. As shown in Figures 10.18 and 10.19, with the increase of resonant controller gain, the fluctuations in the capacitor current and DC voltage decrease. When the controller gain is larger than 20, the fluctuations are almost eliminated. Therefore, by selecting an appropriate controller gain, the impact of the unbalanced grid on the DC bus capacitors in a wind power converter can be suppressed.

It should be noted that, in Figures 10.18 and 10.19, the RSC still adopts the conventional current vector control. No extra controllers are added to deal with the unbalanced grid. Therefore, the aforementioned control strategy can keep the DC voltage constant independently from the GSC side without any dependence on the RSC.

Combining with the GSC DC voltage loop described in Chapter 9, the open-loop frequency characteristics of the DC voltage outer loop with the introduction of

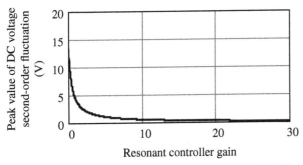

Figure 10.19   Relation between DC voltage second-order fluctuation and resonant controller gain K_r.

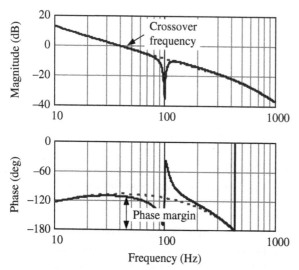

Figure 10.20   Open-loop frequency characteristics of the DC voltage outer loop with the introduction of capacitor current control.

the capacitor current control can be depicted as shown in Figure 10.20. Compared to conventional control method (shown in Figure 10.20 as dashed lines), the phase margin of the DC voltage loop is not significantly reduced, and it is illustrated by solid lines. Therefore, the capacitor current suppression control based on the resonant controller does not impose any impact on the stability of the voltage outer loop.

In general, with respect to the influence of unbalanced grid on the wind power converter, the capacitor current control based on the resonant controller has the following characteristics:

1. The GSC's impedance in negative-sequence voltage is largely increased so that the capacitor current second-order harmonic and DC voltage fluctuation can be suppressed.

2. The stability of GSC control is not influenced and the control system still holds a large stability margin.

3. The modular structure of the converter is retained, and the control of GSC and RSC are independent.

4. No additional hardware is needed as the capacitor current information is obtained from a DC-voltage measurement.

10.4.5   Elimination of Third-Order Harmonic Current Introduced by Capacitor Current Control

10.4.5.1   Analysis of Third-Order Harmonic Current

Under unbalanced grid, a second-order harmonic component is seen in the capacitor current, if the input error of capacitor current control loop can be expressed as

$e_{ic} = E_{ic} \cos(2\omega_s t + \varphi_0)$, and if the gain of the resonant controller $G_R^{(2)}$ at the frequency of $\pm 100\,\text{Hz}$ is K_r and the phase angle is 0, the output voltage command v_{gdq}^R of the resonant controller of capacitor current loop shown in Figure 10.14 can be expressed as

$$v_{gdq}^R = K_r e_{ic} = K_r E_{ic} \cos(2\omega_s t + \varphi_0) = V_{gm}^R \cos(2\omega_s t + \varphi_0) \quad (10.22)$$

where $V_{gm}^R = K_r E_{ic}$ is the magnitude of v_{gdq}^R.

According to Euler's equation, a sinusoidal component can be decomposed into two conjugated vectors rotating in opposite directions, which can be expressed as

$$v_{gdq}^R = V_{gm}^R \cos(2\omega_s t + \varphi_0) = V_{gm}^R \left[\frac{e^{-j(2\omega_s t + \varphi_0)} + e^{j(2\omega_s t + \varphi_0)}}{2} \right] \quad (10.23)$$

Therefore, the output voltage command of the resonant controller of the capacitor current loop contains both positive-sequence and negative-sequence second-order harmonic components.

According to Figure 10.14 and neglecting the feedforward decoupling terms, the output voltage command of the GSC control loop can be written as

$$v_{gdq}^{ref} = v_{gdq}^{PI} + v_{gdq}^R = V_{gdq}^{PI} + 0.5 V_{gm}^R e^{-j(2\omega_s t + \varphi_0)} + 0.5 V_{gm}^R e^{j(2\omega_s t + \varphi_0)} \quad (10.24)$$

Transforming (10.24) from the dq rotating reference frame into $\alpha\beta$ stationary reference frame, the AC voltage command in the $\alpha\beta$ stationary reference frame can be derived as

$$v_{g\alpha\beta}^{ref} = v_{gdq}^{ref} e^{j\omega_s t} = V_{gdq}^{PI} e^{j\omega_s t} + \left(0.5 V_{gm}^R e^{-j\varphi_0} \right) e^{-j\omega_s t} + \left(0.5 V_{gm}^R e^{j\varphi_0} \right) e^{j3\omega_s t} \quad (10.25)$$

As shown in (10.25), the three-phase voltage command of the GSC is composed of a positive-sequence fundamental component $V_{gdq}^{PI} e^{j\omega_s t}$, negative-sequence fundamental component $(0.5 V_{gm}^R e^{-j\varphi_0}) e^{-j\omega_s t}$, and positive-sequence third-order harmonic component $(0.5 V_{gm}^R e^{j\varphi_0}) e^{j3\omega_s t}$. The positive-sequence fundamental voltage command is in charge of controlling the positive-sequence fundamental current of the GSC, and the negative-sequence fundamental voltage command is used to control the negative-sequence fundamental current of the GSC, however the positive-sequence third-order harmonic voltage command may cause the GSC to output third-order harmonic voltage leading to a third-order harmonic current from the GSC.

10.4.5.2 The Extraction of Negative-Sequence Voltage Command Based on dq Transformation

To avoid the introduction of third-order harmonic current by capacitor current control loop, the third-order component $(0.5 V_{gm}^R e^{j\varphi_0}) e^{j3\omega_s t}$ of the output voltage command of capacitor current loop has to be eliminated, extracting only the required negative-sequence voltage command $(0.5 V_{gm}^R e^{-j\varphi_0}) e^{-j\omega_s t}$. Figure 10.21 shows a

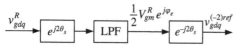

Figure 10.21   Extraction of negative-sequence voltage command based on dq transformation.

method to extract the negative-sequence voltage command based on dq transformation. Since in the positive-sequence dq rotating frame, the negative-sequence voltage command rotates at an angular speed of $-2\omega_s$, it can be transformed into DC component by a dq transformation with the angular speed of $2\omega_s$. Then, the DC component can be extracted by a low pass filter (LPF), and after an inverse dq rotating transformation with an angular speed of $2\omega_s$, the required negative-sequence voltage command $v_{gdq}^{(-2)ref}$ rotating at the angular speed of $-2\omega_s$ can be obtained. After the above operations, the obtained negative-sequence voltage command is $v_{gdq}^{(-2)ref} = (0.5V_{gm}^R e^{-j\varphi_0})e^{-j2\omega_s t}$. In Figure 10.21, θ_s is the phase of the grid voltage, $\theta_s = \omega_s t + \theta_0$, where θ_0 is the initial phase, $\varphi_e = 2\theta_0 - \varphi_0$.

In Figure 10.21, the LPF can be a second-order Butterworth LPF with a transfer function of [20]

$$G_{LPF}(s) = \frac{1}{1 + \sqrt{2}\left(s/\omega_n\right) + \left(s/\omega_n\right)^2} \tag{10.26}$$

where ω_n is the cut-off frequency of the LPF. According to Figure 10.21, the expression of the negative-sequence voltage command $v_{gdq}^{(-2)ref}$ after the extraction algorithm is

$$v_{gdq}^{(-2)ref}(s) = G_{LPF}(s + j2\omega_s)v_{gdq}^R \tag{10.27}$$

Thus, the transfer function of the capacitor current controller becomes

$$G_{R\_LPF}^{(2)}(s) = G_R^{(2)}(s)G_{LPF}(s + j2\omega_s) \tag{10.28}$$

If the cut-off frequency of the LPF is 20 Hz, the frequency characteristics of $G_{LPF}(s + j2\omega_s)$ can be depicted as shown in Figure 10.22a. The parameters of $G_R^{(2)}(s)$ are $K_r = 20$ and $\omega_c = 3$ rad/s. The frequency characteristics of the transfer function $G_{R\_LPF}^{(2)}(s)$ are shown in Figure 10.22b. As shown in Figure 10.22b, the extraction method based on dq rotating transformation is able to effectively eliminate the positive-sequence third-order harmonic component, meanwhile, a large control gain is maintained for the negative-sequence fundamental component.

Figure 10.23 shows the complete control block diagram of the capacitor current control with a negative-sequence voltage extraction method as shown in Figure 10.21. Since the positive-sequence third-order harmonic voltage command is eliminated, the output three-phase voltage v_g of the GSC does not contain the corresponding harmonic voltage and therefore the harmonic distortion in the output current i_g caused by capacitor current control is eliminated.

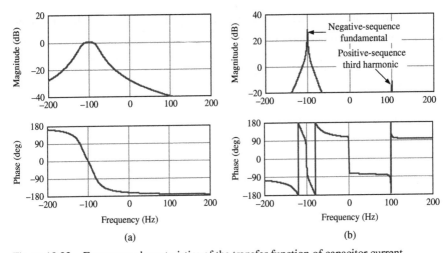

Figure 10.22   Frequency characteristics of the transfer function of capacitor current controller: (a) Frequency characteristics of the transfer function $G_{LPF}(s + j2\omega_s)$; (b) frequency characteristics of the transfer function of capacitor current controller.

10.4.5.3   Improved Capacitor Current Control Scheme Based on Negative-Sequence Resonant Controller

According to the analysis given in Section 10.4.5.1, the root cause of the third-order harmonic in the output of the current control loop is that the second-order harmonic in the capacitor current can be decomposed into conjugated negative-sequence fundamental rotating vector and positive-sequence third-order harmonic rotating vector.

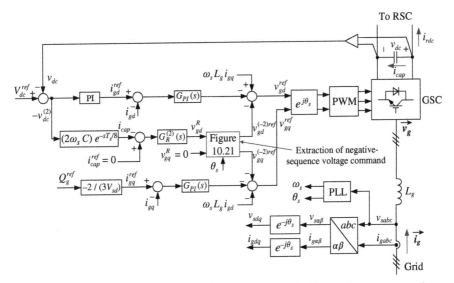

Figure 10.23   Block diagram of the capacitor current control with negative-sequence voltage extraction.

Since the resonant controller $G_R^{(2)}$, as shown in equation (10.1), has the same controller gain $K_r^{(2)}$ at frequencies of ± 100 Hz, the harmonic component has the same amplification as the negative-sequence component, causing a large harmonic in the output command of capacitor current loop.

As illustrated in Section 10.4.5.2, the method of filtering the positive-sequence third harmonic voltage command based on rotating frame transformation can effectively eliminate the harmonic voltage command. However, large computational efforts are needed for a Park transformation, an LPF, and an inverse Park transformation.

If the resonant controller only has a large controller gain at -100 Hz (here the minus sign represents the negative-sequence rotating direction), the output command of the capacitor current control loop would only contain the negative-sequence control command. In [14], a negative-sequence resonant controller is introduced of which the ideal form can be written as

$$G_R^{(-2)}(s) = \frac{K_r}{s + j2\omega_s} = K_r\left(\frac{s}{s^2 + (2\omega_s)^2} - j\frac{2\omega_s}{s^2 + (2\omega_s)^2}\right) \quad (10.29)$$

Similarly, the non-ideal form with a cut-off frequency of ω_c can be expressed as [14]:

$$G_R^{(-2)}(s) = G_{Rd}^{(-2)}(s) + jG_{Rq}^{(-2)}(s) \approx \frac{2K_r\omega_c s}{s^2 + 2\omega_c s + (2\omega_s)^2} - j\frac{2K_r\omega_c(2\omega_s)}{s^2 + 2\omega_c s + (2\omega_s)^2} \quad (10.30)$$

where $G_{Rd}^{(-2)}(s)$ is the d-axis open-loop transfer function of the negative-sequence resonant controller, $G_{Rq}^{(-2)}(s)$ is the q-axis open-loop transfer function of the negative-sequence resonant controller.

Given the parameters of the negative-sequence second-order resonant controller $G_R^{(-2)}(s)$ as $K_r = 25$ and $\omega_c = 3$ rad/s, the frequency characteristics of the transfer function $G_R^{(-2)}(s)$ can be depicted, as shown in Figure 10.24. As shown in the figure, $G_R^{(-2)}(s)$ only has a large control gain at -100 Hz which is the corresponding frequency of the negative-sequence component in the positive-sequence dq reference frame, while having a large damping for the positive-sequence third-order harmonic component. Therefore, the negative-sequence second-order resonant controller can effectively suppress the second-order harmonic current in the capacitor without introducing a third-order harmonic distortion in the output current of the GSC.

Combining equation (10.30) and Figure 10.13, the complete control block diagram of the improved capacitor current control scheme based on the negative-sequence resonant controller is shown in Figure 10.25. As shown in the figure, the error signal of the capacitor current goes into the d-axis controller $G_{Rd}^{(-2)}(s)$ of $G_R^{(-2)}(s)$ obtaining a d-axis voltage command of the output of capacitor current control loop; similarly the error signal of the capacitor current goes into the q-axis controller $G_{Rq}^{(-2)}(s)$ of $G_R^{(-2)}(s)$ obtaining the q-axis voltage command of the output of capacitor current control loop. The output d- and q-axis voltage commands of the capacitor

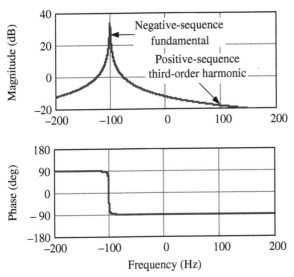

Figure 10.24 Frequency characteristics of the transfer function $G_R^{(-2)}(s)$, $K_r = 25$, and $\omega_c = 3\,\text{rad/s}$.

current loop are added to the d- and q-axis voltage commands given by the fundamental current control loop respectively forming the d- and q-axis voltage commands v_{gd}^{ref} and v_{gq}^{ref} of the GSC. After the SVM modulation, gate drive signals are generated to realize the current control targets of the GSC.

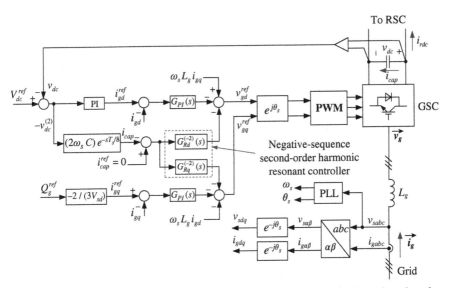

Figure 10.25 Block diagram of the improved capacitor current control scheme based on the negative-sequence resonant controller.

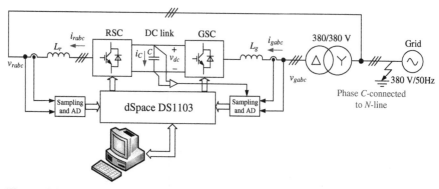

Figure 10.26   Schematic diagram of the back-to-back converter testbed.

10.4.6   Experimental Results

To verify the effectiveness of capacitor current control scheme, experimental tests are done on a back-to-back converter testbed and a small DFIG wind power generation testbed.

10.4.6.1   Back-to-Back Converter Testbed

The schematic diagram of the back-to-back converter testbed is shown in Figure 10.26, where the DFIG is emulated by an inductor L_r. The parameters of the testbed are listed below:

Grid phase voltage, 220 V/50 Hz; rated power of the converter, 2.2 kW; DC bus voltage, 650 V; grid-side inductor L_g, 15 mH; rotor-side inductor L_r, 15 mH; DC bus capacitor C, 360 μF; converter switching frequency, 2 kHz; sampling frequency, 4 kHz.

The GSC controls the DC bus voltage v_{dc}, and the RSC controls the output AC current i_{rabc}. The power of the RSC is the load of the GSC. To emulate unbalanced grid condition, phase C of the testbed is connected to the neutral line of the grid, causing an unbalance factor of 50%. The RSC is used as the fluctuation power source of the GSC, causing second-order fluctuation in the DC current i_{rdc} under the unbalanced grid.

10.4.6.2   Experimental Results from the Back-to-Back Converter Testbed

Figure 10.27 shows the AC current i_{rabc} of the RSC under the unbalanced grid. Figure 10.27a shows the situation with no load. As shown in Figure 10.27a, the three-phase AC current contains only negative-sequence component, causing fluctuations in the DC bus voltage. Figure 10.27b shows the situation under half load and the magnitudes of the three-phase currents are not equal under the unbalanced grid, and fluctuations are contained in the active power. second-order fluctuation is also contained in the DC current flowing from the RSC to the GSC, causing fluctuations in the DC bus voltage. Also shown in Figure 10.27 is that the fluctuation magnitudes under zero load and half load are almost identical, which indicates that the magnitude of active current has a little influence on the DC voltage fluctuation.

When the current loop adopts a PI controller, the suppression capability of the GSC current loop on the negative-sequence current is shown in Figure 10.28. As

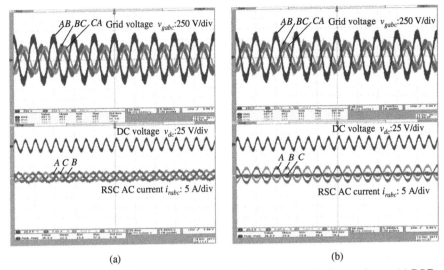

(a)   (b)

Figure 10.27   RSC waveforms under unbalanced grid with the DFIG operating at: (a) RSC AC current $i_r = 0$ pu; (b) RSC AC current $i_r = 0.5$ pu.

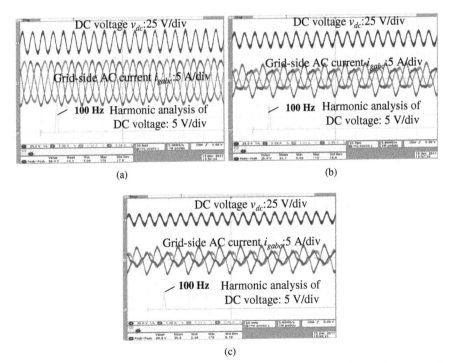

(a)   (b)

(c)

Figure 10.28   Suppression capability of the GSC current loop on negative-sequence current with load condition of $i_r = 0.5$ pu: (a) Current loop open-loop bandwidth 100 Hz; (b) current loop open-loop bandwidth 250 Hz; (c) current loop open-loop bandwidth 500 Hz.

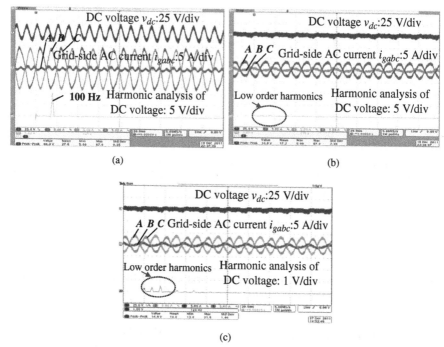

Figure 10.29 Capacitor current control performance under unbalanced grid with the load condition of $i_r = 0.5$ pu: (a) Current PI control; (b) capacitor current control based on negative-sequence voltage extraction; (c) capacitor current control based on the negative-sequence resonant controller.

shown in the figures, with the increase of the bandwidth of the current loop, the GSC damping of negative-sequence current increases, and the second-order fluctuations in the DC voltage decreases. However, although the bandwidth is increased up to 500 Hz, significant fluctuations are still present in the DC voltage. Therefore, the suppression capability of the current PI controller on the negative-sequence current is limited, and the impact of the unbalanced grid on the converter cannot be fully eliminated.

Figure 10.29 shows the test results of the capacitor current control under the unbalanced grid, where Figure 10.29b gives the test waveforms of the capacitor current control based on negative-sequence voltage extraction, and Figure 10.29c shows the test waveforms of capacitor current control based on the negative-sequence resonant controller. As shown in the figures, with the adoption of capacitor current control, the amplitude of DC voltage fluctuation decreases from 10 V to below 1 V, the DC voltage fluctuation is almost eliminated. Besides, due to the elimination of negative-sequence current, the amplitudes of the three-phase AC currents are reduced as well. Therefore, the capacitor current control can effectively eliminate the impact of grid unbalance on the DC bus voltage as well as reduce the current stress on the converter.

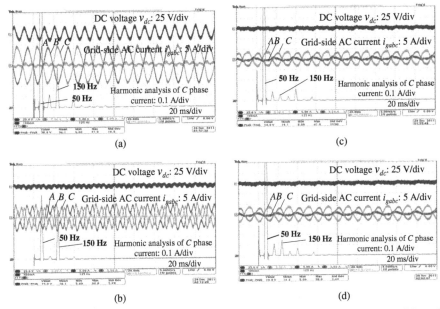

Figure 10.30 Comparison of capacitor current control waveforms with the load condition of $i_r = 0.5$ pu: (a) Current PI control; (b) capacitor current control without third harmonic suppression; (c) capacitor current control based on negative-sequence voltage extraction; (d) capacitor current control based on the negative-sequence resonant controller.

Figure 10.30 gives a comparison of the capacitor current control waveforms with and without third-order harmonic suppression, where the four sets of waveforms represent the situations with PI control, with capacitor current control without third harmonic suppression, with capacitor current control based on negative-sequence voltage extraction and with capacitor current control based on negative-sequence resonant controller, respectively. Compared to the PI control method, no matter with or without third-order harmonic suppression, the second-order fluctuation in DC voltage is largely suppressed. However, as it can be seen from the harmonic analysis of the phase-C current, large third-order harmonic is seen in the AC current of the GSC under PI control and the capacitor current control without third-order harmonic suppression. When the capacitor current control based on the negative-sequence voltage extraction or the capacitor current control based on the negative-sequence resonant controller is adopted, the third-order harmonic component in the AC current is significantly reduced. Therefore, compared to the conventional PI control scheme, on applying the two improved capacitor current control methods, not only the DC voltage fluctuation can be suppressed, but the output current quality of the converter can be largely enhanced as well.

Corresponding to Figure 10.30, the waveforms of the capacitor current and the output of the resonant controller is shown in Figure 10.31. When the capacitor current control is not adopted, large second-order harmonic current flows through the DC capacitor, as shown in Figure 10.31a. When the capacitor current control is

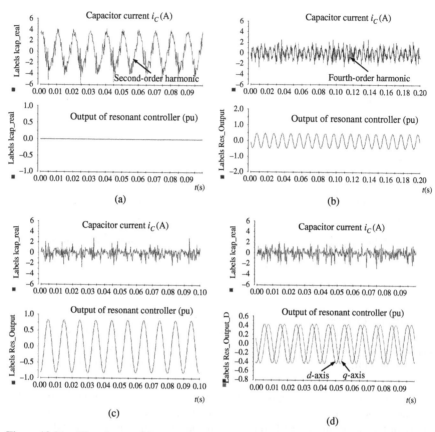

Figure 10.31   Waveforms of the capacitor current and the output of resonant controller: (a) Current PI control; (b) capacitor current control without third harmonic suppression; (c) capacitor current control based on negative-sequence voltage extraction; (d) capacitor current control based on the negative-sequence resonant controller.

adopted without third-order harmonic suppression, fourth-order harmonic current is contained in the capacitor current, which is caused by the fourth-order fluctuation of the active power introduced by the interaction between the positive-sequence third-order harmonic current and the fundamental negative-sequence voltage. When the negative-sequence voltage extraction method and negative-sequence resonant controller are adopted to eliminate the third-order harmonic voltage command, the low order harmonic components are significantly suppressed.

Figure 10.32 shows the transient process when the converter switches from PI control to capacitor current control. As shown in the figure, after enabling the capacitor current control, DC voltage fluctuation as well as the negative-sequence current of the GSC are suppressed. The switching process is smooth, which illustrates that the capacitor current control loop does not affect the stability of the control system.

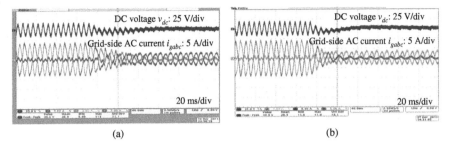

(a)   (b)

Figure 10.32   Transient process when the converter switches from PI control to capacitor current control: (a) Capacitor current control based on negative-sequence voltage extraction; (b) capacitor current control based on negative-sequence resonant controller.

Figure 10.33 shows the waveforms of capacitor current and output of resonant controller when the converter switches from PI control to capacitor current control. As shown in the figure, the capacitor current is significantly reduced with the introduction of capacitor current control, and the transient process of the output command of the resonant controller is smooth.

10.4.6.3   Experimental Results from the DFIG Wind Power Generation Testbed

To further validate the effectiveness of capacitor current control under the unbalanced grid, the experimental analysis is done on the 30 kW DFIG wind power generation testbed. The schematic diagram of the testbed can be found in Figure 8.15. In the experiments, a 11% voltage dip of phase-C voltage is used to emulate grid voltage unbalance with an unbalance factor of 4%. In the steady-state tests, the stator output active power is set to 10 kW, the reactive power is zero, and the rotor speed is kept at 1200 r/min (0.8 pu). In the tests, the PI parameters used for the GSC are $K_p = 1.2$

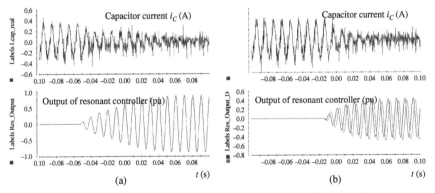

(a)   (b)

Figure 10.33   Capacitor current and the output of resonant controller when the converter switches from PI control to capacitor current control: (a) Capacitor current control based on negative-sequence voltage extraction; (b) capacitor current control based on the negative-sequence resonant controller.

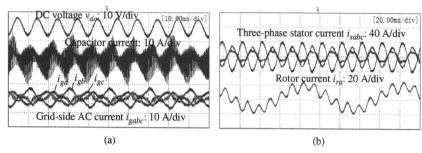

(a)   (b)

Figure 10.34   Waveforms of the wind turbine under unbalanced grid voltage when the conventional PI control is adopted, test conditions: stator active power 10 kW, rotor speed 1200 r/min (0.8 pu). (a) GSC waveforms; (b) DFIG waveforms.

and $K_i = 100$, and the PI parameters used for the RSC are $K_p = 0.36$ and $K_i = 30$, the resonant controller parameters for the capacitor current control loop are $K_r = 15$ and $\omega_c = 3$ rad/s.

Figure 10.34 shows the waveforms of the wind turbine under unbalanced grid voltage when the conventional PI control is adopted. As shown in the figure, 100 Hz fluctuation is seen in the DC voltage, and capacitor current also contains low order harmonics. Meanwhile, the AC current of the GSC and stator current of the DFIG are unbalanced and harmonically distorted, and harmonic components are also present in the rotor current.

When capacitor current control is adopted by the GSC, the GSC injects some fundamental negative-sequence active current into the grid to transmit the second-order power fluctuation generated by the RSC into the DC bus, thus keeping the DC bus voltage almost constant, as shown in Figure 10.35. It should be noted that if the rotor power fluctuation is too large, with the adoption of capacitor current control, severe unbalance may appear in the grid-side AC current, which may exceed the maximum current rating of the GSC. To avoid such a problem, stator current balancing control introduced in Section 10.3 has to be adopted to suppress rotor power fluctuation.

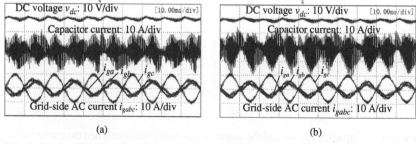

(a)   (b)

Figure 10.35   Waveforms of the wind turbine under unbalanced grid voltage when capacitor current control is adopted. Test conditions: stator active power 10 kW, rotor speed 1200 r/min (0.8 pu). (a) Capacitor current control based on negative-sequence voltage extraction; (b) capacitor current control based on negative-sequence resonant controller.

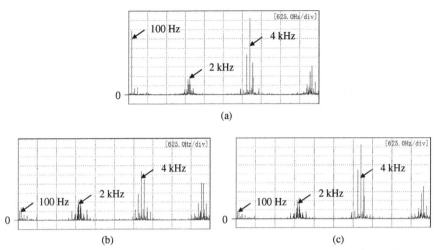

Figure 10.36 Frequency analysis of capacitor current for the test waveforms shown in Figures 10.34 and 10.35: (a) Current PI control; (b) Capacitor current control based on negative-sequence voltage extraction; (c) Capacitor current control based on negative-sequence resonant controller.

The frequency analysis of capacitor current is shown in Figure 10.36. When capacitor current control is not adopted (i.e., current PI control is used), the capacitor current not only contains components of the switching frequency, but also a very large second-order harmonic component. Correspondingly, when the capacitor current control is used, the 100 Hz component of the capacitor current is largely reduced. Therefore, the impact of the grid unbalances on the DC bus capacitor is resolved, which validates the effectiveness of the capacitor current control.

Because of the vaporization of electrolyte, the capacitance of the DC bus capacitor may vary during long-term operation. Usually, the DC bus capacitance varies in the range of ± 20%. Tests are made to evaluate the influence of capacitor current calculation error caused by capacitance change on the control performances and it is shown in Figure 10.37. In the tests, the capacitance factor C in equation (10.10) deviates from the actual capacitance value by − 20% and + 20%. As shown in the figure,

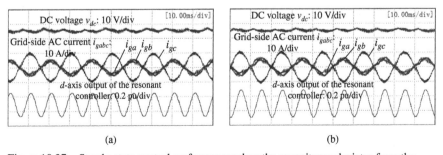

Figure 10.37 Steady-state control performance when the capacitance deviates from the actual capacitance: (a) Capacitance factor C deviates by − 20%; (b) capacitance factor C deviates by + 20%.

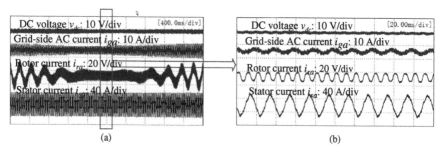

Figure 10.38   Control performance of the capacitor current control under unbalanced grid with the variation of rotor speed, Rotor speed is changing linearly from 0.9 to 1.1 pu and the active power is kept at 10 kW: (a) Rotor speed is changing linearly from 0.9 to 1.1 pu; (b) around synchronous speed, corresponding to the squared area of (a).

the deviation of capacitance has a little influence on the control performance, and the fluctuation in DC voltage is still effectively suppressed. Thus, as long as an appropriate controller gain is set for the resonant controller, the capacitor current control loop can provide a proper negative-sequence voltage command to suppress the DC voltage fluctuation, which illustrates a good robustness of the capacitor current control method.

Figure 10.38 shows the control performance of the capacitor current control under unbalanced grid with the variation of rotor speed. In the tests, the rotor speed changes linearly from 0.9 pu (1350 rpm, sub-synchronous) to 1.1 pu (1650 rpm, super-synchronous). During the rotor speed transition, the active power is kept at 10 kW. As shown in the figure, during the transition of rotor speed, the control of the DC voltage is independent of the rotor speed variation, and the DC voltage fluctuation is largely suppressed. Figure 10.38b shows the waveforms around synchronous speed, where it can be seen that the rotor current contains second-order fluctuation, but the DC bus voltage remains constant. Thus, the capacitor current control method is able to maintain good control performance in the whole operation speed range of the DFIG.

To test the dynamic performance of the GSC, under balanced three-phase grid voltage, the stator active power step changes from 0 to 0.5 pu (15 kW). The dynamic response of the GSC is shown in Figure 10.39. As it is shown in the figure, after

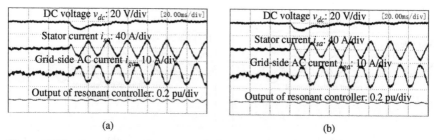

Figure 10.39   System dynamic response during power step change, stator active power step changes from 0 to 0.5 pu (15 kW): (a) Capacitor current control based on negative-sequence voltage extraction; (b) capacitor current control based on the negative-sequence resonant controller.

the step change of active power, the DC bus voltage dips for about 10 V and returns to normal value after two grid cycles, which illustrates that with the introduction of capacitor current control, the GSC remains to have good dynamic performance.

10.5 SUMMARY

Under unbalanced grid, negative-sequence voltage may cause fluctuations in the electromagnetic torque of the DFIG and the DC bus voltage of the back-to-back converter, jeopardizing the safe operation of the wind turbine system. The conventional current control method is not effective in suppressing the impacts of negative-sequence voltage on the DFIG wind turbine. To reduce the second-order fluctuation in the electromagnetic torque, a stator current balancing control method based on the resonant controller is used to suppress the stator negative-sequence fundamental current, solving the problem of lifetime reduction caused by stator current unbalance and the electromagnetic torque fluctuation.

For the GSC, a capacitor current control method is adopted to suppress the second-order harmonic current flowing into the DC bus capacitor, also resulting in the suppression of DC voltage fluctuation. The capacitor current is obtained by a delay algorithm of the DC voltage, avoiding the cost of extra current-sensing circuits. The control scheme is performed by the GSC independently not relying on the power information from the RSC, making such a method suitable for large-scale wind turbine systems adopting a modular design. Since the capacitor current control may cause positive-sequence third harmonic in the output current of GSC, two methods are used to eliminate the positive-sequence third-order harmonic voltage command introduced by the capacitor current control loop which is the negative-sequence voltage extraction based on dq rotating transformation and negative-sequence resonant controller. Steady-state and dynamic experiments validate that the capacitor current control scheme has good steady-state performance without affecting the dynamic performance of the system. The stator current balancing control together with the capacitor current control are able to enhance the performance of DFIG wind turbine system under unbalanced grid condition.

REFERENCES

[1] "Grid code: Issue 4 revision 10," Great Britain N G. 2012. [Online]. Available: http://www2.nationalgrid.com/UK/Industry-information/Electricity-codes/Grid-code/The-Grid-code/

[2] "Technical regulation 3.2.5 for wind power plants with a power output great than 11 kW," Engerinet.dk Denmark. 2010. [Online]. Available: https://en.energinet.dk/Electricity/Rules-and-Regulations/Regulations-for-grid-connection

[3] "Grid connection regulations for high and extra high voltage," E.ON Germany. 2006. [Online]. Available: http://www.nerc.com/docs/pc/ivgtf/German_EON_Grid_Code.pdf

[4] "P.O. 12.3: Response requirements of wind power generation to network voltage dips," Red Electrica Spain. 2006. [Online]. Available: http://www.ree.es/sites/default/files/01_ACTIVIDADES/Documentos/ProcedimientosOperacion/PO_resol_12.3_Respuesta_huecos_eolica.pdf

[5] "Technical rule for connecting wind farm to power system, GB/T 19963-2011," Chinese Standard. 2012. [Online]. Available: http://www.chinesestandard.net/PDF-English-Translation/GBT19963-2011.html

[6] "Voltage characteristics of electricity supplied by public distribution systems, EN 50160," Leonardo Energy. 1999. [Online]. Available: http://admin.copperalliance.eu/docs/librariesprovider5/power-quality-and-utilisation-guide/542-standard-en-50160-voltage-characteristics-in.pdf

[7] F. Iov, R. Teodorescu, F. Blaabjerg, B. Andresen, J. Birk, and J. Miranda, "Grid code compliance of grid-side converter in wind turbine systems," in *Proc. IEEE Power Electron. Spec. Conf.*, 2006, pp. 1–7.

[8] M. Altin, Ö Göksu, R. Teodorescu, P. Rodriguez, B. B. Jensen, and L. Helle, "Overview of recent grid codes for wind power integration," in *Proc. Int. Conf. Optim. Electr. Electron. Equip.*, 2010, pp. 1152–1160.

[9] "Aluminum electrolytic capacitors," Hitachi Ltd. [Online]. Available: http://www.hitachi-aic.com

[10] J. Kolar and S. Round, "Analytical calculation of the RMS current stress on the DC-link capacitor of voltage-PWM converter systems," *IEE Proc. Electr. Power Appl.*, vol. 153, no. 4, pp. 535–543, 2006.

[11] "ACS 800 Low voltage wind turbine converter," ABB. [Online]. Available: http://new.abb.com/power-converters-inverters/wind-turbines/utility-scale/acs800

[12] "Converteam prowind series," [Online]. Available: http://www.gepowerconversion.com/product-solutions/low-voltage-drives/prowind

[13] "PowerModule PM3000 series," AMSC. [Online]. Available: http://www.amsc.com/solutions-products/power_converters.html

[14] X. Yuan, W. Merk, H. Stemmler, and J. Allmeling, "Stationary-frame generalized integrators for current control of active power filters with zero steady-state error for current harmonics of concern under unbalanced and distorted operating conditions," *IEEE Trans. Ind. Appl.*, vol. 38, no. 2, pp. 523–532, 2002.

[15] J. Jung, S. Lim, and K. Nam, "A feedback linearizing control scheme for a PWM converter–inverter having a very small DC-link capacitor," *IEEE Trans. Ind. Appl.*, vol. 35, no. 5, pp. 1124–1131, 1999.

[16] N. Hur, J. Jung, and K. Nam, "A fast dynamic DC-link power-balancing scheme for a PWM converter-inverter system," *IEEE Trans. Ind. Electron.*, vol. 48, no.4, pp. 794–803, 2001.

[17] Y. Suh, Y. Go, and D. Rho, "A Comparative study on control algorithm for active front-end rectifier of large motor drives under unbalanced input," *IEEE Trans. Ind. Appl.*, vol. 47, no. 3, pp. 1419–1431, 2011.

[18] Y. Wang, L. Xu, and B. W. Williams, "Compensation of network voltage unbalance using doubly fed induction generator-based wind farms," *IET Renew. Power Gen.*, vol. 3, no. 1, pp. 12–22, 2009.

[19] A. G. Abo-Khalil, D. C. Lee, and J. I. Jang, "Control of back-to-back PWM converters for DFIG wind turbine systems under unbalanced grid voltage," in *Proc. IEEE Int. Symp. Ind. Electron.*, 2007, pp. 2637–2642.

[20] H. S. Song and K. Nam, "Dual current control scheme for PWM converter under unbalanced input voltage conditions," *IEEE Trans. Ind. Electron.*, vol. 46, no.5, pp. 953–959, 1999.

[21] A. Yazdani and R. Iravani, "A unified dynamic model and control for the voltage-sourced converter under unbalanced grid conditions," *IEEE Trans. Power Deliv.*, vol. 21, no. 3, pp. 1620–1629, Jul. 2006.

[22] S. Alepuz, S. B. Monge, J. Bordonau, J. A. Martinez-Velasco, C. A. Silva, J. Pontt, and J. Rodriguez, "Control strategies based on symmetrical components for grid-connected converters under voltage dips," *IEEE Trans. Ind. Electron.*, vol. 56, no. 6, pp. 2162–2173, 2009.

[23] J. Hu and Y. He, "Reinforced control and operation of DFIG-based wind-power-generation system under unbalanced grid voltage conditions," *IEEE Trans. Energy Convers.*, vol. 24, no. 4, pp. 905–915, 2009.

[24] Y. Suh and T. A. Lipo, "Modeling and analysis of instantaneous active and reactive power for PWM AC/DC converter under generalized unbalanced network," *IEEE Trans. Power Deliv.*, vol. 21, no. 3, pp. 1530–1540, 2006.

[25] J. G. Hwang and P. W. Lehn, "A single-input space vector for control of AC–DC converters under generalized unbalanced operating conditions," *IEEE Trans. Power Electron.*, vol. 25, no. 8, pp. 2068–2081, 2010.

[26] M. H. Bierhoff and F. W. Fuchs, "DC-link harmonics of three-phase voltage-source converters influenced by the pulsewidth-modulation strategy—An analysis," *IEEE Trans. Ind. Electron.*, vol. 55, no. 5, pp. 2085–2092, 2008.

PART IV

GRID FAULT RIDE-THROUGH OF DFIG

CHAPTER 11

DYNAMIC MODEL OF DFIG UNDER GRID FAULTS

In this chapter, the dynamic model of DFIG under grid fault will be analyzed in detail, including the model under voltage dips, during voltage recoveries, and under recurring grid faults. The dynamic model of a DFIG under voltage dips with three operation conditions—with rotor open circuit, with normal vector control, and with the rotor-side crowbar are investigated. Both symmetrical grid fault and asymmetrical grid fault are included. For the dynamic model of the DFIG during voltage recovery and under recurring grid faults, the analysis is based on common fault ride-through strategies of the DFIG, that is, using rotor crowbar under voltage dips but normal control method under voltage recoveries. For DFIG under recurring grid faults, the influences of the recurring fault parameters on the performance of DFIG are analyzed, which includes the voltage dip level and the grid fault angle of the first voltage dip, as well as the duration between two faults. Corresponding simulations and experimental results are provided.

11.1 INTRODUCTION

As introduced in Chapter 3, the fault ride-through (FRT) requirements for wind turbine system (WTS) have been carried out in many countries around the world. The WTSs are required to stay connected to the grid during grid faults and provide reactive power support. The grid faults will normally cause voltage dips at the terminal of the WTS. For a WTS with a full-scale power converter, as the generator is fully isolated from the grid by the power converter, the voltage dips will not be "seen" by the generator directly. So the FRT ability of the system will be determined by the FRT ability of the power converter [16]. However, with respect to the DFIG WTS, besides the grid-side converter, the voltage dips will also influence the generator, as the stator of the DFIG is directly connected to the grid. As a result, the performance of the DFIG under grid faults will be more complicated. Before discussing the advanced FRT strategy of the DFIG, it is essential to pay attention to the dynamic model of the DFIG under grid fault and to estimate the performance of the system under grid faults.

Advanced Control of Doubly Fed Induction Generator for Wind Power Systems, First Edition.
Dehong Xu, Frede Blaabjerg, Wenjie Chen, and Nan Zhu.
© 2018 by The Institute of Electrical and Electronics Engineers, Inc. Published 2018 by John Wiley & Sons, Inc.

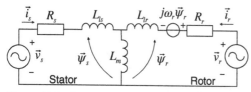

Figure 11.1   Equivalent circuit of the DFIG represented in $\alpha\beta$ frames.

In this chapter, the dynamic model of DFIG under grid fault will be analyzed in detail, including the dynamic model of DFIG under voltage dips, during voltage recoveries, and under recurring grid faults. The dynamic model of the DFIG under voltage dips with three operation conditions—with rotor open circuit, with normal vector control and with rotor-side crowbar are investigated. Both symmetrical grid fault and asymmetrical grid fault are included. For the dynamic model of the DFIG during voltage recovery and under recurring grid faults, the analysis is based on the common FRT strategy of the DFIG, that is, using rotor crowbar under voltage dips but normal control method under voltage recoveries. For DFIG under recurring grid faults, the influences of the recurring fault parameters on the performance of DFIG are analyzed, including the voltage dip level and the grid fault angle of the first voltage dip, as well as the duration between two faults. Corresponding simulations and experimental results are provided.

11.2   BEHAVIOR DURING VOLTAGE DIPS

11.2.1   Equivalent Circuits of DFIG under Voltage Dips

In order to analyze the performance of DFIG, the dynamic model of the DFIG in $\alpha\beta$ frame is rewritten in (11.1), and its equivalent circuit is shown in Figure 11.1 [17].

$$\vec{v}_s = R_s\vec{i}_s + \frac{d}{dt}\vec{\psi}_s \qquad\qquad \vec{\psi}_s = L_s\vec{i}_s + L_m\vec{i}_r$$

$$\vec{v}_r = R_r\vec{i}_r + \frac{d}{dt}\vec{\psi}_r + j\omega_r\vec{\psi}_r \qquad \vec{\psi}_r = L_r\vec{i}_r + L_m\vec{i}_s \qquad (11.1)$$

where $\vec{v}_s$, $\vec{v}_r$ are the stator and rotor voltages, respectively; $\vec{i}_s$, $\vec{i}_r$ are the stator and rotor currents, respectively; $\vec{\psi}_s$, $\vec{\psi}_r$ are the stator and rotor flux linkages, respectively. R_s and R_r are the stator and rotor resistances, respectively; L_s and L_r are the stator and rotor inductances, respectively; and L_m is the mutual inductance.

11.2.1.1   Symmetrical Voltage Dips

The basic principle of a DFIG under voltage dips is that the stator flux of the DFIG cannot be changed abruptly. As a result, a stator natural flux $\vec{\psi}_{sn}$ will be introduced to keep flux continuity. The trajectory of the stator flux under 80% symmetrical voltage dips is shown in Figure 11.2. Before the voltage dips, the stator flux vector $\vec{\psi}_s$ is

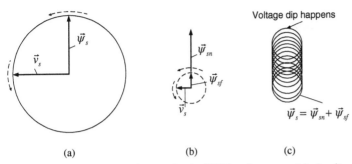

Figure 11.2   Stator flux trajectory for the DFIG under symmetrical voltage dips: (a) before dips; (b) at the dip instant; (c) after dips.

rotating with a grid voltage vector $\vec{v}_s(\vec{v}_s = V_s e^{j\omega_s t})$, when the dips happen at $t_0 = 0$, the voltage amplitude of the grid voltage vector $\vec{v}_s$ is reduced immediately. However, according to the flux conservation principle, the stator flux $\vec{\psi}_s$ cannot change rapidly, so a natural flux $\vec{\psi}_{sn}$ will hereby emerge, as expressed in (11.2).

$$\vec{\psi}_{sn}(t_0) = \frac{pV_s}{j\omega_s} \tag{11.2}$$

where p is the voltage dip level, as also illustrated in Figure 11.2b [18].

As a result, the stator flux after the voltage dip is the sum of two components with different frequencies. One is the stator natural flux $(\vec{\psi}_{sn})$ produced by the voltage dip, which is a time-decayed DC component in the stator $\alpha\beta$ frame. Its decaying speed is related to the operation of the DFIG after voltage dips, which will be introduced in the following parts. The other is the stator-forced flux $(\vec{\psi}_{sf})$ produced by the remaining grid voltage $\vec{v}_s = (1 - p)V_s e^{j\omega_s t}$, which is rotating with the grid voltage vector and with the grid frequency ω_s, which is shown in Figure 11.2b.

According to the superposition principle, the dynamic model of the DFIG under balanced voltage dips can be described with two independent equivalent circuits: the natural machine, whose stator is short-circuited while the natural flux $\vec{\psi}_{sn}$ exists in the stator winding, and the forced machine, where stator voltage is the remaining grid voltage and no transient flux exists, as shown in Figure 11.3 [1].

The equivalent circuit of the forced machine is the same as the DFIG under normal operation, which has been analyzed in the previous chapters. The different equation which describes the relationship between stator natural flux $\vec{\psi}_{sn}$ and rotor natural current $\vec{i}_{rn}$ can be derived from Figure 11.3b, as expressed in (11.3) and (11.4):

$$\frac{d}{dt}\vec{\psi}_{sn} = -\frac{R_s}{L_s}\vec{\psi}_{sn} + \frac{L_s}{L_s}R_s\vec{i}_{rn} \tag{11.3}$$

$$\left(\frac{d}{dt} - j\omega_r\right)\vec{\psi}_{sn} = \vec{v}_{rn} - R_r\vec{i}_{rn} - \left(\frac{d}{dt} - j\omega_r\right)\sigma L_r\vec{i}_{rn} \tag{11.4}$$

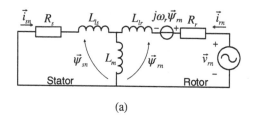

(a)

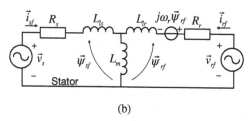

(b)

Figure 11.3 Equivalent circuits of the DFIG under symmetrical voltage dips: (a) the natural machine and (b) the forced machine.

11.2.1.2 Asymmetrical Voltage Dips

According to the symmetrical component method, the asymmetrical grid voltage can be divided into the positive-sequence, negative-sequence, and zero-sequence components. The voltage vectors of the three types of asymmetrical grid faults is shown in Figure 11.4, whereas the positive-sequence (V_s^+), negative-sequence (V_s^-), and zero-sequence (V_s^0) in the grid voltage are listed in Table 11.1.

Besides the stator natural flux $\vec{\psi}_{sn}$ and the forced flux $\vec{\psi}_{sf}$, a stator negative flux $\vec{\psi}_{sne}$ with a frequency of $-\omega_s$ will be introduced by the negative-sequence grid voltage as well. Another equivalent circuit, the negative machine, can be used to analyze the performance of the DFIG, together with the natural machine and the forced machine, which is illustrated in Figure 11.5 [2].

The rotor equation of the negative machine can be there by derived as (11.5).

$$\left(\frac{d}{dt} - j\omega_r\right)\vec{\psi}_{sne} = \vec{v}_{me} - R_r\vec{i}_{rn} - \left(\frac{d}{dt} - j\omega_r\right)\sigma L_r\vec{i}_{me} \qquad (11.5)$$

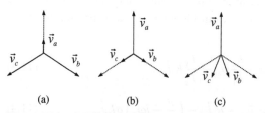

(a)   (b)   (c)

Figure 11.4 Grid voltage during asymmetrical grid faults: (a) single-phase-to-ground faults; (b) two-phase-to-ground faults; and (c) two-phase faults.

TABLE 11.1   1 Symmetrical components in the grid voltage under asymmetrical grid faults

Fault type	V_s^+	V_s^-	V_s^0
Single-phase to ground	$1 - p/3$	$p/3$	$p/3$
Two-phase to ground	$1 - p/2$	$p/2$	0
Two phase	$1 - 2p/3$	$p/3$	0

With the existence of the forced flux and the negative-sequence flux, the steady-state stator flux trajectory after asymmetrical grid fault will be an ellipse. The stator flux trajectory under a single-phase-to-ground fault is shown in Figure 11.6a.

In this case, the stator natural flux produced during the asymmetrical grid faults not only depend on the fault types and voltage dip levels, but also on the time when the fault happens, which is described by the fault angle θ in Figure 11.6. If the fault happens at $\theta = 90°$ when the positive and negative-sequence grid voltage have opposite directions, the sum of the stator force and negative flux after fault will be the same as the stator flux before the fault, and thus the stator natural flux will be zero, which is shown in Figure 11.6b. When the fault happens at $\theta = 0$, the stator natural flux will be maximum, as shown in Figure 11.6c, and in this case, it can be given as

$$\vec{\psi}_{sn}(t_0)|_{\theta=0} = \frac{2}{3} p \frac{V_s}{j\omega_s} \tag{11.6}$$

11.2.2   With Rotor Open Circuit

11.2.2.1   Under Symmetrical Voltage Dips
If the rotor terminals are open-circuited under voltage dips, the rotor current is zero. This situation may happen when the rotor-side converter (RSC) is disabled under voltage dips. In this case, (11.3) and (11.4) can be rewritten as

$$\frac{d}{dt}\vec{\psi}_{sn} = -\frac{R_s}{L_s}\vec{\psi}_{sn} \tag{11.7}$$

$$\left(\frac{d}{dt} - j\omega_r\right)\vec{\psi}_{sn} = \vec{v}_{rn} \tag{11.8}$$

...er voltage dips can be

By solving (11.7), the stator natural current

derived as

$$\frac{?}{j\omega_s}e^{-\frac{t}{\tau_s}}$$

$$\vec{\psi}_{sn}(t) = \vec{\psi}_{s}. \tag{11.9}$$

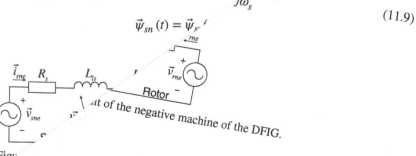

...t of the negative machine of the DFIG.

Figu

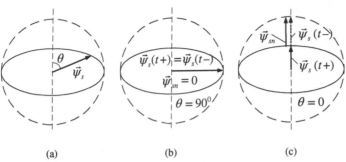

Figure 11.6   Trajectory of the stator flux during asymmetrical grid faults: (a) steady state after faults; (b) situation with zero natural flux; and (c) situation with the maximum natural flux.

where $\tau_s = L_s/R_s$ is the stator time constant. It can be concluded from (11.9) that the stator natural flux is time-decaying with the stator time constant τ_s after voltage dips. As in a large-scale DFIG, the stator resistance R_s is normally very small, so the stator time constant τ_s will be normally be larger than 1 s.

The rotor voltage in this situation can be defined as the rotor electromotive force (EMF) [3], so the rotor EMF $\vec{E}_{rn}$ of the natural machine in this case can be found as

$$\vec{E}_{rn}(t) \approx -j\omega_r\vec{\psi}_{sn} = -\frac{\omega_r}{\omega_s}pV_se^{-\frac{t}{\tau_s}} \tag{11.10}$$

The stator current can in the natural machine $\vec{i}_{sn}$ be calculated from (11.9) and (11.1), as expressed in (11.11):

$$\vec{i}_{sn}(t) = \frac{pV_s}{j\omega_sL_s}e^{-\frac{t}{\tau_s}} \tag{11.11}$$

The
operation, as EMF $\vec{E}_{rf}$ and stator current $\vec{i}_{sf}$ is similar to the DFIG under normal that the rotor messed in (11.12) and (11.13). It can be seen from (11.10) and (11.11) forced EMF $\vec{E}_{rf}$ is EMF $\vec{E}_{rn}$ is proportional to the rotor speed ω_r, while the rotor-operating around syn-tional to the slip speed $\omega_r - \omega_s$. As the DFIG is normally under serious voltage d speed ω_s, ω_r is much larger than $\omega_r - \omega_s$. As a result, larger than in forced mach ctor EMF in the natural machine $\vec{E}_{rn}$ will be much operation as well. much larger than the rotor EMF under normal

$$\vec{E}_{rf}(t) \approx$$

$$\vec{i}_{sf}(t) = \frac{(1 - p)V_se^{j\omega_st}}{j\omega_sL_s} \tag{11.12}$$

$$\tag{11.13}$$

The simulation of a 1.5 MW DFIG und shown in Figure 11.7. The rotor speed is 1.2 pu open-circuited. It can be found that the stator flux i trical voltage dip is rotor of DFIG is stator natural

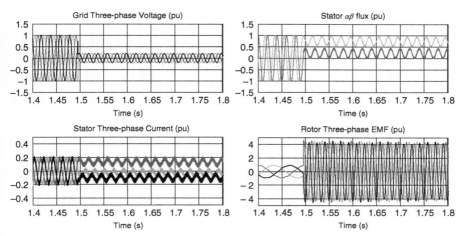

Figure 11.7   Simulation of a 1.5MW DFIG under an 80% symmetrical voltage dip.

flux and an AC forced flux. The natural flux decays very slowly, as the stator time constant is about 1.6 s in this case. The stator current also contains the DC natural current and AC forced current.

It should be noted that the rotor natural EMF under voltage dips (which is with the rotor frequency in the rotor reference) is much larger than that under normal operations (which is with the slip frequency in rotor reference). It reaches more than 4 pu under voltage dips, compared to about 1 pu under normal operation and as analyzed in the previous paragraph.

11.2.2.2 Asymmetrical Voltage Dips

The negative-sequence stator flux will be introduced under asymmetrical voltage dip, and it can be expressed in (11.14).

$$\vec{\psi}_{sne}(t) = \frac{\vec{v}_{sne}(t)}{j\omega_s} = \frac{V_s^-}{j\omega_s}e^{-j\omega_s t} \tag{11.14}$$

The amplitude of the negative-sequence grid voltage V_s^- with different fault types can be found from Table 11.1. The stator current $\vec{i}_{sne}$ and rotor EMF $\vec{E}_{rne}$ in the negative machine can be calculated from the equivalent circuit in Figure 11.5 and (11.5). As shown in (11.15) and (11.16),

$$\vec{i}_{sne}(t) = -\frac{V_s^-}{j\omega_s L_s}e^{-j\omega_s t} \tag{11.15}$$

$$\vec{E}_{rne}(t) \approx j\frac{\omega_r + \omega_s}{\omega_s}(1-p)V_s^-e^{j(\omega_r+\omega_s)t} \tag{11.16}$$

It can be concluded that as $\omega_r + \omega_s$ is much larger than $\omega_r - \omega_s$, so under serious asymmetrical grid faults, the rotor negative EMF $\vec{E}_{rne}(t)$ is also much larger compared to the EMF under normal operation.

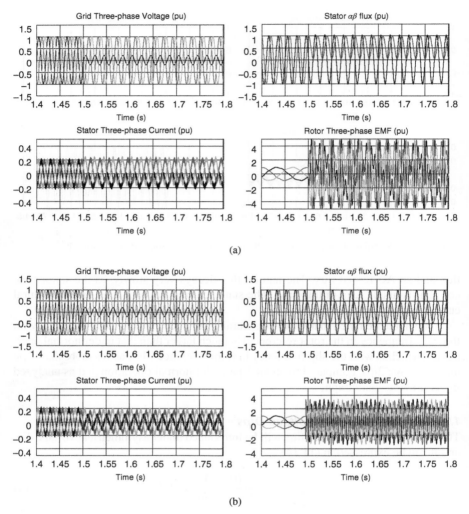

Figure 11.8   Simulation of a 1.5 MW DFIG under one-phase-to-ground asymmetrical voltage dip with (a) grid fault angle $\theta = 0°$ and with (b) grid fault angle $\theta = 90°$.

The forced and natural fluxes introduced by the unbalanced voltage dips have been introduced in the previous analysis, and the corresponding stator current and rotor EMF can be calculated with the same method used in the balanced voltage dips. Notice that the natural flux introduced in unbalanced voltage dips is related to the grid fault angle θ.

The simulation results of a 1.5 MW DFIG under one-phase-to-ground asymmetrical grid faults are shown in Figure 11.8. The rotor speed is 1800 rpm and the voltage dip level of phase A is 80%. In Figure 11.8a, the grid fault angle is 0 and the voltage dip will introduce the largest stator natural flux; it can be found that a DC

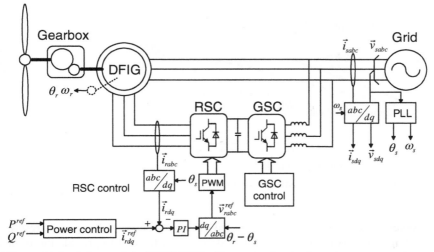

Figure 11.9   Scheme of the DFIG WTS with stator voltage oriented vector control.

component is introduced in stator flux and stator current. At the same time, the stator fluxes and currents are unbalanced, which indicate that there are negative-sequence flux and current. The rotor EMF contains three components as well and the amplitude of rotor EMF is close to 5 pu. In Figure 11.8a, the grid fault angle is 90° and the voltage dip will not introduce any natural flux. As a result, no DC component can be found in the stator flux and current. However, the amplitude of rotor EMF is still very large because of the negative-sequence flux.

11.2.3   With Normal Vector Control

Vector control [4] is often applied when the DFIG WTS is working under normal conditions. If no special solution is applied after the voltage dips, the DFIG will continue to work with vector control. The stator natural and negative-sequence flux introduced by the grid voltage dips will influence the performance of the system. However, it has been analyzed that as the voltage rating of the RSC is limited, the performance of the DFIG will be totally different under small dips compared to that under serious and deep dips [5, 6].

11.2.3.1   Control Limit of RSC with Vector Control

The stator voltage oriented vector control scheme of the DFIG has been introduced in Chapter 5, and it is again represented in Figure 11.9.

Normally, the current controllers try to suppress the rotor natural current $\vec{i}_{rn}$ after voltage dips. However, it can be seen from (11.4) and (11.10) that if the rotor natural current $\vec{i}_{rn}$ is controlled to be zero, the output RSC voltage $\vec{v}_{rn}$ should be equal to the rotor EMF $\vec{E}_{rn}$ in the natural machine. Under asymmetrical grid faults, the RSC needs to provide corresponding negative-sequence output voltage as well.

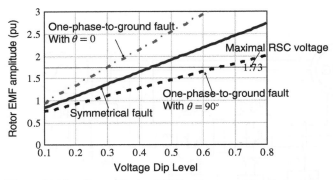

Figure 11.10   Control limit of RSC with vector control.

On the other hand, the RSC voltage is limited by the DC-bus voltage v_{dc}. As a result, if the voltage dip level is too large, the RSC may not be able to provide the rotor voltage, and the RSC will be saturated and the DFIG will be out of control. The rotor EMF amplitude of a 1.5 MW DFIG under different kinds of grid faults are shown in Figure 11.10. The rotor speed is 1800 rpm. The maximum RSC output voltage of the power converter is presented as well; the DC-bus voltage of the RSC is 1200 V, so the maximum RSC output voltage using SVM control can be calculated and transformed to per unit, resulting in 1.73 pu as in Figure 11.10.

For symmetrical faults, the RSC will be saturated when the voltage dip level is larger than 0.44. For one-phase-to-ground fault with $\theta = 90°$, as there is no stator natural flux introduced under voltage dips, the RSC will not be saturated until the voltage dip level is larger than 0.65. However, if $\theta = 0°$, the RSC is able to control the DFIG only if the voltage dip level is smaller than 0.3, as the rotor EMF introduced by the rotor natural and negative-sequence EMF are all very large.

11.2.3.2   Under Small Dips

When the voltage dip level is small and the RSC is still capable of controlling the DFIG, the current controller will try to suppress the rotor natural current. However, as the rotor natural current has a frequency of ω_s in the dq frame, the steady-state error cannot be avoided. Small amount of rotor and stator natural currents may still exist. The fluctuations in the electromagnet torque and DC-bus voltage will be thereby introduced. As the rotor natural current is not large, the stator time constant is similar to that with rotor open circuit, which is $\tau_s \approx L_s/R_s$ [5].

The simulation of a 1.5 MW DFIG WTS under a 20% symmetrical voltage dip is shown in Figure 11.11; the DFIG is generating 1 pu active power before fault and the rotor speed is 1800 rpm. The DC stator natural flux is introduced and it decays slowly and a DC component in stator current also exists, similar to the situation with rotor open circuit. The rotor current is distorted as the rotor natural current with the frequency of ω_r is introduced. The DC-bus voltage and torque fluctuations can be observed as well.

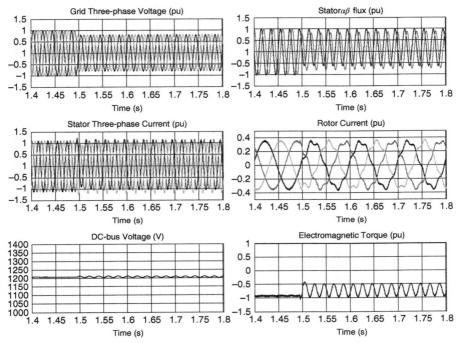

Figure 11.11   Simulation of the DFIG under 20% symmetrical voltage dips.

11.2.3.3 Under Serious Dips

The RSC may be saturated under serious voltage dips. In this case, the DFIG will be out of control, a large rotor current will be produced and will flow into the DC bus of the RSC, the converters may be overloaded, and the DC voltage will be too high. Besides, large electromagnetic torque fluctuation will be introduced, and this may reduce the life time of the gearbox.

The simulation of a 1.5 MW DFIG WTS under an 80% symmetrical voltage dip is shown in Figure 11.12. In this case, the RSC is saturated and the DFIG is out of control. It can be seen that the stator current under voltage dip is larger than 3 pu, and the rotor current is larger than 1.1 pu (about 3 pu refers to the stator side), almost three times larger than under normal operation. Besides, the DC voltage rises from 1200 V to more than 1600 V. The RSC is having overcurrent and DC overvoltage. A large electromagnetic torque fluctuation is introduced as well. The DFIG cannot ride through such serious voltage dips using normal vector control.

11.2.4   With Rotor-Side Crowbar

The rotor-side crowbar is widely used for the low voltage ride-through (LVRT) of a DFIG. The basic control scheme of a DFIG with rotor-side crowbar is shown in Figure 11.13 [19].

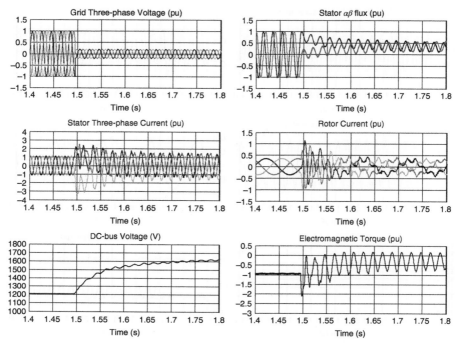

Figure 11.12   Simulation of a 1.5 MW DFIG under 80% symmetrical voltage dips operations at 1800 rpm.

Under grid faults, the crowbar is enabled to protect the RSC after the voltage dips and (or) the over rotor current is detected [7]. The rotor of the DFIG is short-circuited and the RSC is disabled. The crowbar will be introduced in detail in Chapter 12. The active crowbar with a diode rectifier and IGBT switch is usually used. However, the linearized crowbar model is used in the analysis, as shown in Figure 11.13.

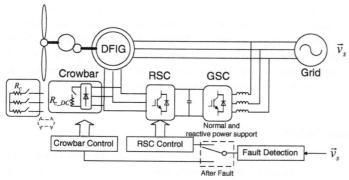

Figure 11.13   Scheme of the DFIG WTS with rotor-side crowbar for handling LVRT.

The equivalent relationship between R_c and R_{c_DC} can be expressed as in (11.17), if the both circuit has the same rotor current amplitude.

$$R_c \approx \frac{\sqrt{3}}{3} R_{c\_DC} \tag{11.17}$$

As the crowbar is active after voltage dips, the rotor natural voltage $\vec{u}_{rn}$ in (11.4) can be represented as

$$\vec{v}_{rn} = -R_c \vec{i}_{rn} \tag{11.18}$$

So, (11.4) can be rewritten as (11.19).

$$\left(\frac{d}{dt} - j\omega_r\right) \frac{L_m}{L_s} \vec{\psi}_{sn} = -(R_c + R_r)\vec{i}_{rn} - \left(\frac{d}{dt} - j\omega_r\right)\sigma L_r \vec{i}_{rn} \tag{11.19}$$

It has been reported in [8] and [9] that the stator natural flux after symmetrical voltage dips is a DC component that decays with the new stator time constant τ_{sc}, and its initial value has been expressed as (11.9). So the stator natural flux $\vec{\psi}_{sn}$ is expressed as (11.20).

$$\vec{\psi}_{sn}(t) \approx p\frac{V_s}{j\omega_s} e^{-\frac{t}{\tau_{sc}}} \tag{11.20}$$

Substituting (11.20) in (11.19) gives

$$\left(-\frac{1}{\tau_{sc}} - j\omega_r\right)\frac{L_m}{L_s}\vec{\psi}_{sn}(t) = -(R_c + R_r)\vec{i}_{rn} - \left(\frac{d}{dt} - j\omega_r\right)\sigma L_r \vec{i}_{rn} \tag{11.21}$$

Normally, in a large-scale DFIG, $\frac{1}{\tau_s} \ll \omega_r$ [1], and $R_c \gg R_r$, so (11.21) can be rewritten as

$$-j\omega_r \frac{L_m}{L_s}\vec{\psi}_{sn}(t) = -R_c \vec{i}_{rn} - \left(\frac{d}{dt} - j\omega_r\right)\sigma L_r \vec{i}_{rn} \tag{11.22}$$

So, with (11.20) and (11.22), the rotor natural current can be

$$\vec{i}_{rn}(t) \approx -\frac{\omega_r}{\omega_s}\frac{p_1 V_s}{R_c + j\omega_r \sigma L_r} e^{-\frac{t-t_0}{\tau_{sc}}} + \frac{\omega_r}{\omega_s}\frac{p_1 V_s}{R_c + j\omega_r \sigma L_r} e^{-j\omega_r t}e^{-\frac{t-t_0}{\tau_r}} \tag{11.23}$$

where τ_r is the rotor time constant and $\tau_r \approx \sigma L_r/R_c$. In a large-scale DFIG $R_c \gg \sigma L_r$, the second term in (11.23) will be decaying very fast after the voltage dips [8], so the approximate expression of (11.23) can be found as

$$\vec{i}_{rn}(t) \approx -\frac{\omega_{sl}}{\omega_s}\frac{p_1 V_s}{R_c + j\omega_r \sigma L_r} e^{-\frac{t-t_0}{\tau_s}} \tag{11.24}$$

By substituting (11.24) and (11.20) in (11.3), the new stator time constant τ_{sc} can be expressed as

$$\frac{1}{\tau_{sc}} \approx \frac{R_s}{L_s}\frac{R_c^2 + (\omega_r \sigma L_r)^2 + \omega_r^2 \sigma L_r L_m}{R_c^2 + (\omega_r \sigma L_r)^2} + \frac{R_s}{L_s}\frac{R_c \omega_r L_m}{R_c^2 + (\omega_r \sigma L_r)^2}j \tag{11.25}$$

The imaginary part indicates that the natural flux and current may rotating slowly while decaying [8]; it will not change the amplitude of $\vec{\psi}_{sn}$ and $\vec{i}_{rn}$, so the stator time constant τ_{sc} can be regarded as

$$\tau_{sc} \approx \frac{L_s}{R_s + R_s L_m \frac{\omega_r^2 \sigma L_r}{R_c^2 + (\omega_r \sigma L_r)^2}} \tag{11.26}$$

It can be found from (11.24) and (11.26) that the rotor current is limited with rotor-side crowbar. Besides, the stator time constant is reduced compared to the vector control, which means the damping of stator natural flux $\vec{\psi}_{sn}$ is accelerated.

The rotor-forced current $\vec{i}_{rf}$ and stator-forced current $\vec{i}_{sf}$ stator with rotor-side crowbar can be represented as

$$\vec{i}_{rf}(t) = \frac{\omega_r - \omega_s}{\omega_s} \frac{(1 - p_1)V_s}{R_c + j(\omega_s - \omega_r)\sigma L_r} e^{j\omega_s t} \tag{11.27}$$

$$\vec{i}_{sf}(t) = \left[\frac{1}{L_s} \frac{(1 - p_1)V_s}{j\omega_s} - \frac{\omega_r - \omega_s}{\omega_s} \frac{(1 - p_1)V_s}{R_c + j(\omega_s - \omega_r)\sigma L_r} \right] e^{j\omega_s t} \tag{11.28}$$

It can be seen from (11.28) that the DFIG will absorb reactive power from the grid if the crowbar is enabled. With respect to the active power, the DFIG will operate as a motor under sub-synchronous speed and will absorb active power ($\omega_r - \omega_s > 0$), while under super-synchronous speed, the DFIG will generate active power ($\omega_r - \omega_s < 0$).

Under asymmetrical faults, the rotor and stator negative-sequence currents $\vec{i}_{rne}$ and $\vec{i}_{sne}$ can be calculated from (11.5), as given in (11.29) and (11.30).

$$\vec{i}_{rne}(t) = \frac{\omega_r + \omega_s}{\omega_s} \frac{V_s^-}{R_c + j(\omega_r + \omega_s)\sigma L_r} e^{j(\omega_r + \omega_s)t} \tag{11.29}$$

$$\vec{i}_{sne}(t) = \left[\frac{1}{L_s} \frac{V_s^-}{j\omega_s} - \frac{\omega_r + \omega_s}{\omega_s} \frac{V_s^-}{R_c + j(\omega_r + \omega_s)\sigma L_r} \right] e^{j(\omega_r + \omega_s)t} \tag{11.30}$$

The rotor and stator negative-sequence currents can be limited by the rotor-side crowbar as well.

The simulation of a 1.5 MW DFIG WTS under an 80% symmetrical voltage dip is shown in Figure 11.14. It can be seen that the damping of stator natural flux is accelerated compared to Figure 11.12 and the rotor and stator current is limited as well. Besides, the DC voltage will not rise and the electromagnet torque fluctuations are relatively small.

The simulated results of the DFIG WTS under symmetrical grid fault are shown in Figure 11.15 (solid lines), compared to the calculated results (dotted lines) as a comparison. The amplitude of the rotor current (refers to the stator side) and stator current, as well as the stator time constant with different crowbar resistance values under different voltage dip levels are simulated. The simulated results are basically in accordance with the calculated results.

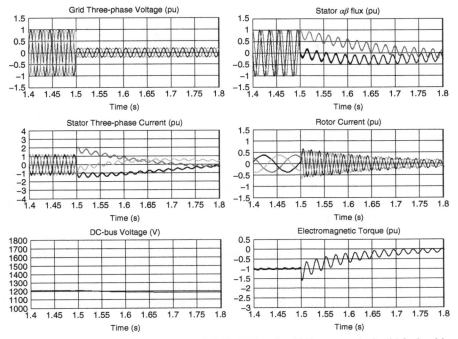

Figure 11.14   Simulation of a 1.5 MW DFIG WTS under 80% symmetrical grid fault with rotor-side crowbar, operating at 1800 rpm.

11.2.5   Non-Instant Voltage Dips

In the previous analysis, the voltage dip is assumed to have happened instantly. However, in a real power system, the voltage dips may be multiple-staged [10, 11]. Also, during the LVRT test in China, a transient time smaller than 20 ms is permitted for the voltage dip generator [12]. The grid voltage dip generated by a grid emulator is shown in Figure 11.16. The transient time of the voltage dip is about 15 ms.

Although the transients of the voltage dips will be different under different grid situations, the non-instant voltage dip in Figure 11.16 is taken as an example to

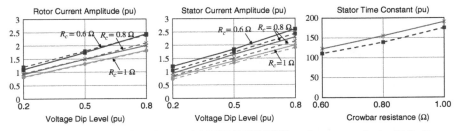

Figure 11.15   Simulated results of a 1.5 MW DFIG WTS under symmetrical grid fault (solid line) compared to the calculated results (dotted line) operating at 1800 rpm.

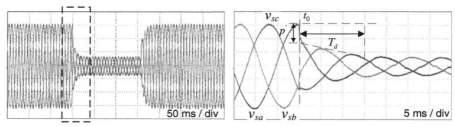

Figure 11.16   Non-instant voltage dip generated by a grid emulator.

investigate its influence on the DFIG operation. The voltage dips generated by the AC source can be regarded as two steps: the grid voltage amplitude instantly falling to $(1 - p')V_s$, and reduced linearly to $(1 - p)V_s$ in a duration of T_d. The stator natural flux introduced in the first step can easily be obtained as

$$\vec{\psi}_{sn\_1} = \frac{(1 - p')V_s}{j\omega_s} \tag{11.31}$$

It has been assumed that the voltage amplitude is reduced linearly in step 2. During voltage dips, at time t, in a smaller enough time Δt, the stator natural flux introduced by the voltage dips can be regarded as

$$\Delta \Psi_{sn} = \frac{(p - p')}{T_d} \frac{V_s}{\omega_s} \Delta t \tag{11.32}$$

As illustrated in Figure 11.17, the x-axis is aligned to $\vec{\psi}_{sn\_2}(t_0)$. So the stator natural flux introduced during T_d in x-axis and y-axis can be derived as

$$|\vec{\psi}_{sn\_2}|_x = \int_0^{T_d} \frac{(p - p')}{\omega T_d} \frac{V_s}{\omega} \cos \omega t dt = \frac{(p - p')V_s}{\omega T_d} \sin \omega T_d \tag{11.33}$$

$$|\vec{\psi}_{sn\_2}|_y = \int_0^{T_d} \frac{(p - p')}{\omega T_d} \frac{V_s}{\omega} \sin \omega t dt = \frac{(p - p')V_s}{\omega T_d}(1 - \cos \omega T_d) \tag{11.34}$$

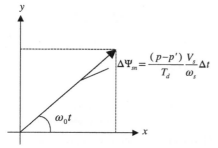

Figure 11.17   Stator flux differential of the DFIG under a non-instant fault.

So, the amplitude of the stator flux $|\vec{\psi}_{sn\_2}|$ can be derived as

$$|\vec{\psi}_{sn\_2}| = \sqrt{|\vec{\psi}_{sn\_2}|_x^2 + |\vec{\psi}_{sn\_2}|_y^2} = \frac{2(p - p')V_s}{\omega T_d} \sin \frac{\omega T_d}{2} \qquad (11.35)$$

If the damping of $|\vec{\psi}_{sn\_1}|$ during T_d is taken into account, the amplitude of the stator natural flux introduced in non-instant fault can be approximately derived as

$$\vec{\psi}_{sn\_d} \approx \vec{\psi}_{sn\_1} e^{-\frac{T_d}{\tau_s}} + \vec{\psi}_{sn\_2} \qquad (11.36)$$

Especially when $T_d \rightarrow 0$, $|\vec{\psi}_{sn\_2}| \approx \frac{(p_1 - p'_1)V_s}{\omega}$, which also represents the case of the instant fault. Also notice that when $T_d = 20\,\text{ms}$, $|\vec{\psi}_{sn\_2}| \approx 0$. The performance of the DFIG system will be greatly influenced by the transients of the voltage dips.

11.3 DFIG BEHAVIOR DURING VOLTAGE RECOVERY

11.3.1 During Instant Voltage Recovery

After grid faults, the voltage will recover to normal operation. If the grid voltage rises from lower fault voltage to normal voltage instantly, the performance of the DFIG can be regarded as a reverse operation of what is happening under voltage dips. The stator natural flux will be introduced by the voltage swell as well, and under symmetrical grid fault, it can also be expressed as given in (11.2). However, the stator-forced flux $\vec{\psi}_{sf}$ after voltage recovery is generated by the normal voltage, it can be expressed as

$$\vec{\psi}_{sf}(t) = \frac{V_s}{j\omega_s} e^{j\omega_s t} \qquad (11.37)$$

Large transient currents may also be introduced under voltage recovery after serious voltage dips as the same stator natural flux is introduced. The situation may be even worse as the stator-forced flux is larger after voltage recovery.

11.3.2 Voltage Recovery in Power Systems

It has been reported in [13, 14] that in power systems, the voltage recovery does not happen instantly, but will be finished in a few steps if the operation of the isolation breaker is taken into account. A simplified three-phase grid fault model is shown in Figure 11.18. The fault started at t_0 and the voltage dip is introduced. At t'_2, the breakers begin to isolate the faults. As most of the breakers in power system can only be open at current zero, the voltage recovery will take place in two steps. At t'_2, assuming the fault current in phase A i_{fA} crosses zero first, the breaker in phase A is opened and the grid voltage in phase A comes back to normal voltage. The time when the fault started is influenced by the grid fault angle θ between the fault current and the voltage. It is determined by the line impedance between the fault location and the WTS terminal. It is usually between 75–85° for the grid fault happening in the transmission system and 45–60° in the distributed system [13]. However, the phases

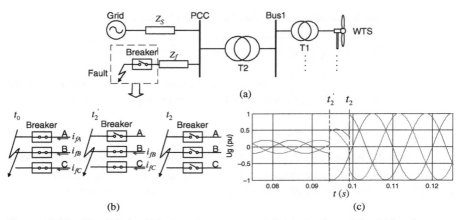

Figure 11.18 Simplified grid fault and recovery model: (a) circuit module; (b) breaker operation; and (c) voltage waveforms.

B and C are still under fault condition and as a result, the three-phase fault becomes a two-phase-short-circuit fault. After about 5 ms, the fault current in phase B and phase C also cross zero, the breaker in phases B and C is opened, and all the three-phase voltages are back to normal. The voltage waveforms during the voltage recovery of a three-phase fault are shown in Figure 11.18.

With different fault types, the operation of voltage recovery will also be different. For the three-phase-to-ground fault, the fault recovery will take place in three steps: three-phase-to-ground fault → two-phase-to-ground fault → one-phase-to-ground fault → normal grid.

For one-phase-to-ground faults, as there is only one circuit breaker to be open, the fault clearance will be finished in one step.

11.3.3 During Three-Phase Fault Recovery

Before fault recovery, the DFIG is normally operating with vector control to provide reactive power support under voltage dips. The voltage recovery begins at t_2' as shown in Figure 11.18. The trajectory of the stator flux $\vec{\psi}_s$ from t_2' to t_2 is shown in Figure 11.19. Assuming the DFIG operates in steady state under lower voltage before t_2', the initial value of stator flux before t_2' $\vec{\psi}_s(t_{2-}')$ can be expressed as

$$\vec{\psi}_s(t_{2-}') = (1-p)\frac{V_s}{j\omega_s}e^{j\omega_s t_2'} \tag{11.38}$$

At t_2', the three-phase fault will become a two-phase fault if the operation of isolation breaker is taken into account. So the forced flux after t_2' $\vec{\psi}_{sf}(t_{2+}')$ will be the sum of the positive and negative sequence as represented in (11.39).

$$\vec{\psi}_{sf}(t_{2+}') = (1-\frac{p}{2})\frac{V_s}{j\omega_s}e^{-j(\frac{\pi}{2}-\theta)} - \frac{p}{2}\frac{V_s}{j\omega_s}e^{j(\frac{\pi}{2}-\theta)} \tag{11.39}$$

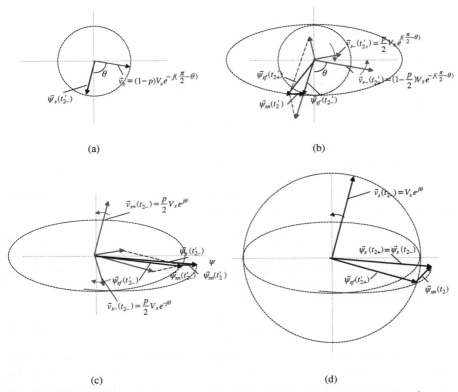

(a)   (b)

(c)   (d)

Figure 11.19   Stator flux trajectory of the DFIG during voltage recovery (a) before t_2'; (b) after t_2'; (c) before t_2; and (d) after t_2.

where θ is the grid fault angle. As the stator flux $\vec{\psi}_s$ cannot be changed abruptly $(\vec{\psi}_s(t_{2-}') = \vec{\psi}_s(t_{2+}'))$, the stator natural flux $\vec{\psi}_{sn}(t_2')$ introduced at t_2' can be derived as

$$\vec{\psi}_{sn}(t_2') = \vec{\psi}_s(t_{2+}') - \vec{\psi}_{sf}(t_{2+}') = p\frac{V_s}{\omega_s}\cos\theta \qquad (11.40)$$

as shown in Figure 11.19b. The grid voltage will back to normal at t_2, and the duration between t_2 and t_2' is normally 5 ms [13]. The stator natural flux $\vec{\psi}_{sn}(t)$ will decay within 5 ms as well. However, as the duration between t_2 and t_2' is small enough to be ignored, it can be assumed that $\vec{\psi}_{sn}(t)$ will not change in this short duration, which means $\vec{\psi}_{sn}(t_2) \approx \vec{\psi}_{sn}(t_2')$. Before t_2, the stator flux $\vec{\psi}_s(t_2)$ will be expressed as (11.41).

$$\vec{\psi}_s(t_{2-}) = \left(1 - \frac{p}{2}\right)\frac{V_s}{j\omega_s}e^{j\theta} - \frac{p}{2}\frac{V_s}{j\omega_s}e^{-j\theta} + p\frac{V_s}{\omega_s}\cos\theta \qquad (11.41)$$

which is illustrated in Figure 11.19c. The stator-forced flux after t_2 $\vec{\psi}_{sf}(t_{2+})$ is the normal stator-forced flux as shown in (11.42).

$$\vec{\psi}_{sf}(t_{2+}) = \frac{V_s}{j\omega_s}e^{j\theta} \qquad (11.42)$$

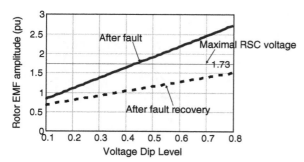

Figure 11.20   Amplitudes of rotor EMF after fault recovery and after the fault occurrence.

The stator natural flux introduced at t_2 can be derived from the flux continuity principle as well, and it is represented in (11.43):

$$\vec{\psi}_{sn}(t_2) = \vec{\psi}_s(t_{2+}) - \vec{\psi}_{sf}(t_{2+}) = \sqrt{2}p\frac{V_s}{\omega_s}\cos\theta e^{j\frac{\pi}{4}} \qquad (11.43)$$

which is the initial value of $\vec{\psi}_{sn}(t)$ after the voltage recovery. The corresponding flux trajectory is represented in Figure 11.19d.

The amplitude of stator natural flux after three-phase-fault recovery is

$$\left|\vec{\psi}_{sn}(t_2)\right| = \sqrt{2}\cos\theta p\frac{V_s}{\omega_s} \qquad (11.44)$$

As the wind farms are normally connected to transmission lines, θ is usually between 75° and 85°. So, compared to (11.2), the stator natural flux introduced by the voltage recovery will be smaller than under voltage dip. The amplitude of the rotor EMF after voltage dips and voltage recovery under a three-phase fault in a 1.5 MW DFIG is shown in Figure 11.20. The rotor speed is 1800 rpm and the grid fault angle θ is 75°. It can be seen that the rotor EMF introduced by the voltage recovery is smaller, and it is always under the maximum RSC output voltage. Consequently, the RSC is capable of controlling the DFIG with normal control after fault recovery. If the fault happens in a distribution line, the fault also influences the DFIG where θ will be smaller and the DFIG may be out of the controllable range. In this case, the rotor-side crowbar may need to be active again during the voltage recovery. This will be discussed in detail in the next chapter.

It should be noticed that during t_2' to t_2, the DFIG is operated under unbalanced fault, and the RSC may be saturated in this period. However, as this duration is short (usually 5 ms), its influence on the DC-bus voltage will normally be acceptable. On the other hand, the RSC is normally capable of handling up to 2 pu current (referred to the stator) for a short time.

After the voltage recovery, the stator flux introduced by the voltage recovery will also decay with the stator time constant τ_s, so the stator natural flux after voltage recovery can be expressed as

$$\vec{\psi}_{sn}(t) = \sqrt{2}p\frac{V_s}{\omega_s}\cos\theta e^{\frac{\pi}{4}j}e^{-\frac{t}{\tau_s}} \qquad (11.45)$$

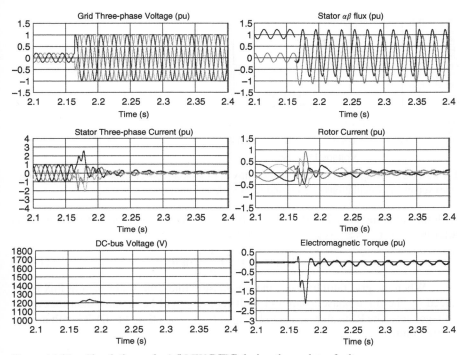

Figure 11.21 Simulations of a 1.5 MW DFIG during three-phase fault recovery.

The DFIG is normally required to restore the active power output with a rating of 0.1 pu/s, and that can be easily achieved with the vector control.

Simulations have been performed on a 1.5 MW DFIG under fault recovery from an 80% symmetrical three-phase fault, which are shown in Figure 11.21. The rotor speed is 1800 rpm and the grid fault angle θ is 75°. Before the voltage recovery, the DFIG is generating 1.0 pu reactive current under lower voltage. The reactive current generation is stopped when voltage recovery happens, but the DFIG is still operating with normal vector control. The DC-stator natural flux after voltage recovery is smaller compared to Figure 11.12, so the DC-bus voltage only rises about 40 V after the voltage recovery, and the transient stator and rotor current are smaller than the case shown in Figure 11.12 as well.

The simulated results of the stator natural flux amplitude after voltage recovery are compared with calculated results from (11.43), as shown in Figure 11.22. With the different voltage dip levels and grid fault angle, the simulated results are basically in accordance with the calculated results.

11.3.4 During Three-Phase-To-Ground Fault Recovery

During the fault recovery from three-phase-to-ground faults, the voltage recovery will normally take place in three steps. The performance of the DFIG can also be analyzed

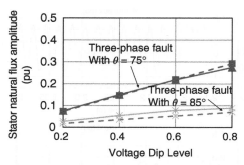

Figure 11.22 Amplitude of stator natural flux during fault recovery: the simulations (solid line) versus the calculations (11.44) and results (dotted line).

with the same method. The details will be neglected here and the stator natural flux after the fault recovery is given as (11.456).

$$|\vec{\psi}_{sn}(t_2)| = \frac{4}{3}p\frac{V_s}{\omega_s}\cos\theta \qquad (11.46)$$

When the grid fault happens in transmission lines, the stator natural flux introduced by the three-phase-to-ground fault recovery will be smaller than that under voltage dips. The RSC also seems to be capable of controlling the DFIG with normal vector control.

11.3.5 During Asymmetrical Fault Recovery

The one-phase-to-ground fault is taken as an example, as most of the faults happening in power systems are one-phase-to-ground faults [15]. It has been analyzed that under grid fault, the stator natural flux amplitude ψ_{sn} can be expressed as

$$|\vec{\psi}_{sn}| = \frac{2}{3}p\frac{\vec{v}_s}{j\omega_s}\cos\theta \qquad (11.47)$$

The fault angle θ is a random value between 0 and 2π under the voltage dips. However, during the voltage recovery at t_2, as the isolated breakers normally open when fault current crosses zero to isolate the faults, θ is no longer a random value but represents the angle between the fault current and the voltage, and it is usually between 75° and 85° in transmission systems.

After the grid fault recovery, the performance of the DFIG will basically be the same as in the case of the three-phase or three-phase-to-ground voltage recovery. A simulation of the DFIG under a 80% one-phase-to-ground fault recovery is made and the results are shown in Figure 11.23. The DFIG is working with rotor-side crowbar before fault recovery. The stator natural flux introduced by the recovery is no more than 0.1 pu, and the transient current is very small.

11.4 UNDER RECURRING GRID FAULTS

In some standards, it is specified that the WTS should be able to handle recurring grid faults, in order to make the system more robust.

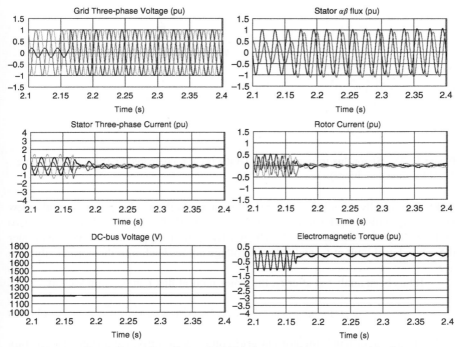

Figure 11.23 Simulations of a 1.5 MW DFIG during one-phase-to-ground fault recovery operating at 1800 rpm.

11.4.1 During Symmetrical Recurring Fault

The three-phase symmetrical fault is taken as an example here. The grid voltage of the recurring grid fault is shown Figure 11.24. The first fault occurs at t_0, a voltage dip with the dip level of p_1 is produced, and the reactive power support is required by the grid code after t_1, the duration between t_0 and t_1 is usually about 100 ms. After a few hundred milliseconds, the protection device is tripped and the circuit breaker isolates the fault at t_2'. The voltage recovery will take place in two steps. The grid voltage returns to normal at t_2. For the three-phase fault, the duration between t_2

$$\vec{v}_s(s) = (1 - p_1)\,\vec{V}_s \qquad \vec{v}_s(s) = (1 - p_2)\,\vec{V}_s$$

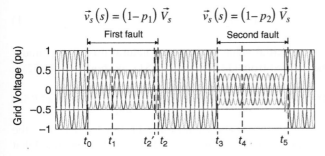

Figure 11.24 Grid voltage waveforms introduced by the recurring grid faults.

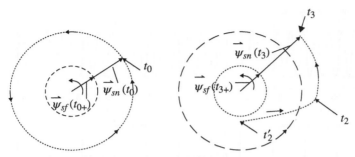

Figure 11.25 Stator flux trajectory of the DFIG: (a) at t_0 compared to; (b) at t_3 (see Figure 11.24 for time definitions).

and t_2' is normally 5 ms [13]. Another three-phase fault may happen at t_3, and a new voltage dip with a dip level of p_2 will be introduced. The duration between the two faults is defined to be $T_{re} = t_3 - t_2$. The reactive power support is also required under the second grid fault at t_4, and the grid voltage comes back into the normal situation at t_5.

It has been analyzed after t_2, the stator natural flux can be expressed as

$$\vec{\psi}_{sn}(t) = \vec{\psi}_{sn}(t_2)e^{-\frac{t-t_2}{\tau_s}} = \sqrt{2}p_1 \frac{V_s}{\omega_s}\cos\theta e^{\frac{\pi}{4}j}e^{-\frac{t-t_2}{\tau_s}} \tag{11.48}$$

On the other hand, the normal grid voltage will introduce a stator-forced flux $\vec{\psi}_{sf}(t)$, as shown in (11.49).

$$\vec{\psi}_{sf}(t) = \vec{\psi}_{sf}(t_{2+})e^{j\omega_s(t-t_2)} = \frac{V_s}{j\omega_s}e^{j[\omega_s(t-t_2)+\theta]} \tag{11.49}$$

As a result, before the second fault happens at t_3, the stator flux $\vec{\psi}_s(t_{3-})$ can be derived as (11.50).

$$\vec{\psi}_s(t_{3-}) = \vec{\psi}_{sn}(t_{3-}) + \vec{\psi}_{sf}(t_{3-}) = \sqrt{2}p_1 \frac{V_s}{\omega_s}\cos\theta e^{\frac{\pi}{4}j}e^{-\frac{T_{re}}{\tau_s}} + \frac{V_s}{j\omega_s}e^{j(\omega_s T_{re}+\theta)} \tag{11.50}$$

When the second grid fault happens at t_3, a new voltage dip with a voltage dip level of p_2 will be introduced. So, the stator-forced flux $\vec{\psi}_{sf}(t_{3+})$ at t_3 after the voltage dip can be expressed as

$$\vec{\psi}_{sf}(t_{3+}) = (1 - p_2)\frac{V_s}{j\omega_s}e^{j(\omega_s T_{re}+\theta)} \tag{11.51}$$

As $\vec{\psi}_s(t_{3+}) = \vec{\psi}_s(t_{3-})$, the initial value of the stator natural flux $\vec{\psi}_{sn}(t_3)$ under the second grid fault can be derived from (11.50) and (11.51), as shown in (11.52).

$$\vec{\psi}_{sn}(t_3) = \vec{\psi}_s(t_{3-}) - \vec{\psi}_{sf}(t_{3+}) = \sqrt{2}p_1 \frac{V_s}{\omega_s}\cos\theta e^{\frac{\pi}{4}j}e^{-\frac{T_{re}}{\tau_s}} + p_2\frac{V_s}{j\omega_s}e^{j(\omega_s T_{re}+\theta)} \tag{11.52}$$

The corresponding flux trajectory is shown in Figure 11.25b. Compared to the situation with the first grid fault, as shown in Figure 11.25a, it can be concluded that

$\vec{\psi}_{sn}(t_3)$ consists of two parts: the first part is introduced by the voltage recovery of the first grid fault and its amplitude depends on the voltage dip level of the first grid fault (p_1), the fault angle of the first grid fault (θ), and the duration between two faults (T_{re}). The second part is introduced by the second grid fault and its amplitude is only influenced by the voltage dip level of the second voltage dip (p_2).

The rotor-side crowbar will be active again at t_3. With (11.52), the stator natural flux $\vec{\psi}_{sn}(t)$ after t_3 can be represented as (11.53).

$$\vec{\psi}_{sn}(t) = \vec{\psi}_{sn}(t_3)e^{-\frac{t-t_3}{\tau_{sc}}} = \left[\sqrt{2}p_1\frac{V_s}{\omega_s}\cos\theta e^{\frac{\pi}{4}j}e^{-\frac{T_{re}}{\tau_s}} + p_2\frac{V_s}{j\omega_s}e^{j(\omega_s T_{re}+\theta)} \right] e^{-\frac{t-t_3}{\tau_{sc}}} \quad (11.53)$$

If the rotor current introduced by the forced flux $\vec{\psi}_{sf}(t)$ is neglected, the rotor current $\vec{i}_r(t)$ under the second voltage dip can also be derived as

$$\vec{i}_r(t) \approx -\frac{j\omega_r}{R_c + j\omega_r\sigma L_r} \left(\sqrt{2}p_1\frac{V_s}{\omega_s}\cos\theta e^{\frac{\pi}{4}j}e^{-\frac{T_{re}}{\tau_s}} + p_2\frac{V_s}{\omega_s}e^{j[\omega_s T_{re}+\theta-\frac{\pi}{2}]} \right) e^{-\frac{t-t_3}{\tau_{sc}}} \quad (11.54)$$

and the stator current $\vec{i}_s$ under the second voltage dip will be

$$
\begin{aligned}
\vec{i}_s(t) &= \frac{1}{L_s}\vec{\psi}_s(t) - \frac{L_m}{L_s}\vec{i}_r(t) \\
&\approx \frac{1}{L_s}\left(\sqrt{2}p_1\frac{V_s}{\omega_s}\cos\theta e^{\frac{\pi}{4}j}e^{-\frac{T_{re}}{\tau_s}} + p_2\frac{V_s}{\omega_s}e^{j[\omega_s T_{re}+\theta-\frac{\pi}{2}]} \right) \\
&\quad \times \left(1 + \frac{j\omega_r L_m}{R_c + j\omega_r\sigma L_r} \right)e^{-\frac{t-t_3}{\tau_{sc}}} + \frac{1}{L_s}\frac{(1-p_2)V_s}{j\omega_s}e^{j\omega_s(t-t_3)} \quad (11.55)
\end{aligned}
$$

The RSC needs to be active 100 ms later as well, during t_4–t_5 and the DFIG will repeat the operation during t_1–t_2.

The rotor current $\vec{i}_r(t)$ after t_3 can be represented as (11.54). If it is regarded as a function of time (t), it is time-decaying with τ_s and it can be found that when $t = t_3$, the largest rotor current can be expressed as (11.56).

$$\vec{i}_r(t_3) \approx -\frac{\omega_r}{R_c + j\omega_r\sigma L_r} \left(\sqrt{2}p_1\frac{V_s}{\omega_s}\cos\theta e^{\frac{\pi}{4}j}e^{-\frac{T_{re}}{\tau_s}} + p_2\frac{V_s}{\omega_s}e^{j[\omega_s T_{re}+\theta-\frac{\pi}{2}]} \right) \quad (11.56)$$

It should be noticed that (11.56) consists of two terms: the first one is the stator natural flux introduced by the voltage recovery of the first fault (at t_2), its amplitude depends on the voltage dip level of the first grid fault (p_1), the grid fault angle of the first grid fault (θ), and the duration between two faults (T_{re}). The second one is produced by the second grid fault and it is related to the voltage dip level of the second grid fault (p_2), which is the same with the situation under single grid fault.

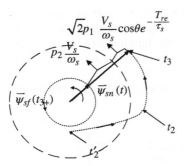

Figure 11.26   Maximum stator natural flux $\vec{\psi}_{sn}$ under the second grid fault.

The amplitude of $\vec{i}_r(t_3)$ can be expressed as (11.57).

$$\left|\vec{i}_r(t_3)\right| \approx \frac{\omega_r}{\sqrt{R_c^2 + (\omega_r \sigma L_r)^2}}$$

$$\times \sqrt{\left(\sqrt{2}p_1 \frac{V_s}{\omega_s}\cos\theta e^{-\frac{T_{re}}{\tau_s}}\right)^2 + 2p_2\frac{V_s}{\omega_s}\sqrt{2}p_1\frac{V_s}{\omega_s}\cos\theta e^{-\frac{T_{re}}{\tau_s}}\cos\left(\omega_s T_{re} + \theta - \frac{3}{4}\pi\right) + \left(p_2\frac{V_s}{\omega_s}\right)^2}$$

$$\tag{11.57}$$

where

$$\omega_s T_{re} + \theta - \frac{3}{4}\pi = 2\pi n, n \in N \tag{11.58}$$

The maximum amplitude I_{rm} of (11.57) is expressed as (11.59). It is also the situation that the stator natural flux produced by the voltage recovery of the first fault and by the voltage dips of the second fault are in the same direction, so that their amplitude is superimposed, and the rotor natural current produced by them are also superimposed, as shown in Figure 11.26.

$$I_{rm} \approx \frac{\omega_r}{\sqrt{R_c^2 + (\omega_r \sigma L_r)^2}}\left(\sqrt{2}p_1\frac{V_s}{\omega_s}\cos\theta e^{-\frac{T_{re}}{\tau_s}} + p_2\frac{V_s}{\omega_s}\right) \tag{11.59}$$

It should be noticed that (11.29) describes the time when the largest rotor current and stator natural flux can appear with an ideal PI controller and some error could be introduced in a real case. However, it can be found that the largest rotor current and stator natural flux always occur when the stator natural flux produced by the voltage recovery of the first fault and by the voltage dips of the second fault are in the same direction, as shown in Figure 11.26.

With the same method, it can be found that when $t = t_3$ and when the relationship between T_{re} and θ can be expressed as (11.58), the stator current under the second grid fault will reach its maximum amplitude I_{sm}, as represented in (11.60).

$$I_{sm} \approx \frac{1}{L_s}\left(\sqrt{2}p_1\frac{V_s}{\omega_s}\cos\theta e^{-\frac{T_{re}}{\tau_s}} + p_2\frac{V_s}{\omega_s}\right)\left(1 + \frac{\omega_r L_m}{\sqrt{R_c^2 + (\omega_r \sigma L_r)^2}}\right)$$

$$+ \frac{1}{L_s}\frac{(1-p_2)V_s}{\omega_s} \tag{11.60}$$

The electromagnetic torque of the DFIG can be expressed as

$$T_{em} = \frac{3}{2}P_s L_m(i_{sa}i_{r\beta} - i_{s\beta}i_{ra}) \tag{11.61}$$

With (11.60) and (11.59), the maximum torque fluctuations under the second grid fault T_{em} (in pu) can be obtained as (11.62).

$$T_{em} \approx \left[\frac{\omega_r}{\omega_s}\frac{(p_2 V_s + \sqrt{2}\cos\theta p_1 V_s e^{-\frac{T_{re}}{\tau_s}})R_c}{R_c^2 + (\omega_r \sigma L_r)^2} + \frac{\omega_r - \omega_s}{\omega_s}\frac{(1-p_2)V_s R_c}{R_c^2 + ((\omega_r - \omega_s)\sigma L_r)^2}\right]$$

$$\times \frac{p_2 V_s + \sqrt{2}\cos\theta p_1 V_s e^{-\frac{T_{re}}{\tau_s}}}{\omega_s}$$

$$+ \frac{(1-p_2)V_s}{\omega_s}\frac{\omega_r}{\omega_s}\frac{p_2 V_s + \sqrt{2}\cos\theta p_1 V_s e^{-\frac{T_{re}}{\tau_s}}}{\sqrt{R_c^2 + (\omega_r \sigma L_r)^2}} \tag{11.62}$$

With (11.60), (11.61), and (11.62), the influence of the recurring fault parameters (p_1, θ, and T_{re}) on the transient rotor current, stator current, and torque fluctuations can be analyzed in detail.

For the DFIG under three-phase-to-ground recurring faults, its performance can be analyzed by (11.47) with the same method. It will not be repeated in this chapter.

11.4.2 Influence of the First Dip Level

The influence of p_1 on the performance of DFIG under the second grid fault in a 1.5 MW DFIG WTS is shown in Figure 11.27, and the parameters of the DFIG is shown in Table 11.2. The rotor speed is 1650 rpm and the duration between two faults $T_{re} \approx$ 500 ms. The grid fault angle θ is 60°.

It can be seen from Figure 11.27a that the maximum rotor current I_{rm} is increased with the increase of p_1 and it reaches 3.5 pu if $p_1 = 80\%$ and $p_2 = 80\%$, compared to 2.3 pu at most under the first grid fault (the same situation with single grid fault). It can be found in Figures 11.27a–11.27c that the maximum stator current I_{sm} is also increased to more than 3.6 pu with $p_1 = 80\%$, and the maximum torque fluctuation T_{em} will be about 3 pu, much larger than that under single grid fault (about 1.5 pu). When p_1 is reduced to 50% and 20%, the maximum rotor current I_{rm}

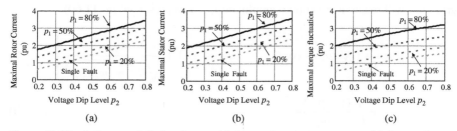

Figure 11.27 Influence of dip level p_1 on (a) the maximum rotor current; (b) the maximum stator current; and (c) the maximum torque fluctuations under the second grid fault for a 1.5 MW DFIG operating at 1650 rpm.

is reduced to 3.0 pu and 2.4 pu with $p_2 = 80\%$; the maximum stator current I_{sm} and the maximum torque fluctuation T_{em} are also reduced with the decrease of p_1, as the stator natural flux $\vec{\psi}_{sn}$ after the voltage recovery from the first grid fault is decreased.

The corresponding simulation results are shown in Figure 11.28. The performances of the DFIG when the two faults happen at $T_{re} \approx 500$ ms are simulated, the voltage dip level of the second fault $p_2 = 80\%$, the rotor speed is 1650 rpm, and the grid fault angle of the first grid fault $\theta = 60°$, the voltage dip levels of the first grid fault p_1 are 80, 50, and 20%. The rotor current, stator current, stator flux, and electromagnetic torque of the DFIG with different p_1 are shown in Figure 11.28. The maximum rotor current I_{rm} reaches 3.6 pu at most under the second grid fault, compared to 2.2 pu under the first grid fault (the same situation as under single grid fault). The maximum stator current I_{sm} and the maximum torque fluctuation T_{em} are also increased to 4 pu and to about 3.3 pu, as shown in Figure 11.28a. In Figures 11.28b and 11.28c, the voltage dip level of the first grid fault p_1 is reduced to 50% and 20%, respectively. The maximum rotor current I_{rm}, the maximum stator current I_{sm}, and the maximum torque fluctuation T_{em} are decreasing with the decrease of p_1. This is also in accordance with the calculated results in Figure 11.27a.

More simulations are made for different p_1 to verify the mathematical model introduced in (11.58), (11.59), and (11.61). The maximum rotor current I_{rm}, maximum stator current I_{sm}, and the maximum torque fluctuation T_{em} are shown in

TABLE 11.2   Parameters of the DFIGs

Rated power	1.5 MW	30 kW
Rated voltage (L-L) (V_{rms})	690	380
Stator resistance R_s (mΩ)	2.139	92
Stator inductance L_s (mH)	4.05	26.7
Rotor resistance R_r (mΩ)	2.139	56
Rotor inductance L_r (mH)	4.09	26.2
Mutual inductance L_m (mH)	4.00	25.1
Crowbar resistance R_c (Ω)	0.55	21
Turns ratio n_{sr}	0.369	0.32
Stator time constant τ_s (s)	1.89	0.28

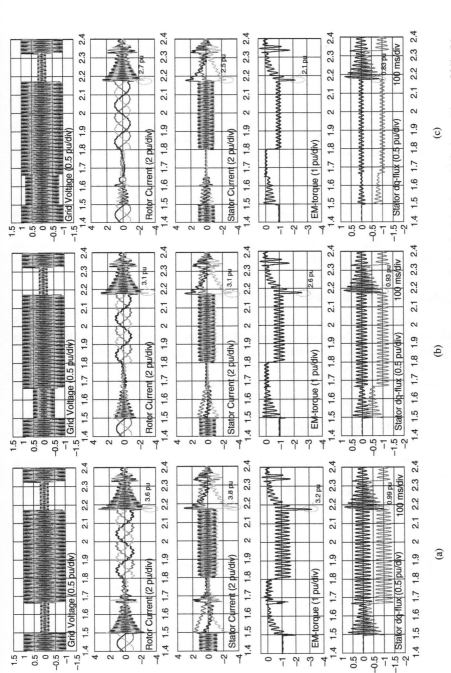

Figure 11.28 Simulated performance of a 1.5 MW DFIG with different voltage dip level of the first grid fault p_1: (a) $p_1 = 80\%$; (b) $p_1 = 50\%$; and (c) $p_1 = 20\%$.

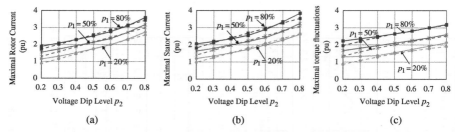

Figure 11.29 Simulated performance of a 1.5 MW DFIG at 1450 rpm (solid lines) with different voltage dip level of the first grid fault p_1 compared to calculated results (dotted lines): (a) maximum rotor current I_{rm}; (b) maximum stator current I_{sm}; and (c) maximum torque fluctuation T_{em}.

Figure 11.29 (solid lines) and compared with the calculated result which is shown in Figure 11.29 (dotted lines). Some error may exist but it can be found from Figure 11.29 that the simulated results are basically in accordance with the calculated results from the mathematical model introduced in (11.58), (11.59), and (11.61).

11.4.3 Influence of the Grid Fault Angle

The influence of θ on the performance of the DFIG under the second grid fault in a 1.5 MW DFIG WTS is shown in Figure 11.30, the rotor speed is 1650 rpm, the duration between two faults is $T_{re} \approx 500$ ms, and the voltage dip level $p_1 = 80\%$.

It can be found from Figure 11.30 that the maximum rotor current I_{rm} is increased with the decrease of θ, as well as the maximum stator current I_{sm} and the torque fluctuation T_{em} are increased. A smaller θ will lead to larger stator natural flux $\vec{\psi}_{sn}$ after the voltage recovery from the first grid fault, as shown in (11.43). If θ is smaller than 60°, the RSC may not be able to control the DFIG under voltage recovery with an 80% voltage dip and the crowbar may need to be activated again during voltage recovery. However, this will normally not happen in the transmission system [13].

The performance of the DFIG under two faults with different grid fault angles θ are also simulated. The duration between two faults is $T_{re} \approx 500$ ms, the voltage dip level of the first and second fault are $p_1 = 80\%$ and $p_2 = 80\%$ and the rotor speed

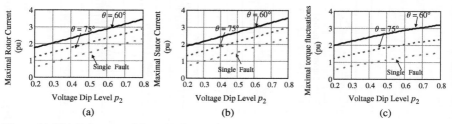

Figure 11.30 Influence of θ on (a) the maximum rotor current; (b) the maximum stator current; and (c) the maximum torque fluctuations under the second grid fault.

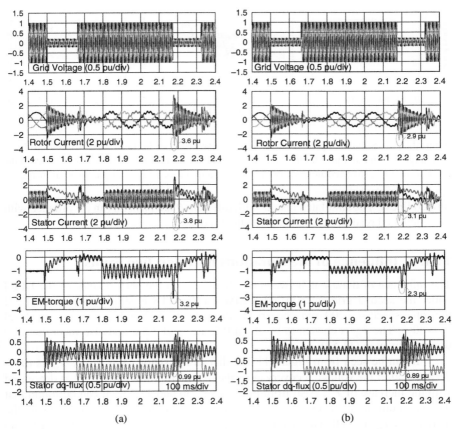

Figure 11.31   Simulated performance of a 1.5 MW DFIG at 1650 rpm with different grid fault angles of the first grid fault θ: (a) $\theta = 60°$; (b) $\theta = 75°$.

is 1650 rpm. The grid fault angle of the first grid fault θ are $\theta = 60°$ in Figure 11.31a and $\theta = 75°$ in Figure 11.31b, respectively. The rotor current, stator current, stator flux, and electromagnetic torque of the DFIG with different voltage dip levels of the first grid fault p_1 is shown in Figure 11.31. It can be concluded from Figures 11.31a and 11.31b that with the same T_{re}, p_1, and p_2, the maximum rotor current I_{rm} is decreased with the increase of θ, as the stator natural flux introduced by the voltage recovery of the first grid fault is decreased. The maximum stator current I_{sm} and the maximum torque fluctuation T_{em} are also decreased. It is in accordance with the calculated results in Figure 11.31.

More simulations are made to verify the mathematical model introduced in (11.58), (11.59), and (11.61). The maximum rotor current I_{rm}, maximum stator current I_{sm}, and the maximum torque fluctuation T_{em} are shown in Figure 11.32 (solid lines) and compared with the calculated results which are shown in Figure 11.32

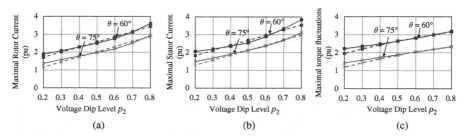

Figure 11.32 Simulated performance of a 1.5 MW DFIG (solid line) with different grid fault angles of the first grid fault θ compared to calculated results (dotted line): (a) maximum rotor current I_{rm}; (b) maximum stator current I_{sm}; and (c) maximum torque fluctuation T_{em}

(dotted lines). The simulated results with different θ are basically in accordance with the calculated result from the mathematical model.

11.4.4 Influence of the Durations between Two Faults

The influence of durations between two faults T_{re} on the performance of the DFIG under the second grid fault in a 1.5 MW DFIG WTS is shown in Figure 11.33: the rotor speed is 1200 rpm, the grid fault angle $\theta = 60°$ and the voltage dip level $p_1 = 80\%$. The T_{re} is chosen to be about 0.5, 1.5, and 2.5 s, and also meet the relationship given in (11.58).

It can be seen from Figure 11.33a that as the stator natural flux $\vec{\psi}_{sn}$ is time-decaying after the voltage recovery at t_3, the maximum rotor current I_{rm} is decreased with the increase of T_{re}. Also from Figures 11.33b and 11.33c, it can be found that the maximum stator current I_{sm} and the torque fluctuation T_{em} are decreased with the increase of T_{re} as well. The stator time constant $\tau_s \approx 1.6$ s in the 1.5 MW DFIG, so after 2.5 s, the influence from the voltage recovery from the first fault on the maximum rotor current I_{rm}, maximum stator current I_{sm}, and the torque fluctuation T_{em} of the second grid fault will be relatively smaller.

The performance of the DFIG under two faults with different durations between two faults (T_{re}) are shown in Figure 11.34, where the voltage dip level of the first and second faults are $p_1 = 80\%$ and $p_2 = 80\%$ and the rotor speed is 1650 rpm. The

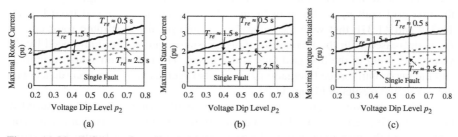

Figure 11.33 Influence from T_{re} on (a) the maximum rotor current; (b) the maximum stator current; and (c) the maximum torque fluctuations under the second grid fault.

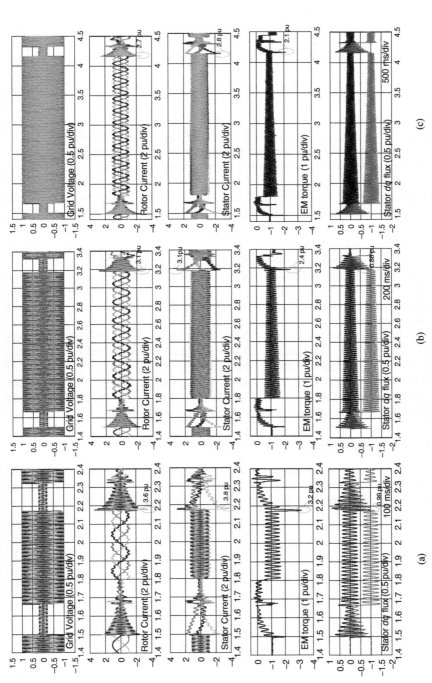

Figure 11.34 Simulated performance of a 1.5 MW DFIG with different durations between two faults T_{re}: (a) $T_{re} \approx 500$ ms; (b) $T_{re} \approx 1.5$ s; and (c) $T_{re} \approx 2.5$ s.

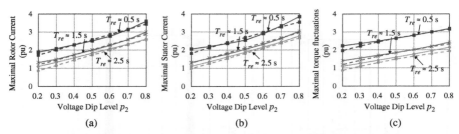

Figure 11.35 Simulated performance of a 1.5 MW DFIG (solid line) with different durations between two faults T_{re} compared to calculated results (dotted line): (a) maximum rotor current I_{rm}; (b) maximum stator current I_{sm}; and (c) maximum torque fluctuation T_{em}.

grid fault angle of the first grid fault θ is $\theta = 60°$. The durations between two faults (T_{re}) are $T_{re} \approx 500$ ms in Figure 11.34a, $T_{re} \approx 1.5$ s in Figure 11.34b, and $T_{re} \approx 2.5$ s in Figure 11.34c. The rotor current, stator current, stator flux, and electromagnetic torque of the DFIG with different T_{re} are shown. It can be concluded that with the same θ, p_1, and p_2, the maximum rotor current I_{rm} is decreased with the increase of T_{re}, as the stator natural flux introduced by the voltage recovery of the first grid fault is further decayed with the increase of T_{re}. The maximum stator current I_{sm} and the maximum torque fluctuation T_{em} are also decreased, which are in accordance with the calculated results.

More simulations are made in order to verify the mathematical model, and they are shown in Figures 11.35 and 11.36 The maximum rotor current I_{rm}, maximum stator current I_{sm}, and the maximum torque fluctuation T_{em} are shown in Figure 11.35 (solid lines) and compared with the calculated result (dotted lines). Some error may exist but the simulated results are basically in accordance with the calculated results from the mathematical model verifying the accuracy of the model.

11.4.5 Asymmetrical Recurring Faults

Also, in this part, the performance of the DFIG under one-phase-to-ground recurring fault is considered as an example to be analyzed, and it is assumed that the two dips happen at the same phase. It has been evaluated that after the fault recovery from

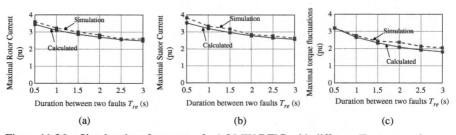

Figure 11.36 Simulated performance of a 1.5 MW DFIG with different T_{re} compared to calculated results, when $p_1 = p_2 = 80\%$: (a) maximum rotor current I_{rm}; (b) maximum stator current I_{sm}; and (c) maximum torque fluctuation T_{em}.

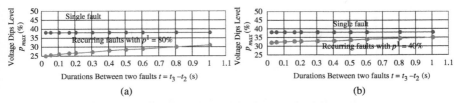

Figure 11.37 Capability of the 1.5 MW DFIG to ride through recurring single-line-to-ground faults with the voltage dip level of the first fault: (a) 80% and (b) 40%.

a one-phase-to-ground fault, the stator natural flux is expressed as (11.47), and it is rewritten here as

$$|\vec{\psi}_{sn}| = \frac{2}{3}p_1 \frac{V_s}{j\omega_s} \cos\theta \tag{11.63}$$

After the voltage recovery, the DFIG is usually operating with a normal vector control method, and hence the stator natural current after t_3 can be expressed as

$$\vec{\psi}_{sn}(t) = \frac{2}{3}p_1 \frac{V_s}{j\omega_s} \cos\theta e^{-\frac{t}{\tau_s}} \tag{11.64}$$

Similar to the analysis for symmetrical recurring faults, it can be found that the maximum stator natural flux under the second voltage dip can be calculated as (11.65).

$$|\vec{\psi}_{sn}(t_3)|_{max} = \frac{2}{3}p_1 \frac{V_s}{j\omega_s} \cos\theta e^{-\frac{T_{re}}{\tau_s}} + \frac{2}{3}p_2 \frac{V_s}{j\omega_s} \tag{11.65}$$

With (11.65), the rotor EMF of the DFIG under recurring asymmetrical faults can be evaluated together with (11.16). The stator-forced and negative-sequence fluxes $\vec{\psi}_{sf}$ and $\vec{\psi}_{sne}$ after the second fault can be expressed in the same way as a single-line-to-ground fault. However, the stator natural flux after the second voltage dip may be superposed with the stator natural flux produced by the voltage recovery. The rotor EMF in natural machine is also related to p_1, τ_s, and T_{re}. The maximum value of the required rotor voltage is also increased.

$$|\vec{v}_{rn}|_{max} = |\vec{e}_{rn}|_{max} \approx j\omega_r \left(\frac{2}{3}p_1 \frac{V_s}{j\omega_s} \cos\theta e^{-\frac{T_{re}}{\tau_s}} + \frac{2}{3}p_2 \frac{V_s}{j\omega_s} \right) \tag{11.66}$$

On the other hand, the rotor voltage is limited by the DC-bus voltage, so the ride-through ability of the DFIG under asymmetrical recurring faults can be derived [20]. The capability of the system to ride through the second grid fault is reduced. It can be described by p_{max} of the system which is established from (11.66) and (11.16), by letting $p_{max} = p_2$. When the first voltage dip levels are 80% and 40%, it is shown in Figures 11.37a and 11.37b, compared to that under one single-line-to-ground fault, it can be seen that p_{max} is reduced by about 12% if the voltage dip level of the first grid fault $p_1 = 80\%$, and by about 7% if the voltage dip level of the first grid fault $p_2 = 40\%$ after t_2, as the stator natural fluxes produced by the voltage recovery and by the second grid fault are superposed. The stator time constant is around 1.5 s with

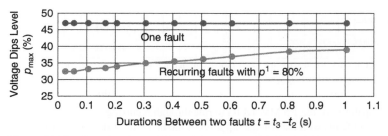

Figure 11.38 Simulated capability of the 1.5 MW DFIG to ride through single-line-to-ground recurring faults with $p_1 = 80\%$.

normal vector control, so this reduction will last for more than 1 s. It is longer than the shortest duration between the two faults that the WTS needs to ride though (500 ms) according to the grid code. When $T_{re} = 500$ ms, the ride-through capability of the system is still about 10% less than that under only one voltage dip with the first voltage dip level $p_1 = 80\%$, and 6% less than that under only one voltage dip with the first voltage dip level $p_1 = 40\%$.

The simulations of a 1.5 MW DFIG WTS under recurring asymmetrical grid faults are shown in Figure 11.38. As it is difficult to observe the saturation of the RSC, the p_{max} is defined to be the voltage dip level when the transient rotor current reaches the safety limit (about 1.8 pu in this case). As the rotor resistance and leakage inductance may limit the current even when the RSC is saturated, the simulated ride-though capability may be larger than the analyzed one. The simulated ride-through capability of the system under recurring single-line-to-ground faults is shown in Figure 11.39, and compared with the simulated ride-through capability of the system

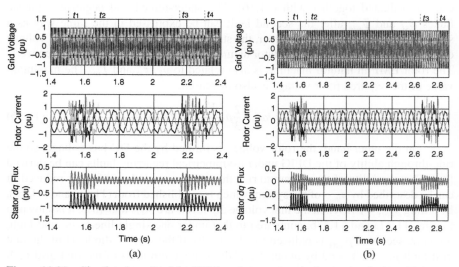

Figure 11.39 Simulated result of the DFIG under asymmetrical recurring faults with durations between the two faults of (a) 500 ms and (b) 1 s.

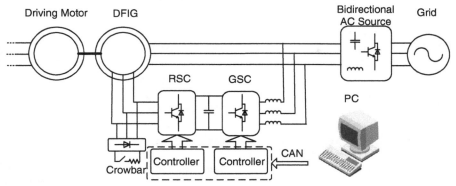

Figure 11.40   Scheme of the DFIG test system.

under one single-line-to-ground fault. The dip level of the first fault (p_1) is 80% and the rotor speed is 1800 rpm. It can be seen that the p_{max} is reduced from 47% to 32% under recurring faults. Although this reduction is time-decaying, as the stator time constant τ_s is around 1.5 s with the normal vector control, p_{max} is still less than 40% even after 1 s. This phenomenon matches with the analysis. On the other hand, the shortest duration between the two faults that the WTS needs to ride though is only 500 ms. As a result, a larger transient rotor current may occur.

Another simulation is made with the dip level of the first fault $p_1 = 40\%$, and the dip level of the second grid fault $p_2 = 40\%$, the stator voltage $\vec{v}_s$, rotor current $\vec{i}_s$, and the stator flux $\vec{\psi}_s$ of the DFIG under recurring faults are shown in Figure 11.39.

The two faults happen within 500 ms in Figure 11.39a and within 1 s in Figure 11.39b. It can be concluded that due to the influence of the first grid fault, the transient rotor current during the second grid fault is larger than the first one and the ride-through capability of the system under recurring faults is reduced. As the stator time constant τ_s of the system with vector control is about 1.5 s, when the duration between two faults is about 1 s, this reduction still exists. The simulated result matches the analysis in Figure 11.39b.

11.4.6   Experiments of DFIG under Recurring Grid Faults

In the last part of this chapter, experimental results on a 30 kW and a 7.5 kW reduced-scaled DFIG system are given. The DFIG under recurring grid faults, including symmetrical and asymmetrical recurring faults, are tested. And the performance of DFIG under single fault and during fault recovery can also be observed from the test results under recurring faults.

The scheme of the test system is shown in Figure 11.40. The DFIG is driven by a caged motor, and connected to a bidirectional AC source. The voltage dips are generated by the bidirectional AC source to emulate the recurring grid faults. The three-phase voltage can be controlled separately in the AC sources, and the voltage waveform of the grid fault with different fault angles can be emulated as well [21].

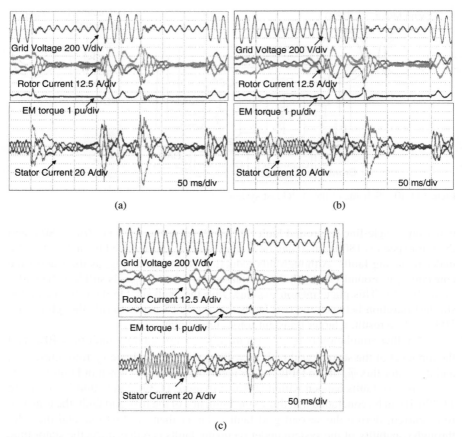

Figure 11.41 Performance of a 30 kW DFIG with different voltage dip level of the first grid fault p_1: (a) $p_1 = 80\%$; (b) $p_1 = 50\%$; and (c) $p_1 = 20\%$.

The RSC and crowbar are controlled by TI F2808 DSP which communicates with the PC by CAN. The parameters of the 30 kW DFIG are shown in Chapter 4. The details of the test bench will be introduced in Chapter 14.

First, the performance of the DFIG under recurring fault with the voltage dip level of the second fault $p_2 = 80\%$, grid fault angle $\theta = 60°$ are tested. As for the small-scaled DFIG, the stator time constant τ_{re} after voltage recovery is only 200–300 ms and the duration between two faults T_{re} is set to be $T_{re} \approx 80$ ms in order to show the influence of the first fault on the second fault. The DFIG was generating 0.2 pu active power under a rotor speed of 1600 rpm before the first fault happened. The crowbar is active after the voltage dip and disabled after 100 ms in order to provide reactive power support under each fault. The rotor current, stator current, and electromagnetic torque (calculated by stator and rotor currents) of the DFIG with the voltage dip level of the first fault $p_1 = 80$, 50, and 20% are shown in Figures 11.41a–11.41c, respectively [22].

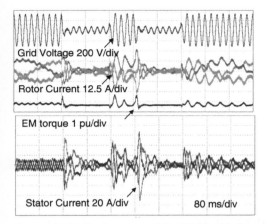

Figure 11.42   Performance of the DFIG with the grid fault angle of $\theta = 75°$.

It can be seen from Figure 11.41a that the maximum rotor transient current I_{rm} is increased to about 30 A under the recurring fault, compared to 18 A under single grid fault. The maximum stator transient current I_{sm} and the torque fluctuation T_{em} are increased as well. The voltage dip level of the first fault p_1 is reduced to 50% and 20% in Figures 11.41b and 11.41c, respectively, and it can be found that the maximum rotor current I_{rm}, stator current I_{sm}, and torque fluctuation T_{em} under the second grid fault are reduced compared to Figure 11.41a. A conclusion can be drawn that the magnitudes of rotor current I_{rm}, stator current I_{sm}, and electromagnetic torque fluctuation T_{em} under the second grid fault are decreased with the decrease of p_1, which is consistent with the analysis.

The performance of the DFIG under recurring faults with grid fault angle $\theta = 75°$ are tested. The voltage dip levels of the first and second faults are $p_1 = p_2 = 80\%$, the duration between two faults T_{re} are set to be $T_{re} \approx 80$ ms. The rotor current, stator current and electromagnetic torque of the DFIG are shown in Figure 11.42. It can be found from Figures 11.42 and 11.41a that the rotor and stator transient currents are decreased with the increase of the grid fault angle θ.

The performance of the DFIG under two faults with different durations between two faults (T_{re}) are tested and the results are shown in Figure 11.43; the voltage dip level of the first and second fault are $p_1 = 80\%$ and $p_2 = 80\%$ and the rotor speed is 1600 rpm. The grid fault angle of the first grid fault θ is $\theta = 60°$. The durations between two faults (T_{re}) are $T_{re} \approx 320$ ms and $T_{re} \approx 480$ ms. It can be concluded that with the same θ, p_1, and p_2, the maximum rotor current I_{rm} is decreased with the increase of T_{re}, as the stator natural flux introduced by the voltage recovery of the first grid fault further decays with the increase of T_{re}. The maximum stator current I_{sm} and the maximum torque fluctuation T_{em} are also decreased.

The experiments of DFIG under asymmetrical grid faults are made based on a 7.5 kW DFIG test system. Two tests are made with two 50% one-phase-to-ground faults, and the duration between two faults are 27 ms and 67 ms. The stator natural

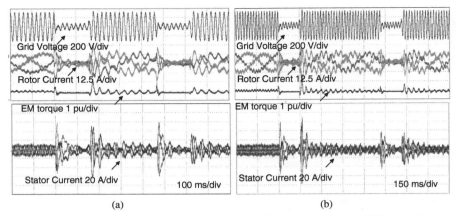

Figure 11.43 Performance of a 30 kW DFIG with different T_{re}: (a) $T_{re} \approx 320$ ms and (b) $T_{re} \approx 600$ ms.

flux produced by the voltage recovery of the first grid fault and by the voltage dips of the second grid fault have the same direction. The grid voltage, rotor current, and stator dq flux of the DFIG under recurring one-phase-to-ground faults are shown in Figure 11.44. The first grid fault happens at t_1, and the voltage recovers at t_2. The second grid fault happens at t_3 and the voltage recovers at t_4. It can be found that after the second fault happens at t_3, as the $\vec{\psi}_{sn}|_{t_3}$ is increased, the transient rotor current after the second grid fault at t_3 is larger than that at t_1, and p_{max} is reduced under recurring grid faults. As the stator time constant τ_s is about 150 ms, the reduction of the ride-through capability (p_{max}) of the system last more than 67 ms.

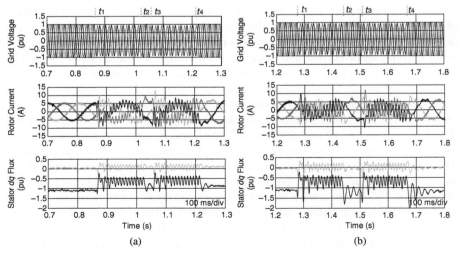

Figure 11.44 Performance of the 7.5 kW DFIG under recurring faults with the duration of (a) 27 ms and (b) 67 ms.

11.5 SUMMARY

In this Chapter, the dynamic model of a DFIG under grid fault is analyzed in detail, including the dynamic models of the DFIG under voltage dips, during voltage recoveries, and under recurring grid faults. For the dynamic model of the DFIG under voltage dips, three operation conditions are studied—with rotor open circuit, with normal vector control, and with rotor-side crowbar are investigated, respectively [23]. The stator natural flux introduced by the voltage dips and voltage recoveries are investigated: both symmetrical grid fault and asymmetrical grid fault are included. The results indicate that the RSC is normally not able to control the DFIG under serious voltage dips, and extra fault ride-through strategies need to be applied under voltage dips [24].

The dynamic model of a DFIG under symmetrical recurring grid faults is also analyzed. The analysis is based on the common FRT strategies of the DFIG. That is using rotor crowbar under voltage dips but normal control method under voltage recoveries [25]. For the DFIG under recurring grid faults, the influences of the recurring fault parameters on the performance of the DFIG are analyzed, which includes the voltage dip level and the grid fault angle of the first voltage dip, as well as the duration between two faults. The stator natural flux under recurring faults may be superposed on the stator natural flux introduced by the voltage recovery, so a larger transient voltage and current will be introduced [26]. The FRT ability of the DFIG under asymmetrical recurring faults are also investigated, and it is found that the FRT ability is reduced under recurring faults compared to that under single faults. Corresponding simulations and experiment results are provided.

REFERENCES

[1] L. Jesus, S. Pablo, S. Xavier, and M. Luis, "Dynamic behavior of the doubly fed induction generator during three-phase voltage dips," *IEEE Trans. Energy Convers.*, vol. 22, no. 3, pp. 709–718, Sep. 2007.

[2] J. Lopez, E. Gubia, P. Sanchis, and X. Roboam, "Wind turbines based on doubly fed induction generator under asymmetrical voltage dips," *IEEE Trans. Energy Convers.*, vol. 23, no. 1, pp. 321–330, Mar. 2008.

[3] R. Cardenas, R. Pena, S. Alepuz, and G. Asher, "Overview of control systems for the operation of DFIGs in wind energy applications," *IEEE Trans. Ind. Electron.*, vol. 60, no. 7, pp. 2776–2798, Jul. 2013.

[4] X. Kong, Z. Zhang, X. Yin, and M. Wen, "Study of fault current characteristics of the DFIG considering dynamic response of the RSC," *IEEE Trans. Energy Convers.*, vol. 29, no. 2, pp. 278–287, Jun. 2014.

[5] S. Xiao, G. Yang, H. Zhou, and H. Geng, "Analysis of the control limit for rotor-side converter for doubly fed induction generator-based wind energy conversion system under various voltage dips," *IET Renew. Power Gener.*, vol. 7, no. 1, pp. 71–81, 2013.

[6] J. Morren and S. W. H. de Haan, "Ride through of wind turbines with doubly-fed induction generator during a voltage dip," *IEEE Trans. Energy Convers.*, vol. 20, no. 2, pp. 435–441, Jun. 2005.

[7] J. Morren and S. W. H. de Haan, "Short-circuit current of wind turbines with doubly fed induction generator," *IEEE Trans. Energy Convers.*, vol. 22, no. 1, pp. 174–180, Mar. 2007.

[8] P. Zhou, Y. He, "Control strategy of an active crowbar for DFIG based wind turbine under grid voltage dips," in *Proc. Int. Conf. Electr. Mach. Syst.*, Seoul, Korea, 2007, pp. 259–264.

[9] E. Styvaktakis and M. H. J. Bollen, "Signatures of voltage dips: transformer saturation and multistage dips," *IEEE Trans. Power Deliv.*, vol. 18, no. 1, pp. 265–270, Jan. 2003.

[10] L. Zhang and M. H. J. Bollen, "Characteristic of voltage dips in power systems," *IEEE Trans. Power Deliv.*, vol. 15, pp. 822–832, Apr. 2000.

[11] Y. Huang, "The LVRT test of Gold Wind 1.5 MW wind turbine," *Electr. Manuf.*, vol. 2, pp. 70–71, 2012 (in Chinese).

[12] M. H. J. Bollen, "Voltage recovery after unbalanced and balanced voltage dips in three-phase systems," *IEEE Trans. Power Deliv.*, vol. 18, no. 4, pp. 1376–1381, Oct. 2003.

[13] M. Mohseni, S. M. Islam, and M. A. S. Masoum, "Impacts of symmetrical and asymmetrical voltage sags on DFIG-based wind turbines considering phase-angle jump, voltage recovery, and sag parameters," *IEEE Trans. Power Electron.*, vol. 26, no. 5, pp. 1587–1598, May 2011.

[14] W. Chen, F. Blaabjerg, N. Zhu, M. Chen, and D. Xu, "Doubly fed induction generator wind turbine system subject to recurring grid faults," in Proc. IEEE Appl. Power Electron. Conf. Exposition, 2014, pp. 3097–3104.

[15] W. Chen, F. Blaabjerg, N. Zhu, M. Chen, and D. Xu, "Comparison of control strategies for doubly fed induction generator under recurring grid faults," in Proc. IEEE Appl. Power Electron. Conf. Exposition, 2014, pp. 398–404.

[16] W. Chen, F. Blaabjerg, N. Zhu, M. Chen, and D. Xu, "Capability of DFIG WTS to ride through recurring asymmetrical grid faults" in Proc. ECCE, 2014.

[17] M. H. J. Bollen, G. Olguin, and M. Martins, "Voltage dips at the terminals of wind power installations," *Wind Energy*, vol. 8, no. 3, pp. 307–318, Jul. 2005.

[18] M. Mohseni, S. M. Islam, and M. A. S. Masoum, "Impacts of symmetrical and asymmetrical voltage sags on DFIG-based wind turbines considering phase-angle jump, voltage recovery, and sag parameters," *IEEE Trans. Power Electron.*, vol. 26, no. 5, pp. 1587–1598, May 2011.

[19] D. Xiang, L. Ran, P. J. Tavner, and S. Yang, "Control of a doubly fed induction generator in a wind turbine during grid fault ride-through," *IEEE Trans. Energy Convers.*, vol. 21, no. 3, pp. 652–662, Sep. 2006.

[20] J. Morren and S. W. H. de Haan, "Ridethrough of wind turbines with doubly-fed induction generator during a voltage dip," *IEEE Trans. Energy Convers.*, vol. 20, no. 2, pp. 435–441, Jun. 2005.

[21] F. K. A. Lima, A. Luna, P. Rodriguez, E. H. Watanabe, and F. Blaabjerg, "Rotor voltage dynamics in the doubly fed induction generator during grid faults," *IEEE Trans. Power Electron.*, vol. 25, no. 1, pp. 118–130, Jan. 2010.

[22] C. Wessels, F. Gebhardt, and F. W. Fuchs, "Fault ride-through of a DFIG wind turbine using a dynamic voltage restorer during symmetrical and asymmetrical grid faults," *IEEE Trans. Power Electron.*, vol. 26, no. 3, pp. 807–815, Mar. 2011.

[23] S. Seman, J. Niiranen, and A. Arkkio, "Ride-through analysis of doubly fed induction wind-power generator under unsymmetrical network disturbance," *IEEE Trans. Power Syst.*, vol. 21, no. 4, pp. 1782–1789, Nov. 2006.

[24] A. Petersson, T. Thiringer, L. Harnefors, and T. Petru, "Modeling and experimental verification of grid interaction of a DFIG wind turbine," *IEEE Trans. Energy Convers.*, vol. 20, no. 4, pp. 878–886, Dec. 2005.

[25] I. Erlich, H. Wrede, and C. Feltes, "Dynamic behavior of DFIG-based wind turbines during grid faults," in Proc. Power Convers. Conf., 2007, pp. 1195–1200.

[26] G. Abad, J. Lopez, M. Rodriguez, L. Marroyo, and G. Iwanski, *Doubly Fed Induction Machine: Modeling and Control for Wind Energy Generation*. John Wiley & Sons, 2011.

CHAPTER 12

GRID FAULT RIDE-THROUGH OF DFIG

In this chapter, the commonly used fault ride-through strategies for the DFIG WPS are introduced, which include the improved control strategies and also new enhanced hardware solutions. The improved control strategies normally use software changes in order to apply the methods, for realizing the FRT of the DFIG, while enhanced hardware solutions use additional circuits to assist the DFIG WPS to withstand the voltage dips. Their effects have been compared and the corresponding simulations and test results are shown. It should be also noticed in many cases that the hardware solutions and improved control strategies need to be applied at the same time. At the same time, for recurring fault ride-through, the challenges are also analyzed and the corresponding FRT strategy is evaluated.

12.1 INTRODUCTION

The performance of a DFIG under grid faults has been analyzed in Chapter 11. The voltage dips introduced by the grid fault will produce stator natural flux in the DFIG windings. As this stator natural flux is rotating with the rotor speed referring to the rotor windings, and a large rotor EMF may be introduced. The RSC may be saturated by this large EMF and lose control to the DFIG. Furthermore, a transient rotor overcurrent and over DC-bus voltage may be brought as well, which may harm the RSC. Under asymmetrical grid fault, the unbalanced voltage will introduce negative-sequence stator flux, together with the stator natural flux, which makes the situation even worse. Under recurring faults, the stator natural flux generated during the voltage recovery of the first grid fault may be superposed with the stator natural flux generated under the second fault, so larger transient current and voltage may be introduced on the second fault.

On the other hand, it has been introduced in Chapter 3 that the grid codes in many countries have required the WPS to ride through the grid faults, including symmetrical grid faults, asymmetrical grid faults, and even recurring grid faults. As a result, many efforts have been made on the fault ride-through of the WPS,

Advanced Control of Doubly Fed Induction Generator for Wind Power Systems, First Edition.
Dehong Xu, Frede Blaabjerg, Wenjie Chen, and Nan Zhu.
© 2018 by The Institute of Electrical and Electronics Engineers, Inc. Published 2018 by John Wiley & Sons, Inc.

especially the fault ride-through of DFIG WPS. In this chapter, different fault ride-through strategies are introduced, respectively, including the improved control strategies and the hardware solutions. The improved control strategies normally use software changes to apply the improved control for realizing the FRT of the DFIG, while hardware solutions using additional circuit is used to help the DFIG WPS to withstand the voltage dips. Their effects are compared with simulation and test results. In many cases, the hardware solutions and improved control strategies need to be applied at the same time. For the recurring fault ride-through, its challenges are analyzed and the corresponding FRT strategy is also evaluated.

12.2 PLL UNDER GRID FAULTS

An accurate grid synchronous is a prerequisite of the control under grid faults, and this is typically done by a PLL like a synchronization reference frame PLL. However, the SRF-PLL introduced in Chapter 5 is designed only for the ideal grid voltage. Under grid fault, the performance of the SRF-PLL needs to be re-evaluated.

12.2.1 SRF-PLL under Grid Faults

The SRF-PLL has been widely used in three-phase grid-connected system to provide the grid synchronous information; the scheme of the SRF-PLL is shown in Figure 12.1.

In SRF-PLL, as long as the grid voltage in the dq reference frame is controlled to be zero, the output phase angle will always track the grid phase angle. Under ideal grid voltage, the q components of the grid voltage v_{gq} is controlled and remains at zero as long as the output of the PLL θ_s tracks the grid angle. Under asymmetrical grid faults, the three-phase voltage will be reduced symmetrically. It has been introduced that the amplitudes of the balanced three-phase voltages are related to the d components of the grid voltage v_{gd}; so SRF-PLL will normally not be influenced by the symmetrical grid faults. Notice that there may be grid phase jumps together with the symmetrical voltage dips, in this case, and the PLL will operate at a new steady state, and a transient process may be introduced.

Nevertheless, under asymmetrical grid faults, the three-phase grid voltages are heavily unbalanced. It has been discussed in Chapter 11 that in this case, the grid

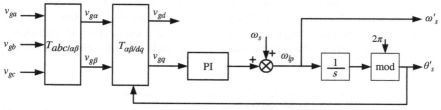

Figure 12.1   Scheme of an SRF-PLL for DFIG synchronization.

voltage $\vec{v}_{gabc}$ contains the positive sequence $\vec{v}^{+}_{gabc}$, negative sequence $\vec{v}^{-}_{gabc}$, and the zero sequence $\vec{v}^{0}_{gabc}$ at the same time, as expressed in (12.1).

$$\vec{v}_{gabc} = v_{ga} + \alpha v_{gb} + \alpha^2 v_{gc} = \vec{v}^{+}_{gabc} + \vec{v}^{-}_{gabc} + \vec{v}^{0}_{gabc} \tag{12.1}$$

Normally, the DFIG WPS are three-phase, three-line system, so the zero sequence can be neglected and then (12.1) can be rewritten as

$$\vec{v}_{gabc} = v_{ga} + \alpha v_{gb} + \alpha^2 v_{gc} = \vec{v}^{+}_{gabc} + \vec{v}^{-}_{gabc} \tag{12.2}$$

where

$$\vec{v}^{+}_{gabc} = V_g^{+} \begin{bmatrix} \cos(\omega_s t + \phi_g{}^{+}) \\ \cos\left(\omega_s t - \frac{2}{3}\pi + \phi_g{}^{+}\right) \\ \cos\left(\omega_s t + \frac{2}{3}\pi + \phi_g{}^{+}\right) \end{bmatrix} \begin{bmatrix} 1 & \alpha & \alpha^2 \end{bmatrix} \tag{12.3}$$

is the positive-sequence component, with amplitude $V_g{}^{+}$ and phase angle $\phi_g{}^{+}$, and

$$\vec{v}^{-}_{gabc} = V_g^{-} \begin{bmatrix} \cos(\omega_s t + \phi_g{}^{-}) \\ \cos\left(\omega_s t + \frac{2}{3}\pi + \phi_g{}^{-}\right) \\ \cos\left(\omega_s t - \frac{2}{3}\pi + \phi_g{}^{-}\right) \end{bmatrix} \begin{bmatrix} 1 & \alpha & \alpha^2 \end{bmatrix} \tag{12.4}$$

is the negative-sequence component with amplitude $V_g{}^{-}$ and phase angle $\phi_g{}^{-}$. If the unbalanced grid voltage $\vec{v}_{gabc}$ is transferred into the synchronous dq reference frame, the grid voltage synchronous with dq reference end can be found as

$$\begin{aligned} \vec{v}_{gdq} &= \begin{bmatrix} v_{gd} \\ v_{gq} \end{bmatrix} \begin{bmatrix} 1 & j \end{bmatrix} = \mathbf{T_p}\vec{v}_{gabc} = \mathbf{T_p}\left(\vec{v}^{+}_{gabc} + \vec{v}^{-}_{gabc}\right) \\ &= V_g^{+} \begin{bmatrix} 1 \\ 0 \end{bmatrix} \begin{bmatrix} 1 & j \end{bmatrix} + V_g^{-} \begin{bmatrix} \cos(-2\omega_s t - \phi_g{}^{-} - \phi_g{}^{+}) \\ \sin(-2\omega_s t - \phi_g{}^{-} - \phi_g{}^{+}) \end{bmatrix} \begin{bmatrix} 1 & j \end{bmatrix} \end{aligned} \tag{12.5}$$

The corresponding voltage vector trajectory is shown in Figure 12.2. In the stationary abc and $\alpha\beta$ reference frames, the trajectory of the positive-sequence voltage $\vec{v}^{+}_{gabc}$ is a circle rotating anti-clockwise, while the trajectory of the negative-sequence voltage $\vec{v}^{-}_{gabc}$ is also a circle, but it is rotating clockwise. So the trajectory of the resultant vector is an ellipse. In the dq reference frame, the positive-sequence voltage $\vec{v}^{+}_{gabc}$ is a fixed vector lying on the d-axis, while the trajectory of the positive-sequence voltage $\vec{v}^{-}_{gabc}$ is a circle rotating clockwise with an angular speed of $2\omega_s$. Consequently, the angular speed of the resultant grid voltage ω_g is not equal to ω_s anymore, but it contains fluctuations components introduced by the negative sequence. As a result, the grid angle contains also a fluctuation component, as expressed in (12.6), and also

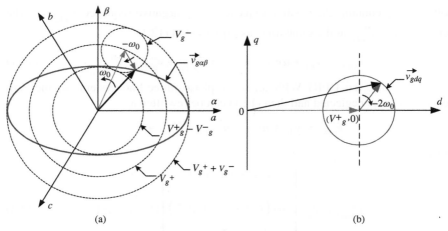

(a)   (b)

Figure 12.2   Unbalanced grid voltage trajectory: (a) In stationary abc and $\alpha\beta$ reference frames and (b) in the synchronous dq reference frame.

as illustrated in Figure 12.3.

$$\theta_g = \theta_s + \arctan\left[\frac{V_g^- \sin(-2\omega_s t - \phi_g^- - \phi_g^+)}{V_g^+ + V_g^- \cos(-2\omega_s t - \phi_g^- - \phi_g^+)}\right] \tag{12.6}$$

As for the SRF-PLL, the grid angle of the negative sequence is only needed for the dq transformation, and the fluctuation components in the grid angle will be introduced as an error in SRF-PLL, as it is tracking the grid angle and angular speed of the resultant grid voltage. In Figure 12.1, the q-axis grid current v_{gq} will contain an AC fluctuation component with a frequency of $2\omega_s$, as it is also the input of the PI controller and this AC fluctuation component will be found in the output of the PI controller as well. If the PI controller is designed with a crossover frequency much smaller than $2\omega_s$, the AC fluctuation in the output of the SRF-PLL can be suppressed. However, this will lead to a poor dynamic performance of the SRF-PLL.

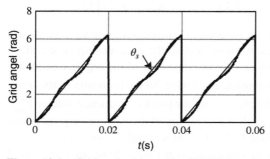

Figure 12.3   Grid angle detected by SRF-PLL under unbalanced grid voltage.

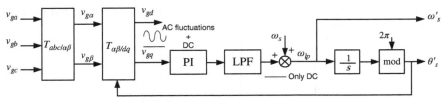

Figure 12.4   Scheme of the SRF-PLL using an LPF.

12.2.2   SRF-PLL with a Low Pass Filter

Generally, there are two ways to ensure the PLL gets the phase information of the positive-sequence grid voltage precisely under the unbalanced grid. One commonly used method is to add an LPF after the PI controller if the crossover frequency of the LPF is much smaller than $2\omega_0$ and the output of the LPF will only contain DC components, and the AC fluctuation component is suppressed. Its scheme is shown in Figure 12.4.

However, the LPF in the control loop will also lead to a slower dynamic response of the system, similar to the method to reduce the crossover frequency of the PI controller. Under grid fault, sometimes the voltage dip will come together with the phase angle jump. In this case, if the PLL cannot respond to the phase jump fast enough, a phase error will be introduced during grid faults.

12.2.3   SRF-PLL with Negative/Positive-Sequence Separation

Another method is to separate the negative-sequence and positive-sequence grid voltages, and only use the positive sequence as the input of the PLL, as shown in Figure 12.5.

The sequence separation is the key to achieve the phase tracking for the positive-sequence grid voltage under the unbalanced grid. The output of the sequence-separation block contains only the positive-sequence grid voltage, so that the SRF-PLL can work in the same way under ideal grid voltage. The sequence-separation blocks are normally designed using a time-delayed method and one of the methods is introduced below.

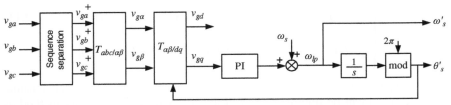

Figure 12.5   Scheme of the SRF-PLL with voltage-sequence separation.

The relationship among the positive, negative, zero components, and the grid *abc* voltage can be found as

$$
\begin{bmatrix} \vec{v}_{gabc}{}^+ \\ \vec{v}_{gabc}{}^- \\ \vec{v}_{gabc}{}^0 \end{bmatrix} = \frac{1}{3} \begin{bmatrix} 1 & 1 & 1 \\ 1 & \alpha & \alpha^2 \\ 1 & \alpha^2 & \alpha \end{bmatrix} \begin{bmatrix} v_{sa} \\ v_{sb} \\ v_{sc} \end{bmatrix}
\tag{12.7}
$$

So, the positive-sequence three-phase grid voltage $v_{ga}{}^+$, $v_{gb}{}^+$, $v_{gc}{}^+$ can be expressed as (12.8).

$$
\begin{bmatrix} v_{ga}^+ \\ v_{gb}^+ \\ v_{gc}^+ \end{bmatrix} = \begin{bmatrix} 1 \\ \alpha^2 \\ \alpha \end{bmatrix} \vec{v}_{gabc}{}^+ = \frac{1}{3} \begin{bmatrix} 1 & \alpha & \alpha^2 \\ \alpha^2 & 1 & \alpha \\ \alpha & \alpha^2 & 1 \end{bmatrix} \begin{bmatrix} v_{sa} \\ v_{sb} \\ v_{sc} \end{bmatrix}
\tag{12.8}
$$

The corresponding time domain relationship of (12.8) can be expressed as

$$
\begin{cases}
v_{ga}^+(t) = \frac{1}{3} \left[v_{ga}(t) + v_{gb}\left(t - 2T_s/3\right) + v_{gc}\left(t - T_s/3\right) \right] \\
v_{gb}^+(t) = \frac{1}{3} \left[v_{ga}\left(t - T_s/3\right) + v_{gb}(t) + v_{gc}\left(t - 2T_s/3\right) \right] \\
v_{gc}^+(t) = \frac{1}{3} \left[v_{ga}\left(t - 2T_s/3\right) + v_{gb}\left(t - T_s/3\right) + v_{gc}(t) \right]
\end{cases}
\tag{12.9}
$$

With (12.9), the positive components in three-phase grid voltage can then be calculated. T_s is the period of the three-phase voltage. As it needs prior information on the three-phase grid voltage $2T_s/3$, the response time of this method cannot be faster than $2T_s/3$. For the AC components with the period of T_s, it satisfies the following relationship:

$$
v_{ga}\left(t - 2T_s/3\right) = -v_{ga}\left(t - 2T_s/3 + T_s/2\right) = -v_{ga}\left(t - T_s/6\right)
\tag{12.10}
$$

This relationship can also be applied to u_{gb} and u_{gc}, so (12.9) can be rewritten as

$$
\begin{cases}
v_{ga}^+(t) = \frac{1}{3} \left[v_{ga}(t) - v_{gb}\left(t - T_s/6\right) + v_{gc}\left(t - T_s/3\right) \right] \\
v_{gb}^+(t) = \frac{1}{3} \left[v_{ga}\left(t - T_s/3\right) + v_{gb}(t) - v_{gc}\left(t - T_s/6\right) \right] \\
v_{gc}^+(t) = \frac{1}{3} \left[-v_{ga}\left(t - T_s/6\right) + v_{gb}\left(t - T_s/3\right) + v_{gc}(t) \right]
\end{cases}
\tag{12.11}
$$

Compared to (12.9), the calculation of (12.11) only needs $T_s/3$; so the dynamic response speed can be improved. As a result, the sequence-separation method as expressed in (12.11) is used in this book. Other parts of the SRF-PLL is the same as that introduced in Chapter 5, and it will not be introduced again.

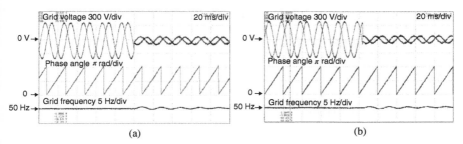

Figure 12.6   Test performance of (a) convention SRF-PLL and (b) SRF-PLL with sequence separation under symmetrical grid faults.

12.2.4   Test Results of PLL with Sequence Separation

Experiments are carried out on the 30 kW DFIG test system to show the performance of the SRF-PLL and the SRF-PLL with sequence separation under grid faults. The grid faults are generated by the AC-source connected to the stator of the DFIG. The performance of the conventional SRF-PLL and the SRF-PLL with sequence separation under symmetrical grid faults are shown in Figure 12.6. The three-phase grid voltage dips with a dip level of 80% is introduced by the grid faults (only two phases are shown in Figure 12.6). As for the symmetrical grid fault, no negative-sequence grid voltage is introduced. Both the conventional SRF-PLL and the SRF-PLL with sequence separation can track the grid-phase angle successfully. Only some fluctuations in the grid frequency are introduced, with an amplitude no more than 1 Hz, and it is decaying after about 80 ms.

The performance of the conventional SRF-PLL and the SRF-PLL with sequence separation under asymmetrical grid faults are tested and shown in Figure 12.7. One phase voltage is reduced to 50% of the normal value while the other phases remain at the normal grid voltage. It can be found from Figure 12.7a that the grid frequency fluctuations are introduced by the unbalanced grid voltage for the conventional SRF-PLL. As for the SRF-PLL with sequence separation, it can be seen from Figure 12.7b that the grid frequency and phase angle remains steady after the

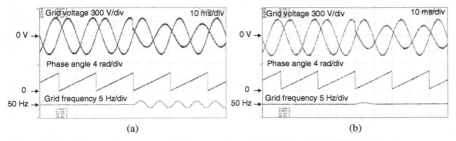

Figure 12.7   Test performance of (a) convention SRF-PLL and (b) SRF-PLL with sequence separation under asymmetrical grid faults.

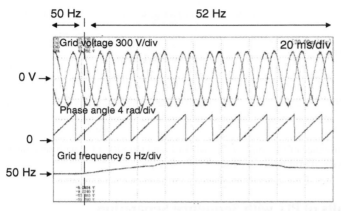

Figure 12.8   Test performance of SRF-PLL with sequence separation under grid frequency changes.

unbalanced voltage dips, the transient process last for about 6 ms. The SRF-PLL with sequence separation is able to track the phase angle of the positive-sequence grid voltage under unbalanced grid faults.

In Figure 12.8, the grid frequency changes from 50 to 52 Hz abruptly. It can be found that the SRF-PLL with sequence separation responds to the frequency change at once, and the steady state is restored in about 100 ms. The dynamic performance of the SRF-PLL with sequence separation for frequency changes is acceptable. Notice that in the real power system, the frequency will not change so fast, as the inertia of the power system typically is very large.

12.3   FRT STRATEGIES FOR DFIG BASED ON IMPROVED CONTROL

It has been studied in Chapter 11 that the conventional vector control is not able to control the DFIG under grid faults properly. As with the conventional vector control, the large rotor EMF introduced by the voltage dips may introduce large transient current and voltage in the RSC and the DFIG. Also, the stator natural flux introduced by the grid voltage dips decays slowly with the conventional vector control, which makes the transient process longer, and more difficult to overcome. The improved control strategies aim to suppress the overcurrent and overvoltage of the DFIG and RSC under grid faults, as well as accelerate the decaying of the stator natural flux, and make the transient process introduced by the voltage dips shorter. The merits of the FRT strategies based on the improved control include the lower cost and ease of implementation, as normally only software changes are necessary. Also, the DFIG is under control all the time, and it is much easier to switch back to normal operation, compared to the hardware solutions which will be introduced in the next section. However, as the capacity of the RSC is limited, the FRT strategies based on the improved control is also limited due to the controllable range. If the DFIG is operating outside the controllable range, for example, if the voltage dip level is too high, the RSC may also be saturated and lose control of the DFIG.

Many improved control strategies have been evaluated for the FRT of the DFIG WPS. However, in this section, three different FRT strategies based on the improved control are introduced as examples. They are the demagnetizing current control, the flux linkage tracking control, and the feedforward transient compensation control. Their controllable range, the damping time constant of stator flux, and the torque fluctuation are analyzed at different rotor speeds and voltage dip levels. Corresponding simulation and test results are given too.

12.3.1   Demagnetizing Current Control

Similar to an induction motor, the demagnetizing current control of the DFIG applies a rotor current reference which is opposite to the stator natural flux, so that the damping of the stator natural flux can be accelerated [1]. This method has been introduced in Chapter 11 under grid fault and for the natural machine; the relationship between the rotor natural current $\vec{i}_{rn}$ and the stator natural flux $\vec{\psi}_{sn}$ can be expressed as given (11.3) and it is rewritten here in (12.12).

$$\frac{d}{dt}\vec{\psi}_{sn} = -\frac{R_s}{L_s}\vec{\psi}_{sn} + \frac{L_m}{L_s}R_s\vec{i}_{rn} \qquad (12.12)$$

If the rotor natural current $\vec{i}_{rn}$ is controlled to be in the opposite direction of the stator natural flux $\vec{\psi}_{sn}$, which means

$$\vec{i}_{rn} = -k\vec{\psi}_{sn} \qquad (12.13)$$

where $k > 0$ is the demagnetizing coefficient. By substituting (12.13) in (12.12), the stator natural flux can be derived as

$$\vec{\psi}_{sn} = \vec{\psi}_{sn0}e^{-\frac{t}{\frac{L_s}{R_s + kL_mR_s}}} \qquad (12.14)$$

where $\vec{\psi}_{sn0}$ is the initial value of the stator natural flux, under symmetrical dip, $\vec{\psi}_{sn0} \approx p\frac{\vec{v}_s}{j\omega_s}$. The stator time constant with demagnetizing current control becomes

$$\tau_s = \frac{L_s}{R_s + kL_mR_s} \qquad (12.15)$$

Compared with the rotor open circuit, where

$$\tau_s = \frac{L_s}{R_s} \qquad (12.16)$$

It can be found that as long as $k > 0$, the stator time constant with demagnetizing current control will always be smaller than that with the rotor open circuit, which means that the damping of the stator natural flux is accelerated by the demagnetizing current control; so the transient process brought by the voltage dip can be finished

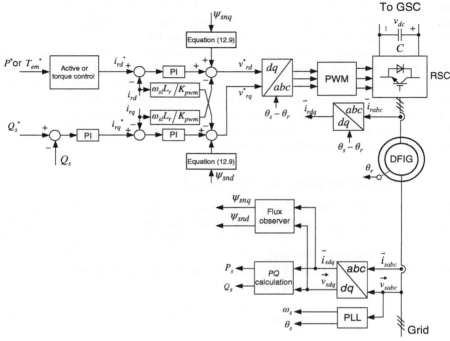

Figure 12.9   Scheme of the demagnetizing current control.

much faster. The control scheme of the demagnetizing current control for the FRT of the DFIG is shown in Figure 12.9.

The stator natural fluxes ψ_{snd} and ψ_{snq} are observed by the flux observer, and the rotor natural current references i_{rnd}^{ref} and i_{rnq}^{ref} are calculated according to (12.13). Then, they are calculated with the rotor-forced current references i_{rfd}^{ref} and i_{rfq}^{ref} from the power control loop, and the sum of them is the current reference of the current inner loop. The rest part of the control scheme is similar to that in the conventional vector control. Noticed that if the DFIG is in normal operation, the stator natural flux components ψ_{snd} and ψ_{snq} are zero, and so the current reference only contains the rotor-forced current reference i_{rfd}^{ref} and i_{rfq}^{ref}. The demagnetizing current control does not need to be enabled and disabled during and after these grid faults.

The demagnetizing coefficient k is a control parameter to be designed. The relationship between the demagnetizing coefficient k and the stator time constant τ_s after the demagnetizing control is applied is as shown in Figure 12.10. The 1.5 MW DFIG WPS is taken as an example. It is easy to conclude that the stator time constant τ_s is reduced with the increase of k, which means the damping of the stator natural flux $\vec{\psi}_{sn}$ is faster with larger k. So, from this point of view, the demagnetizing coefficient k should be chosen as large as possible.

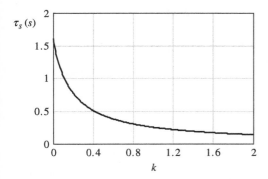

Figure 12.10   Relationship between stator time constant τ_s and demagnetizing coefficient k for a 1.5 MW DFIG.

On the other hand, as the control is achieved by the RSC, the hardware capacity of the RSC must be taken into account too. With (12.13) and (12.14), the rotor current can be found as

$$\vec{i}_{rn} = -kp\frac{\vec{v}_s}{j\omega_s}\vec{\psi}_{sn0}e^{-\frac{t}{\tau_s}} \tag{12.17}$$

Then the rotor voltage can be derived from the rotor equations, as given in (11.4). It is rewritten here in (12.18).

$$\left(\frac{d}{dt} - j\omega_r\right)\vec{\psi}_{sn} = \vec{v}_{rn} - R_r\vec{i}_{rn} - \left(\frac{d}{dt} - j\omega_r\right)\sigma L_r\vec{i}_{rn} \tag{12.18}$$

The relationship between the rotor current amplitude $\left|\vec{i}_r\right|$ and demagnetizing coefficient k, as well as between the rotor voltage amplitude $\left|\vec{v}_r\right|$, and the demagnetizing coefficient k are shown in Figure 12.11. The 1.5 MW DFIG WPS is taken as an example, the grid fault is a symmetrical fault with the voltage dip level of 60%,

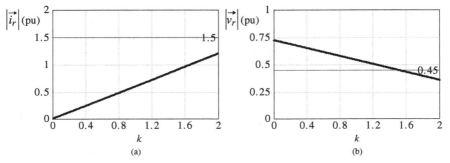

Figure 12.11   Relationship between (a) the amplitude of rotor current; (b) rotor voltage, and demagnetizing coefficient k, when the voltage dip level is 60%.

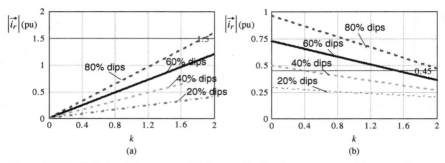

Figure 12.12   Relationship between (a) the amplitude of rotor current; (b) rotor voltage, and demagnetizing coefficient k, with different symmetrical voltage dip level in the 1.5 MW DFIG.

and the rotor speed is 1.2 pu. The rotor current amplitude $\left|\vec{i}_r\right|$ is increased with the increase of k, while the rotor voltage amplitude $\left|\vec{v}_r\right|$ is decreased with the increase of k. Noticed they all have a limitation as labeled in Figure 12.11. Normally, the transient rotor current cannot exceed 1.5–2.0 times the rated value (horizontal dark line in Figure 12.11a). With respect to the rotor voltage, as it is the output voltage of the RSC, it is limited by the DC-bus voltage of the RSC. In the 1.5 MW DFIG WPS, it is 1200 V; so the rotor voltage is limited to be about 692 V at the rotor side, which is about 0.45 pu referring to the stator side (horizontal dark line in Figure 12.11b). If the actual rotor voltage amplitude is larger than this value, the RSC may be saturated and the DFIG may be out of control for a short time. So in this case, k should be chosen larger than 1.6 to ensure the DFIG is in the controllable range.

It should be noticed that the relationship shown in Figure 12.11 will be different for different voltage dip levels p and rotor speed ω_r. With different voltage dip levels p, the relationship between the rotor current amplitude and demagnetizing coefficient k, as well as between the rotor voltage amplitude and demagnetizing coefficient k, are shown in Figure 12.12. The rotor speed ω_r is fixed to be 1.2 pu. With different rotor speeds ω_r, their relationship is shown in Figure 12.13; the voltage dip level p is 80%.

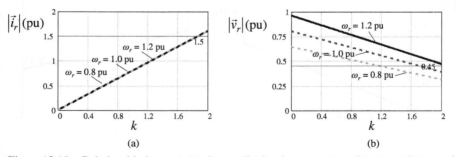

Figure 12.13   Relationship between (a) the amplitude of rotor current; (b) rotor voltage, and demagnetizing coefficient k, with different rotor speeds.

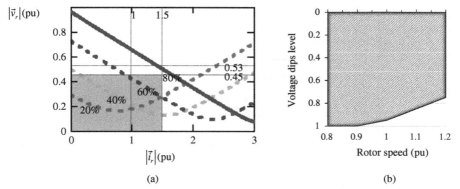

(a)   (b)

Figure 12.14   (a) Relationship between the rotor voltage amplitude $|\vec{v}_r|$ and rotor current amplitude $|\vec{i}_r|$ with different voltage dip levels and (b) the controllable range of the demagnetizing control (shadowed area).

In general, the rotor current amplitude $|\vec{i}_r|$ with the same k increases with an increase of the voltage dip level, and it is about the same with different rotor speeds. Regarding the rotor voltage amplitude $|\vec{v}_r|$ with the same k, it is increased with the increase of voltage dip level and the rotor speed.

Notice that when the rotor speed ω_r is 1.2 pu, and the voltage dip level is 80%, the rotor current amplitude $|\vec{i}_r|$ will exceed 1.5 pu when k is larger than 1.8, while the rotor voltage amplitude $|\vec{v}_r|$ will not be smaller than 0.45 pu if k is smaller than 2. This means that no matter what k is, the rotor current amplitude $|\vec{i}_r|$ or the rotor voltage amplitude $|\vec{v}_r|$ is going to exceed the limited value. So, the 80% voltage dip with the rotor speed of 1.2 pu is beyond the controllable range of the demagnetizing current control. To make it more clear, the relationship between the rotor voltage amplitude $|\vec{v}_r|$ and rotor current amplitude $|\vec{i}_r|$ with different voltage dip levels can be derived from (12.17) and (12.18). By eliminating k, as shown in Figure 12.14a, the rotor speed is 1.2 pu. The shadowed area indicates the safe operation area of the RSC if the corresponding line is covered by the safe operation area of the RSC; the RSC is able to control the DFIG with demagnetizing current control under grid fault with the corresponding voltage dip level and rotor speed. For 80% voltage dips, the corresponding line is out of the controllable range, which means the DFIG cannot work in the safety range with demagnetizing current control under 80% voltage dips; it is out of the controllable range. With this method, the controllable range of the demagnetizing control can be evaluated for a 1.5 MW DFIG, under symmetrical voltage dips, as shown in Figure 12.14b (the shadowed area). With demagnetizing control, the DFIG is able to ride through full voltage dips when the rotor speed is smaller than 0.9 pu. When the rotor speed is 1.2 pu, the largest voltage dip level that the DFIG is able to ride though is about 75%. If the DFIG is out of the controllable range, the RSC may be saturated, and the DFIG may be out of control for a short time.

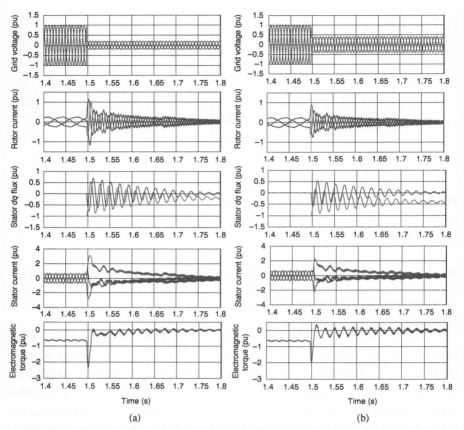

Figure 12.15 Simulation results of the 1.5 MW DFIG with demagnetizing current control under symmetrical voltage dips: (a) With the voltage dip level of 80% and (b) with the voltage dip level of 60%.

The simulation results of a 1.5 MW MATLAB simulation model, as well as the experiment results on a 15 kW DFIG test bench, is shown in Figures 12.15 and 12.16, respectively. The DFIG is controlled by the RSC with demagnetizing current control under symmetrical voltage dips. The waveforms of the three-phase grid voltage, rotor current, stator current, the stator flux in dq reference frames, as well as the torque fluctuations, are shown in Figures 12.15 and 12.16. The rotor speed is 1.2 pu in Figure 12.15, and 1.1 pu in Figure 12.16. The voltage dip level is 80% in Figures 12.15a and 12.16a. After the voltage dip, the demagnetizing current control generates rotor current in the opposite direction of the stator natural flux, which is the 60 Hz frequency components in the rotor current in Figure 12.15a, and a 60 Hz frequency component in the rotor current in Figure 12.15a. As the stator natural flux is a DC component in stator *abc* reference frame, it is rotating with the rotor speed referring to the rotor reference frame. The stator natural flux components, which are the AC components with a frequency of f_s in the stator dq flux, is damping very fast after

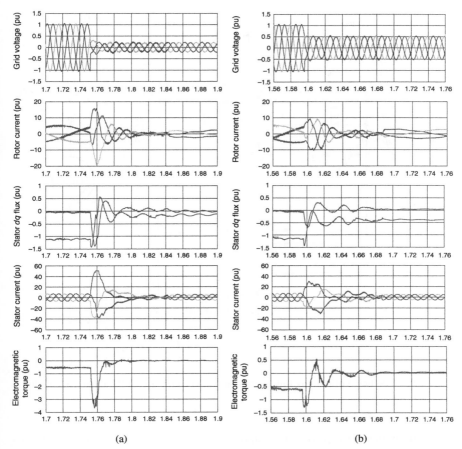

Figure 12.16   Test results of the 1.5 MW DFIG with demagnetizing current control under symmetrical voltage dips: (a) With the voltage dip level of 80% and (b) with the voltage dip level of 60%.

the voltage dips. Compared to the performance of the DFIG with conventional vector control (Figure 11.11), the decay of the stator natural flux is obviously accelerated by the demagnetizing current control. As has been discussed above, the 80% voltage dips with 1.2 pu rotor speed is out of the controllable range of the demagnetizing control; so the RSC is saturated for a short time after the voltage dips and large transient rotor current and stator current is introduced, but only last for a short time, as the stator natural flux is decaying fast. A transient electromagnetic torque is also introduced after the voltage dips; besides that, the AC electromagnetic torque fluctuations can be found after the voltage dips, which is introduced by the rotor natural current and stator-forced flux.

In Figures 12.15b and 12.16b, the voltage dip levels are 60% and 50%, respectively; they are covered by the controllable range of the demagnetizing control. The

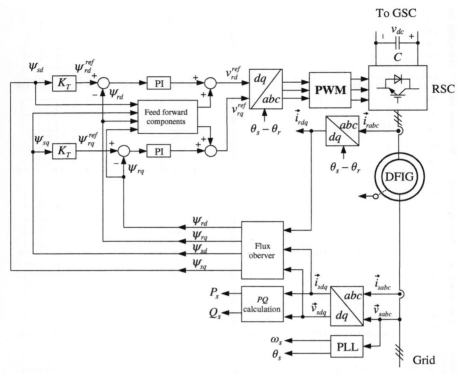

Figure 12.17 Scheme of the flux linkage tracking control for the DFIG under voltage dip operation.

rotor current and stator current are controlled below the limitation and the damping of the stator natural flux is accelerated by the control.

12.3.2 Flux Linkage Tracking Control

The flux linkage tracking control controls the rotor flux linkage to track part of the stator flux linkage by altering the output rotor voltage of the RSC, in order to keep the rotor current under control and to accelerate the damping of the stator natural flux. The control scheme of the stator flux linkage tracking control is shown in Figure 12.17. The stator flux linkage $\vec{\psi}_s$ and rotor flux linkage $\vec{\psi}_r$ in dq reference frames are observed by the flux observer. Then the rotor flux linkage in the dq reference frame (ψ_{rd}, ψ_{rq}) is controlled to track part of the stator flux linkage in the dq reference frame $(K_T \psi_{sd}, K_T \psi_{sq}$, and $0 < K_T < 1)$ [2]. The corresponding rotor flux references are expressed as

$$\begin{cases} \psi_{rd}^{ref} = K_T \psi_{sd} \\ \psi_{rq}^{ref} = K_T \psi_{sq} \end{cases} \tag{12.19}$$

If the controllers are assumed to be ideal controllers, then the rotor flux linkage $\vec{\psi}_r$ follows the reference perfectly, which means

$$\vec{\psi}_{rdq} = K_T \vec{\psi}_{sdq} \tag{12.20}$$

Substituting (12.20) in the flux linkage function of the DFIG in the dq reference frame (12.21), (12.22):

$$\vec{\psi}_{sdq} = L_m \vec{i}_{rdq} + L_s \vec{i}_{sdq} \tag{12.21}$$

$$\vec{\psi}_{rdq} = L_m \vec{i}_{sdq} + L_r \vec{i}_{rdq} \tag{12.22}$$

The relationship between the rotor current and the stator flux can be found as

$$\vec{i}_{rdq} \approx \frac{K_T - 1}{L_{lr} + L_{ls}} \vec{\psi}_{sdq} \tag{12.23}$$

As $0 < K_T < 1$, the coefficient $\frac{K_T - 1}{L_{lr} + L_{ls}}$ in (12.23) is actually a negative value, which means the rotor current is in the opposite direction of the stator flux. The stator flux under voltage dips contains the stator natural flux, stator-forced flux, and stator negative flux under asymmetrical dip. As a result, the corresponding rotor natural current is in the opposite direction of the stator natural flux, similar to that in demagnetizing control, and it is given in (12.24).

$$\vec{i}_{rn} \approx \frac{K_T - 1}{L_{lr} + L_{ls}} \vec{\psi}_{sn} \tag{12.24}$$

Therefore, the decaying of the stator natural flux can be also accelerated in this case. Furthermore, as the electromagnetic torque T_{em} of the DFIG can be expressed as

$$T_{em} = \frac{3}{2} \frac{P_s L_m}{L_s L_r - L_m^2} (\psi_{sq}\psi_{rd} - \psi_{sd}\psi_{rq}) \tag{12.25}$$

Substituting (12.19) in (12.25), it can be found the electromagnetic torque of the DFIG is close to zero if the flux linkage tracking control is used. So, the electromagnetic torque fluctuations of the DFIG are suppressed under voltage dip by the flux linkage tracking control.

Simulation results of the 1.5 MW DFIG WPS with flux linkage tracking control under symmetrical grid faults are shown in Figure 12.18. The rotor speed is 1.2 pu, and the DFIG is generating 1 pu active power before the fault happens. In Figure 12.18a, the voltage dip level is 80%, and the flux linkage tracking control is enabled after the grid fault. It can be found the decay of the stator natural flux (AC components in stator dq flux) is accelerated after the voltage dips, compared with the normal vector control, and the electromagnetic torque fluctuations are much smaller compared to the demagnetizing control. A transient over rotor current up to 1pu and stator current up to 2.7 pu can be found after the voltage dip, as the RSC may be saturated for a short time. In Figure 12.18b, the voltage dip level is reduced to 60%, and the flux linkage tracking control is able to limit the rotor current below 0.6 pu and limit the stator current to be below 2 pu. The decay of the stator natural flux

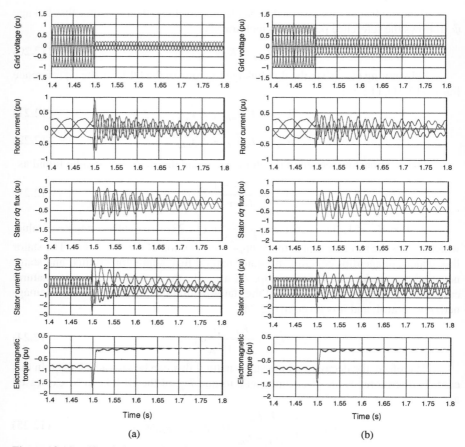

Figure 12.18   Simulation results of a 1.5 MW DFIG with flux linkage tracking under symmetrical voltage dip: (a) With the voltage dip level to be 80% and (b) with the voltage dip level to be 60%.

(AC components in stator dq flux) is also accelerated and the electromagnetic torque fluctuations are suppressed.

Test results on the 15 kW DFIG test system with flux linkage tracking control under symmetrical grid faults are shown in Figure 12.19. When the rotor speed is 1.06 pu, the DFIG is generating 0.2 pu active power before the fault happens. In Figure 12.19a, the voltage dip level is 80%. It can be found that the decay of stator natural flux (AC components in stator dq flux) is accelerated after the voltage dips, compared to the normal vector control. Except for a large transient torque pulse after the voltage dip, the electromagnetic torque fluctuations are much smaller compared to demagnetizing control. Large transient rotor over currents up to 20 A and stator currents up to about 50 A can be found after the voltage dips, as under 80% voltage dips with super-synchronous speed, the RSC may be saturated for a short time after the voltage dips.

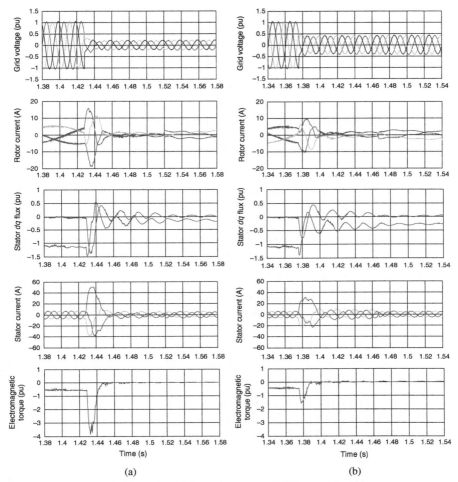

Figure 12.19   Test results of the 15 kW DFIG with flux linkage tracking under symmetrical voltage dips: (a) With a voltage dip level of 80% and (b) with a voltage dip level of 60%.

In Figure 12.19b, the voltage dip level is reduced to 60%, the flux linkage tracking control is able to limit the rotor current under 10 A, and limit the stator current under about 30 A. The decay of stator natural flux (AC components in stator dq flux) is also accelerated and the electromagnetic torque fluctuations are suppressed.

12.3.3   Feedforward Control

The feedforward control during FRT tries to suppress the transient rotor current introduced by the voltage dips, to ensure safe operation of the DFIG. The rotor equation of the DFIG under voltage dip have been introduced in (12.18). The rotor equivalent

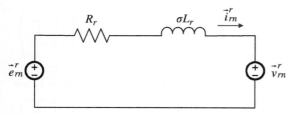

Figure 12.20   Rotor equivalent circuit of the DFIG under voltage dips.

circuit of the DFIG under voltage dip can then be derived as shown in Figure 12.20. In this case

$$\vec{e}^{\,r}_{rn} = \left(\frac{d}{dt} - j\omega_r\right)\vec{\psi}_{sn}e^{-j\omega_r t} \approx -\frac{\omega_r}{\omega_s}pV_se^{-j\omega_r t} \tag{12.26}$$

is the rotor EMF introduced by the stator natural flux. The rotor voltage $\vec{v}^{\,r}_{rn}$ is the RSC output voltage. It can be seen that when the rotor EMF $\vec{e}^{\,r}_{rn}$ is much larger than the RSC output voltage $\vec{v}^{\,r}_{rn}$, large transient rotor current $\vec{i}^{\,r}_{rn}$ will be introduced. This may be caused by the following reasons:

- The rotor EMF $\vec{e}^{\,r}_{rn}$ introduced by the voltage dip occurs abruptly, while the response speed of the RSC output voltage $\vec{v}^{\,r}_{rn}$ is limited by the dynamic performance of the control loop; so $\vec{v}^{\,r}_{rn}$ is not able to respond timely.
- The rotor amplitude EMF $\vec{e}^{\,r}_{rn}$ introduced by the voltage dip is too large, while the RSC output voltage $\vec{v}^{\,r}_{rn}$ is limited by the DC-bus voltage, so the RSC is saturated and a transient rotor current is introduced.

The proposed feedforward control uses the feedforward terms to enhance the dynamic performance of the RSC in order to make the RSC response quickly against the rotor EMF introduced by the voltage dips, so the rotor transient current can be suppressed [3, 4].

The scheme of the feed forward control of the DFIG under voltage dips is shown in Figure 12.21. The stator natural flux can be calculated from the flux observer and the feedforward term calculated from (12.26) is added to the rotor reference; so the RSC is able to response quickly against grid voltage dips. The rest of the control is similar to that in a conventional vector control method.

However, the feedforward control cannot suppress the transient current introduced by the RSC saturation. As the control target is to control the rotor natural current to be zero, the required rotor voltage can be found as (12.27).

$$\vec{v}^{\,r}_{rn} = \vec{e}^{\,r}_{rn} \approx -\frac{\omega_r}{\omega_s}pV_se^{-j\omega_r t} \tag{12.27}$$

On the other hand, the maximum amplitude of the RSC output voltage is limited by the DC-bus voltage, as given by

$$\left|\vec{v}^{\,r}_{rn}\right| \le \frac{\sqrt{3}}{3}v_{dc} \tag{12.28}$$

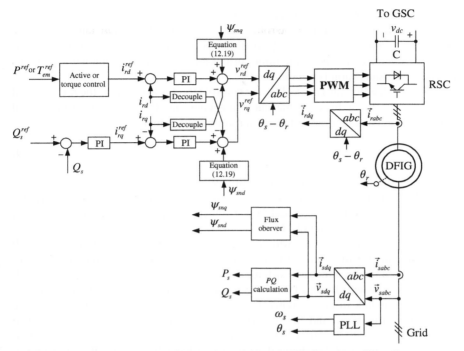

Figure 12.21   Scheme of the feedforward control for the DFIG under voltage dips.

From (12.27) and (12.28), the controllable range of the feedforward control for the DFIG can be calculated. Take the 1.5 MW DFIG as an example, the controllable range is shown in Figure 12.22. The DFIG with feedforward control is able to ride through about 50% voltage dips under a rotor speed of 1.2 pu, and 65% voltage dips under a rotor speed of 0.8 pu. Generally speaking, the controllable range of the

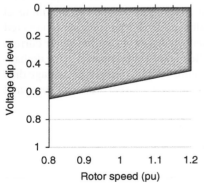

Figure 12.22   Controllable range of the feedforward control of the DFIG under symmetrical grid faults.

feedforward control is smaller than the demagnetizing control, as for the demagnetizing control, the voltage drop at the rotor resistance and leakage inductance will make the required rotor voltage smaller. It should be noticed that even the RSC is saturated out of the controllable range; the rotor current may be still be within the limitations as long as the saturation is not too deep.

For the feedforward control, as the control target is to suppress the rotor natural current to be zero, in this case, the stator time constant after the feedforward control is applied will be similar as that of open rotor circuit, which is

$$\tau_s \approx \frac{L_s}{R_s} \qquad (12.29)$$

The damping of the stator natural flux is not accelerated in theory.

The feedforward control can also be used under asymmetrical grid faults. In this case, the feedforward terms should also include the negative-sequence components rotor EMF, as expressed in (11.16).

Some simulation results from a 1.5 MW MATLAB simulation model, as well as experimental results from a 15 kW DFIG test bench, for the DFIG with feedforward control under symmetrical voltage dips are shown in Figures 12.23 and 12.24, respectively. The waveforms of the three-phase grid voltage, rotor current, stator current, the stator flux in dq reference frames, as well as the torque fluctuations, are shown in Figures 12.23 and 12.24. The rotor speed is 0.8 pu in Figure 12.23a, and 1.2 pu in Figure 12.23b. The voltage dip level is 60% in Figure 12.23. The simulation results in Figure 12.23a shows that after the voltage dips, the feedforward control suppressed the transient rotor current, the transient rotor current is limited to less than 0.5 pu, and the transient stator current is also limited to 1.2 pu at most. However, the damping of the stator natural flux, which is the AC components in the dq reference frame, is much slower compared to the DFIG with demagnetizing control, as shown in Figure 12.15. In Figure 12.23b, the RSC is saturated after voltage dips, as the corresponding rotor speed and voltage dip level are out of the controllable range. Larger transient rotor and stator currents are then introduced.

Some test results are shown in Figure 12.24. The rotor speed is 1.1 pu; in Figure 12.24a, the voltage dip level is 40% while in Figure 12.24b, the voltage dip level is 60%. As for Figure 12.24a, the 40% voltage dips under 1.1 pu rotor speed covered by the controllable range of the feedforward control, the rotor current, stator current is limited. But the damping of the stator natural flux is slower compared to the demagnetizing control, as shown in Figure 12.16. In Figure 12.24b, as the 60% voltage dips under 1.1 pu rotor speed is out of the controllable range, larger transient rotor and stator currents are introduced for a short period.

12.4   FRT STRATEGIES BASED ON HARDWARE SOLUTIONS

The FRT strategies based on improving the control have been introduced in the last section; they can limit the transient current under voltage dips, and some of them are

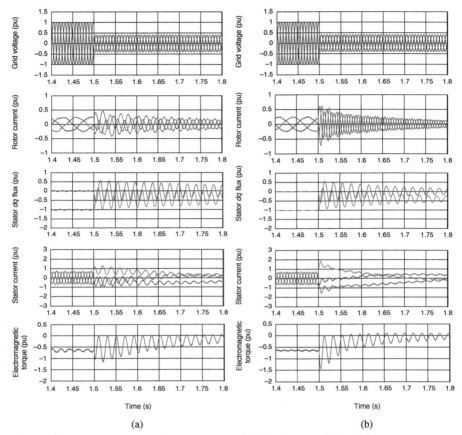

Figure 12.23 Simulation results of the 1.5 MW DFIG with demagnetizing current control under symmetrical voltage dip: (a) With the rotor speed of 0.8 pu and (b) with the rotor speed of 1.2 pu.

able to accelerate the damping of the stator natural flux. However, it has been analyzed that the voltage and current capacities of the RSC may not satisfy the rotor voltage needed for the improved control strategies. As shown in Figures 12.14b and 12.22, for the 80% voltage dip at 1.2 pu rotor speed, the RSC may be saturated with either demagnetizing control or feedforward control. In this case, large transient rotor currents and stator currents may be introduced, as shown in Figure 12.15a, also in Figures 12.24a and 12.24b, up to 1pu rotor transient current and 3pu stator transient currents may be introduced. On the other hand, many countries demand the DFIG WPS to ride through 80% voltage dips, and in some countries, even 100% voltage dip ride-through capabilities are demanded. Consequently, only improved control methods are not enough for the FRT of the DFIG under grid faults [5, 6].

Hardware solutions are use additional circuits, in order to assist the DFIG to ride through serious grid faults. This may lead to additional cost and weight, but they

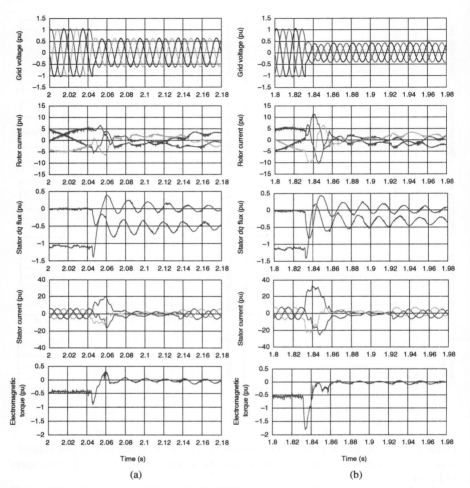

Figure 12.24 Test results of the 1.5 MW DFIG with demagnetizing current control under symmetrical voltage dip: (a) With a voltage dip level of 80% and (b) with a voltage dip level of 60%.

provide a reliable protection for the RSC and DFIG under voltage dips. Many kinds of hardware solutions have been applied or analyzed for the DFIG. In this section, four typical hardware solutions, including the rotor-side crowbar, DC choppers, stator/rotor breaking resistance, and the dynamic voltage restorer (DVR) are going to be introduced as solutions.

12.4.1 Rotor-Side Crowbar

The rotor-side crowbar is one of the commonly used hardware solutions for the DFIG to ride through grid faults. The rotor-side crowbar short-circuits the rotor windings

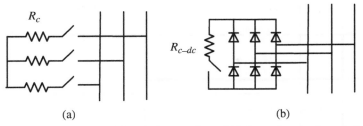

(a) (b)

Figure 12.25 Rotor-side crowbar consists of (a) three-phase resistor and switches; (b) rectifier, and DC-side resistor.

under grid faults, so that the transient rotor current can be limited and the RSC can be protected [7]. The rotor-side crowbar normally consists of a three-phase resistor and the three-phase switches, as shown Figure 12.25. In some cases, the rotor-side crowbar with the rectifier and a DC-side resistor and a DC switch is used, as this concept may be more cost-efficient compared to the three-phase concept. It has been discussed in Chapter 6 that if the current amplitude in the three-phase crowbar and the DC-switch crowbar are the same, the equivalent relationship of the crowbar resistance in the three-phase crowbar R_c and DC-switch crowbar R_{c_dc} can be found as

$$R_c = \frac{\sqrt{3}}{3}R_{c\_dc} \tag{12.30}$$

The scheme of a DFIG WPS with a rotor-side crowbar is shown in Figure 12.26, and its action under voltage dips are illustrated in Figure 12.27. After the grid fault, the voltage dip will be introduced. The crowbar will be enabled according to the fault-detection algorithm [8]. When the voltage dip is detected, the rotor current is larger than the safety value, or the DC-bus voltage is larger than the safety limitation, as illustrated in Figure 12.26. The RSC is then disabled so that the transient

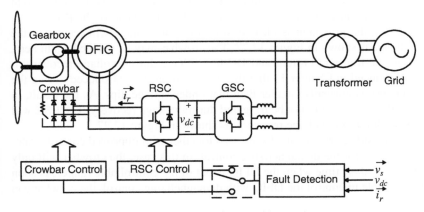

Figure 12.26 Scheme of a DFIG WPS with rotor-side crowbar.

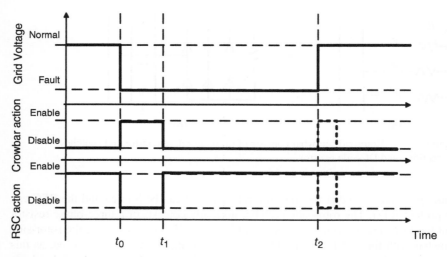

Figure 12.27 Action of the DFIG WPS with rotor-side crowbar activation under voltage dips.

overcurrent are all taken by the crowbar, at the time of t_0 in Figure 12.27. With the crowbar enabled, the DFIG will operate as a cages generator, and reactive power will be absorbed from the grid. The dynamic model of the DFIG with rotor-side crowbar has been introduced in Chapter 11, so it will not be repeated here. After a few tens of milliseconds, at the time t_1 in Figure 12.27, as the stator natural flux introduced by the voltage dips has been damped fast when the crowbar is enabled, it is relatively smaller at t_1, so the RSC is able to control the DFIG in this case. Then, the crowbar is disabled and the RSC is restarted to control the DFIG. After t_1, the DFIG is controlled by the RSC to generate reactive current to support the grid, as demanded by the grid codes. The restart of the RSC is also an important issue. Normally, the crowbar needs to be disabled first, then after one or two switching periods, the RSC can be restarted. The integrator of the inner current control loop of the RSC should be reset to zero in order to prevent a transient overt current introduced by the restart of the RSC.

Assuming the grid fault is cleared at t_2, the grid voltage will restore to the normal voltage, as shown in Figure 12.27. The crowbar may be enabled again here, according to the transient current and DC-bus voltage during voltage recovery. It has been introduced in Chapter 11 that the stator natural flux introduced during the voltage recovery is normally smaller than that under voltage dips. As the result, in most cases, the crowbar will not be triggered during voltage recovery. The RSC will control the DFIG during voltage recovery, as shown in Figure 12.27.

The rotor-side crowbar can also be used together with improved control strategies, for example, for the demagnetizing control. In this case, the crowbar will only be enabled for a short time when the DFIG is out of the controllable range. As soon as the RSC is able to control the DFIG with demagnetizing control, the crowbar can be switched off.

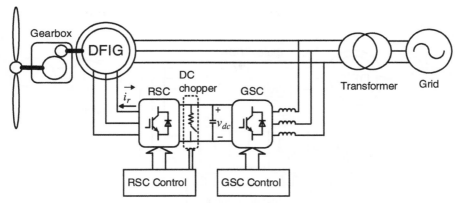

Figure 12.28   Scheme of the DFIG WPS using a DC chopper for FRT.

12.4.2   DC Chopper

The DC chopper is a breaker resistance in the DC bus, as shown in Figure 12.28. It has been analyzed that under voltage dips, the RSC may be saturated as the EMF introduced by the stator natural flux is too large. In this case, the rotor current may flow into the DC bus of the RSC and lead to DC-bus overvoltage, which will destroy the switching devices immediately. The DC chopper is enabled when the DC-bus voltage is over the safety value, then the energy from the RSC can be dissipated on the resistor of the DC chopper thereby the DC-bus voltage is limited [9].

If the RSC is disabled when the DC chopper is activated at the DC link, the DC chopper will act as a rotor-side crowbar from the rotor-side point of view. In this case, the DFIG will be out of control and the stator side will absorb a large amount of reactive power from the grid. However, the DC chopper is usually cooperating with the improved control strategies under voltage dips. When the RSC is saturated and/or when the rotor EMF is too large for the RSC to handle, the DC chopper can absorb the power fed to the DC link and thus avoid an overvoltage of the DC capacitor. The safety operation is guaranteed under serious voltage dips.

The DC chopper can also be used together with a crowbar. In this case, the rotor current can be shared between the DC chopper and the rotor-side crowbar. The activation time of the rotor-side crowbar can somehow be reduced.

12.4.3   Series Dynamic Breaking Resistor

The rotor-side crowbar uses a resistor in parallel with the rotor windings of the DFIG, so the rotor current can be limited. Another approach is to add resistors in series with the DFIG windings, so that the rotor and stator currents can also be limited, as shown in Figure 12.29. It is also mentioned as a series dynamic braking resistor (SDBR), as shown in Figure 12.28. The SDBR can be placed on the stator side or on the rotor side.

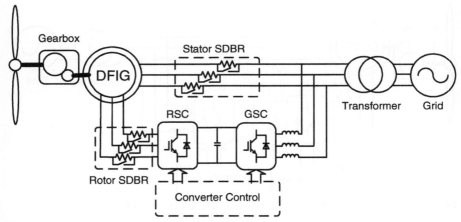

Figure 12.29   Scheme of the DFIG WPS with SDBR.

It is connected to the stator or rotor circuit when the voltage drops, so the stator or rotor resistance is enlarged, then the transient current can be limited [10, 11]. Besides, as the stator time constant with vector control can be found as

$$\tau_s = \frac{L_s}{R_s} \qquad (12.31)$$

The SDBR on the stator side will increase the stator resistance R_s, the stator time constant τ_s will be reduced, which means the damping of the stator natural flux will be accelerated.

When the SDBR is enabled, the RSC does not need to be switched off and the DFIG can be under control all the time. However, when the DFIG is providing reactive power to the grid, the SDBR should be disconnected, otherwise, the power losses on the SDBR will be very large.

12.4.4   Dynamic Voltage Restorer

The DVR is a voltage source converter in series connected to the stator side of the DFIG, as shown in Figure 12.30. With the DVR, the stator voltage of the DFIG is the sum of the grid voltage and the output voltage of the DVR. So under the voltage dip, the DVR is able to compensate for the grid voltage dip and the stator side of the DFIG does not "see" the voltage dips; then no transient stator current, rotor current, or DC-bus voltage will be introduced in the DFIG.

The main disadvantage of the DVR is the higher cost of the converters and the inductor, compared to the rotor-side crowbar or the DC chopper. Also, in some cases, the DVR is used to compensate part of the grid voltage dips, so that the transient stator, rotor current, and DC-bus voltage will be reduced, and the RSC is able to control the DFIG for all voltage dips [12]. As in this case, the capacity of the DVR can be reduced.

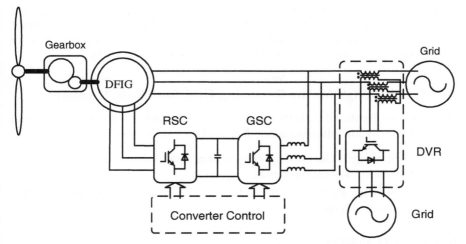

Figure 12.30   Scheme of the DFIG WPS with dynamic voltage restorer (DVR) to handle a complete fault ride though.

A similar strategy can also be used on wind farm terminal, so all the WPS terminal voltage in the wind farm can be compensated. It is normally mentioned as static synchronized serious compensator (SSSC) [13].

12.5   RECURRING FAULT RIDE THROUGH

It has been analyzed in Chapter 11 that under recurring grid fault, the voltage recovery of the first grid fault will also introduce the stator natural flux. If this stator natural flux still exists when the next grid fault happens, the stator natural flux produced by the voltage recovery and by the next voltage dip may be superposed. If vector control is applied after voltage recovery, the stator natural flux decays slowly after the voltage recovery of the first grid fault. In a large-scale DFIG WTS, the typical stator time constant is around 1–2 s. On the other hand, the shortest duration between two faults is only 500 ms according to some grid codes. As a result, the stator natural flux produced by the voltage recovery and by the next voltage dip may be superposed and the transient current and voltage under recurring faults may be larger than that under single fault. The result is that the DFIG may fail to ride through the recurring faults even with rotor-side crowbar. On the other hand, if the crowbar is triggered again during voltage recovery, the rotor current can accelerate the damping of the stator natural flux. But as the rotor current is not under control, large electromagnetic torque (EM-torque) fluctuations may be introduced. It will also influence the reliability of the mechanical system. Consequently, the FRT strategies designed for single grid faults seem not to be the best solution for the FRT of DFIG under the recurring grid faults. The FRT strategy for DFIG WTS to ride through recurring symmetrical grid faults need to be investigated. An improved control strategy is introduced and it is applied during the voltage recovery after the voltage dips. A rotor natural current with rotor rotation frequency is generated in the opposite direction of the stator natural flux, so that the damping of the stator natural flux can be accelerated [14, 15].

12.5.1 Challenge for the Recurring Grid Fault Ride Through

The challenges for the DFIG WPS under recurring grid faults will be discussed based on the DFIG with the following commonly used FRT strategies [16]:

> **FRT Strategy I:** Use the rotor-side crowbar under voltage dips and vector control during voltage recoveries.

> **FRT Strategy II:** Use the rotor-side crowbar both under voltage dips and during voltage recovery.

12.5.1.1 FRT Strategy I

The operation of DFIG WPS under recurring grid faults with FRT strategy I is shown in Figure 12.31. The first grid fault happens at t_0, the voltage drop is detected and

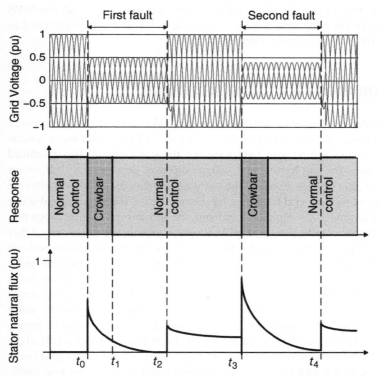

Figure 12.31   Operation of the DFIG WPS under recurring grid faults with FRT strategy I.

the crowbar is enabled. At t_1, as the RSC is able to control the DFIG, the crowbar is disabled and the vector control is applied to provide the reactive power support. The voltage recovery started at t_2' as the stator natural flux introduced during voltage recovery is normally smaller than that under voltage dips and the crowbar does not need to be triggered in most cases, and the DFIG is still operating with vector control. When the next grid fault happens at t_3, the crowbar will be active again, and the response of DFIG WTS will be the same as in the case of the first grid fault.

The dynamic model of the DFIG, in this case, has been introduced in Chapter 11 and the largest rotor current I_{rm} under the second fault can be found as (11.59); it is rewritten here as

$$I_{rm} \approx \frac{\omega_r}{\sqrt{R_c^2 + (\omega_r \sigma L_r)^2}} \left(\sqrt{2} p_1 \frac{V_s}{\omega_s} \cos\theta e^{-\frac{T_{re}}{\tau_s}} + p_2 \frac{V_s}{\omega_s} \right) \qquad (12.32)$$

and the largest stator current I_{sm} under the second fault can be found as (11.60). It is rewritten here as

$$I_{sm} \approx \frac{1}{L_s} \left(\sqrt{2} p_1 \frac{V_s}{\omega_s} \cos\theta e^{-\frac{T_{re}}{\tau_s}} + p_2 \frac{V_s}{\omega_s} \right) \left(1 + \frac{\omega_r L_m}{\sqrt{R_c^2 + (\omega_r \sigma L_r)^2}} \right)$$

$$+ \frac{1}{L_s} \frac{(1 - p_2)V_s}{\omega_s} \qquad (12.33)$$

The maximum rotor current I_{rm} and stator current I_{sm} under recurring faults in a 1.5 MW DFIG WPS are shown in Figures 12.32a and 12.32b and compared with that under single grid fault. The dip level of the first and second fault $p_1 = p_2 = 80\%$ and the grid fault angle θ is 75°. It can be seen that under a single grid fault, the maximum rotor and stator current are limited to about 2.3 pu under 80% voltage dip. However, under recurring grid faults, as the stator natural flux decays slowly after voltage recovery, the maximum rotor and stator current can reach 3.0 pu if the

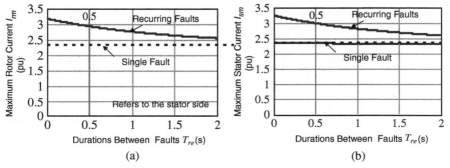

Figure 12.32   Performance of the DFIG under recurring faults with vector control after voltage recovery: (a) Maximum rotor current and (b) maximum stator current.

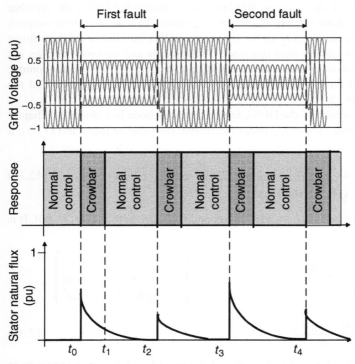

Figure 12.33 Operation of the DFIG WPS under recurring grid faults with FRT strategy II.

duration between the two faults is 500 ms. The current capacity of the crowbar may not be large enough for this transient current under the second voltage dip. So, the DFIG may fail to ride through the recurring fault even with the rotor-side crowbar.

12.5.1.2 FRT Strategy II

In some cases, the crowbar can be active again during the voltage recovery and the operation of the DFIG with this FRT strategy is as illustrated in Figure 12.33. The response of the DFIG after a voltage dip is the same as that in Figure 12.31, the crowbar is enabled and the RSC is disabled. After the voltage recovery is detected after t_2 (also after t_4), the crowbar is active again. As the enabling of crowbar will accelerate the damping of the stator natural flux, the stator natural flux decays much faster compared to what is shown in Figure 12.33 [17].

It has been introduced in Chapter 11 that the stator time constant with rotor-side crowbar can be expressed as (11.26) and it is rewritten here as

$$\tau_{sc} \approx \frac{L_s}{R_s + R_s L_m \frac{\omega_r^2 \sigma L_r}{R_c^2 + (\omega_r \sigma L_r)^2}} \tag{12.34}$$

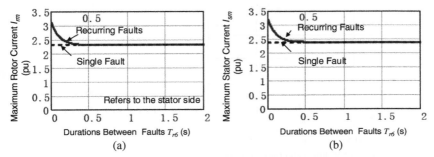

Figure 12.34   Performance of the DFIG under recurring faults with crowbar enabled after voltage recovery: (a) Maximum rotor current and (b) maximum stator current.

So, by replacing τ_s with τ_{sc} in (12.32) and (12.33), the maximal rotor current and stator current under the second fault can be found as

$$I_{rm} \approx \frac{\omega_r}{\sqrt{R_c^2 + (\omega_r \sigma L_r)^2}} \left(\sqrt{2} p_1 \frac{V_s}{\omega_s} \cos \theta e^{-\frac{T_{re}}{\tau_{sc}}} + p_2 \frac{V_s}{\omega_s} \right) \tag{12.35}$$

$$I_{sm} \approx \frac{1}{L_s} \left(\sqrt{2} p_1 \frac{V_s}{\omega_s} \cos \theta e^{-\frac{T_{re}}{\tau_{sc}}} + p_2 \frac{V_s}{\omega_s} \right) \left(1 + \frac{\omega_r L_m}{\sqrt{R_c^2 + (\omega_r \sigma L_r)^2}} \right) + \frac{1}{L_s} \frac{(1 - p_2) V_s}{\omega_s} \tag{12.36}$$

For a large-scale DFIG, the new stator time constant τ_{sc} is usually reduced to around 100 ms. As a result, when the next fault happens at t_3 (in Figure 12.33), the stator natural flux $\vec{\psi}_{sn}$ introduced by the voltage recovery will fully decay and its influence on the next grid fault will be small. In a 1.5 MW DFIG, the maximum rotor current I_{rm} and stator current I_{sm} under recurring faults, when the dip levels of the first and second faults are $p_1 = p_2 = 80\%$, and the grid fault angle θ is 75° are shown in Figures 12.34a and 12.34b. Compared to that under single grid fault, it can be concluded that as the damping of $\vec{\psi}_{sn}$ is accelerated after voltage recovery when the duration between the two faults are larger than 0.5 s, the maximum rotor current and stator current under recurring fault are basically the same as in the case of a single grid fault.

However, the enabling of the rotor-side crowbar may cause other problems for the DFIG WPS. The rotor natural current $\vec{i}_{rn}^r$ and rotor-forced current $\vec{i}_{rf}^r$ will exist at the same time. In the rotor reference frame, they can be derived from the Chapter 11, as given in (12.37) and (12.38).

$$\vec{i}_{rn}^r(t) \approx -\frac{\omega_r}{\omega_s} \frac{\sqrt{2} p_1 V_s \cos \theta_1}{R_c + j\omega_r \sigma L_r} e^{-\frac{t-t_2}{\tau_{sc}}} e^{j\omega_r(t-t_2)} \tag{12.37}$$

$$\vec{i}_{rf}^r(t) = \frac{\omega_s - \omega_r}{\omega_s} \frac{V_s}{R_c + j(\omega_s - \omega_r)\sigma L_r} e^{j(\omega_s - \omega_r)(t-t_2)} \tag{12.38}$$

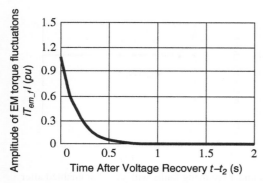

Figure 12.35 Electromagnetic torque fluctuations of the DFIG after voltage recovery.

It has been introduced that the stator natural flux $\vec{\psi}_{sn}$ and the stator-forced flux $\vec{\psi}_{sf}$ after the voltage recovering can be found as (11.48) and (11.49) and they are rewritten here as

$$\vec{\psi}_{sn} = \sqrt{2}p_1 \frac{V_s}{\omega_s} \cos\theta e^{\frac{\pi}{4}j} e^{-\frac{t-t_2}{\tau_s}} \tag{12.39}$$

$$\vec{\psi}_{sf} = \frac{V_s}{j\omega_s} e^{j[\omega_s(t-t_2)+\theta]} \tag{12.40}$$

The electromagnetic torque T_{em} of the DFIG can be found as

$$T_{em} \approx \frac{3}{2}n_p \frac{L_m}{L_s}(\psi_{r\beta}i_{r\alpha} - \psi_{r\alpha}i_{r\beta}) \tag{12.41}$$

With (12.35)–(12.41), the amplitude of electromagnetic torque fluctuations in the DFIG after voltage recovery with the rotor-side crowbar can be derived, and there will be an AC component with a frequency of ω_s in T_{em}, as shown in (12.42).

$$\left|T_{em\_f}\right|(t) \approx \frac{3}{2}n_p \frac{L_m}{L_s}$$

$$\times \left[\frac{V_s}{\omega_s}\frac{\omega_r}{\omega_s}\frac{\sqrt{2}\cos\theta p_1 V_s}{\sqrt{R_c^2 + (\omega_r\sigma L_r)^2}} + \frac{\omega_r}{\omega_s}\frac{V_s}{\sqrt{R_c^2 + ((\omega_s - \omega_r)\sigma L_r)^2}}\sqrt{2}p_1\frac{V_s}{\omega_s}\cos\theta\right] e^{-\frac{t-t_2}{\tau_s}} \tag{12.42}$$

The amplitude of the electromagnetic torque fluctuations after voltage recovery in a 1.5 MW DFIG when the grid fault angle $\theta = 75°$ is shown in Figure 12.35 (in pu). The electromagnetic torque fluctuations will reach 1 pu after the voltage recovery. Such large torque fluctuations may influence the reliability of the mechanical system and may even damage to the gearbox.

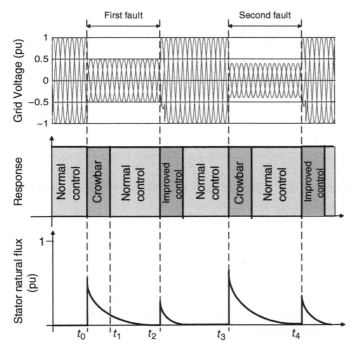

Figure 12.36 Operation of the DFIG WPS under recurring grid faults with a crowbar and improved control.

12.5.2 Control Target for Recurring Fault Ride Through

The basic idea of the improved control is to accelerate the damping of the stator natural flux after voltage recovery so that the influence from the first grid fault on the second grid fault will be relatively smaller. At the same time, the electromagnetic torque fluctuations need to be suppressed, so that the reliability of the mechanical system will be improved. The response of the DFIG with improved control under recurring grid faults is shown in Figure 12.36. The operation of the DFIG under voltage dips will be the same as in the case of single grid fault where the crowbar is active at t_0 and t_3, and the RSC is restarted to generate reactive current after about 100 ms at t_1. The improved control is implied by the RSC after the voltage recovery at t_2. A rotor natural current $\vec{i}_{rn}$ is generated in the opposite direction of stator natural flux $\vec{\psi}_{sn}$, as shown in Figure 12.37. As analyzed in (11.13)–(11.15), the damping of stator natural flux can be accelerated by the rotor natural current after voltage recovery. So, when the next grid fault happens, the stator natural flux $\vec{\psi}_{sn}$ introduced by the voltage recovery will be very small so that the transient stator and rotor currents will basically be the same as in the case of the first dip.

At the same time, in order to suppress the electromagnetic torque fluctuations after voltage recovery, a rotor-forced current $\vec{i}_{rf}$ is generated, and it is in the opposite

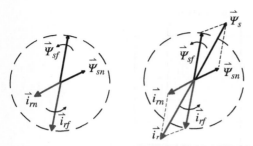

Figure 12.37 Stator flux and rotor current vector of the DFIG with improved control.

direction of stator-forced flux $\vec{\psi}_{sf}$, as long as the relationship between the amplitude of $\vec{i}_{rf}$ and $\vec{\psi}_{sf}$ meet the relationship in (12.43).

$$\frac{\left|\vec{\psi}_{sn}^{s}\right|}{\left|\vec{i}_{rn}^{s}\right|} = \frac{\left|\vec{\psi}_{sf}^{s}\right|}{\left|\vec{i}_{rf}^{s}\right|} \tag{12.43}$$

The angle between the total stator flux $\vec{\psi}_{s}$ and rotor current $\vec{i}_{r}$ is always 180°, as shown in Figure 12.37 and in this case, according to (12.41), the electromagnetic torque of the DFIG can always be controlled nearly to be zero, and the electromagnetic torque fluctuations can be suppressed.

12.5.3 Control Implication

In the improved control, the RSC is required to generate a rotor current with two different frequencies. One is the rotor natural current $\vec{i}_{rn}$, which is a DC component in stator $\alpha\beta$ reference frame, and an AC component with a rotor rotation frequency ω_{r} in the rotor reference frame. The other is the rotor-forced current $\vec{i}_{rf}$, which is an AC component with a grid frequency of ω_{s} in the stator $\alpha\beta$ reference frame, and with a slip frequency of $\omega_{s} - \omega_{r}$ in the rotor reference frame. They are in the opposite direction of the stator natural and forced fluxes $\vec{\psi}_{sn}$ and $\vec{\psi}_{sf}$, respectively; so in the stator $\alpha\beta$ reference frame, they are given as

$$\vec{i}_{rn}^{s} = -k_{1}\vec{\psi}_{sn}^{s} \quad \text{and} \quad \vec{i}_{rf}^{s} = -k_{2}\vec{\psi}_{sf}^{s} \tag{12.44}$$

where k_{1} and k_{2} are proportional coefficients between the rotor current and stator flux. The relationship between the amplitude of $\vec{i}_{rn}^{s}$ and $\vec{i}_{rf}^{s}$ will meet (12.43) as well, so it can be found that

$$k_{1} = k_{2} \tag{12.45}$$

By substituting (12.44) in the stator flux equation, the rotor natural current $\vec{i}_{rn}^{s}$ and forced current $\vec{i}_{rf}^{s}$ can be expressed as

$$\vec{i}_{rn}^{s} = -\frac{L_{s}}{L_{m} + 1/k_{1}}\vec{i}_{sn}^{s} \quad \text{and} \quad \vec{i}_{rf}^{s} = -\frac{L_{s}}{L_{m} + 1/k_{2}}\vec{i}_{sf}^{s} \tag{12.46}$$

After voltage recovery, there are only the natural and forced components in the DFIG in the ideal case, so that the rotor current and stator current can be represented as

$$\vec{i}_r^s \approx \vec{i}_{rn}^s + \vec{i}_{rf}^s \quad \text{and} \quad \vec{i}_s^s \approx \vec{i}_{sn}^s + \vec{i}_{sf}^s \tag{12.47}$$

The rotor current with the improved control can thereby be determined as

$$\vec{i}_r^s = -\frac{L_s}{L_m + 1/k_1(\text{or } k_2)}\vec{i}_s^s \tag{12.48}$$

The stator current can be directly measured. The values of k_1 (or k_2) need to be determined.

With the improved control, the rotor natural current can be expressed as (12.46), by substituting (12.46) in (12.12). The differential equation of the stator natural flux can be found as

$$\frac{d}{dt}\vec{\psi}_{sn}^s = \left(\frac{R_s}{L_s} - \frac{R_s L_m}{L_s}k\right)\vec{\psi}_{sn}^s \tag{12.49}$$

By solving (12.49), the stator time constant τ_{si} can be expressed as (12.50).

$$\tau_{si} = \frac{L_s}{R_s + k_1(\text{or } k_2)R_s L_m} \tag{12.50}$$

On the other hand, the amplitude of the rotor current I_r and stator current I_s with improved control can be calculated from (12.39), (12.40), and (12.44), as given in

$$I_r \approx (1 + \sqrt{2}p_1 \cos\theta)k_1(\text{or } k_2)\frac{V_s}{\omega_s} \tag{12.51}$$

$$I_s = \frac{(1 + \sqrt{2}p_1 \cos\theta)V_s(k_1(\text{or } k_2)L_m + 1)}{\omega_s L_s} \tag{12.52}$$

The maximum voltage dip level p_1 is 80% according to the grid code, while the grid fault angle θ in the transmission line is normally larger than 75°. So the approximated value of I_r and I_s can be represented as (12.53) and (12.54).

$$I_r \approx 1.3k_1(\text{or } k_2)\frac{V_s}{\omega_s} \tag{12.53}$$

$$I_s \approx \frac{1.3V_s\left(k_1(\text{or } k_2)L_m + 1\right)}{\omega_s L_s} \tag{12.54}$$

The relationships between k_1 (or k_2) and τ_{si}, as well as the relationships between k_1 (or k_2) and I_r, I_s in a 1.5 MW DFIG are shown in Figure 12.38. It can be found from Figure 12.38a that τ_{si} is reduced by the improved control compared with vector control and the damping of the stator natural flux $\vec{\psi}_{sn}$ is accelerated. With the increase of k_1 (or k_2), τ_{si} is decreased, so that the damping of $\vec{\psi}_{sn}$ can be faster with larger k_1 (or k_2). On the other hand, it can be concluded from Figures 12.38b and 12.38c that the amplitudes of the rotor current I_r and stator current I_s are increased with the increase of k_1 (or k_2). As I_r and I_s are limited by the safety current I_{rm} of the RSC, and the

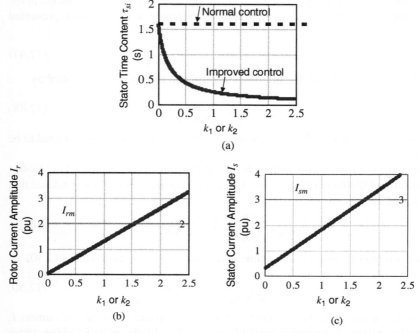

Figure 12.38 Stator flux and rotor current vector of the DFIG with improved control (red dotted lines are the limit).

safety current I_{sm} of the stator, the value of k_1 (or k_2) are limited as well. In this case, the safety current I_{rm} of the RSC is set to be 2 pu and the safety current I_{sm} of the stator is 3 pu.

It can be found with the increase of k_1 (or k_2), the rotor current I_r reaches the safety limit first. So k_1 (or k_2) is chosen to achieve the best acceleration effect in the safety operation area (SOA) of the RSC, as given in (12.55).

$$k_1 = k_2 \approx 0.8 \frac{I_{rm}\omega_s}{V_s} \tag{12.55}$$

So the control reference of the rotor current vector $\vec{i}_r^{s*}$ can be expressed as

$$\vec{i}_r^{s,ref} = -\frac{L_s}{L_m + V_s/0.8I_{rm}\omega_s}\vec{i}_s^s \tag{12.56}$$

Normally, the DFIG is controlled in the synchronous dq reference frame, as given in (12.57).

$$\vec{i}_{rdq}^{ref} = -\frac{L_s}{L_m + V_s/0.8I_{rm}\omega_s}\vec{i}_{sdq} \tag{12.57}$$

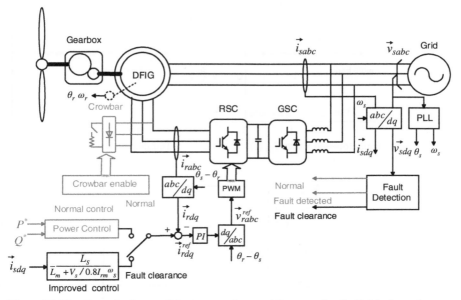

Figure 12.39   Control scheme of the improved control for recurring fault ride through operation.

12.5.4   Control Scheme

Under recurring grid faults, the operation of a DFIG with improved control is shown in Figure 12.40. Before the fault happens, the vector control is used and the DFIG generates active power under normal grid voltage. The first grid fault happens at t_0, the crowbar is enabled after the voltage dip is detected. Then the crowbar is switched off at t_1. The voltage recovers at t_2 from the first grid fault. When the voltage recovery is detected, the improved control is applied. The control diagram of the improved control is shown in Figure 12.39. The rotor current reference $\vec{i}_{rdq}^{\,ref}$ is switched from the output of the power control loop to the value calculated from (12.57). As analyzed in the previous paragraph, the damping of the stator natural flux introduced by the voltage recovery can be accelerated and the electromagnetic torque fluctuations will be suppressed at the same time. When the stator natural flux introduced by the voltage recovery is small enough, the DFIG will go back into normal operation with vector control. When the next grid fault happens, the crowbar is active again and the transient current will be nearly the same as in the case of a single grid fault.

The control performance of the improved control in a 1.5 MW DFIG is shown in Figure 12.40. Compared with Figures 12.32 and 12.34, it can be seen that when the next fault happens 500 ms later, the maximum rotor and stator current are about the same as the case with a single fault. At the same time, the electromagnetic torque fluctuations will be suppressed after the voltage recovery.

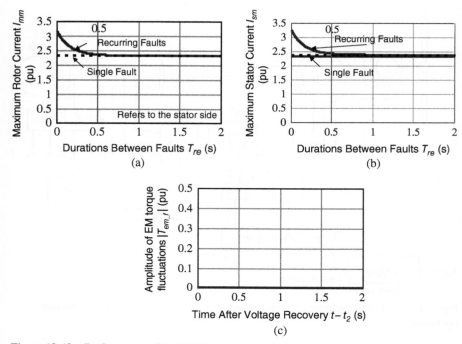

Figure 12.40 Performance of the DFIG under recurring faults with improved control: (a) Maximum rotor current; (b) maximum stator current; and (c) amplitude of electromagnetic torque fluctuations after voltage recovery.

12.5.5 Simulation and Test Results

Simulations are made based on a 1.5 MW DFIG model in Simulink/MATLAB. The parameters of the DFIG are the same as the ones used in the analysis. Firstly, the performance of DFIG under recurring grid faults with different control strategies are simulated and the results are shown in Figure 12.41. The voltage dip levels of the first and second grid faults (p_1 and p_2) are both 80%, the grid fault angle θ of the first faults is 75°. The duration between two faults (T_{re}) is 500 ms. The DFIG generates 1.0 pu active power before the fault happens (at $t_0 = 1.5$ s) and the rotor speed is 1.2 pu (1800 rpm). It can be seen from Figure 12.41a that if the vector control method is applied after the voltage recovery (at $t_2 = 1.67$ s), the stator natural flux introduced by the voltage recovery decays slowly. So when the next voltage dip occurs at $t_3 = 2.18$ s, the stator natural flux introduced by the voltage recovery and by the next voltage dip may be superposed and larger rotor current, stator current, and torque fluctuations can be seen. The rotor transient current under the second faults (at $t_3 = 2.18$ s) reaches 1.5 pu, compared to about 1.0 pu under the first grid fault (at $t_0 = 1.5$ s), while the and stator transient current under the second faults (at $t_3 = 2.18$ s) reaches 3.8 pu, compared to about 2.3 pu under the first grid fault. In Figure 12.41b, the crowbar is active again for about 250 ms after the voltage recovery (at $t_2 = 1.67$ s), and the damping of

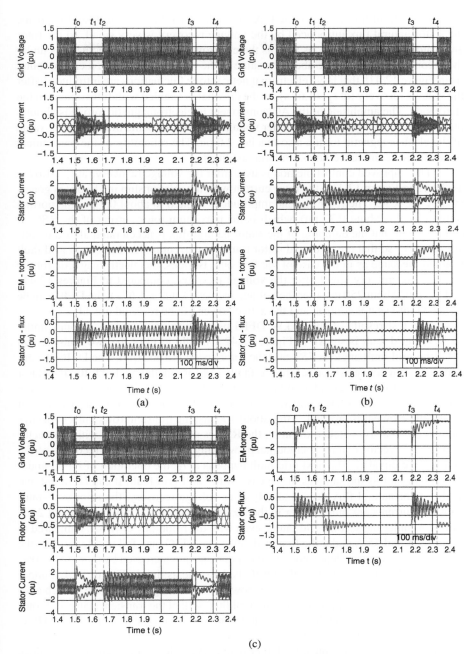

Figure 12.41   Simulation of a 1.5 MW DFIG under recurring grid faults with a crowbar and (a) vector control; (b) crowbar active again; and (c) improved control after voltage recovery.

the stator natural flux is accelerated. As a result, when the next grid fault happens, the stator natural flux introduced by the voltage recovery is very small, so the rotor and stator transient current, as well as the electromagnetic torque fluctuations under the second grid fault will be about the same as the first one. However, it can be found from Figure 12.41b that large electromagnetic torque fluctuations are introduced after the voltage recovery (at $t_2 = 1.67$ s), as the crowbar is enabled with full voltage on the stator side. The amplitude of the AC component in electromagnetic torque is about 1 pu and it is even larger than that under voltage dips. At the same time, the peak value of the electromagnetic torque also reaches 2.0 pu. The reliability of the mechanical system may be influenced by such large loadings.

The improved control method is used after the voltage recovery (at $t_2 = 1.67$ s) as shown in Figure 12.41c, the rotor current with the relationship in (12.56) is generated by the RSC. The rotor current amplitude is limited to about 0.7 pu (the normal rotor current is about 0.37 pu), which is the maximum rotor current of the RSC. It can be found that the damping of the stator natural flux is accelerated after the voltage recovery so that the rotor and stator transient current under the second grid fault is about the same as in the case of the first one. Besides, the electromagnetic torque fluctuations are very small after the voltage recovery, compared to Figures 12.41a and 12.41b. About 250 ms after the voltage recovery (at $t_2 = 1.67$ s), the DFIG can be switched into vector control, and the active power can be restored quickly.

The test results of the 15 kW DFIG using different control strategies under recurring grid fault is shown in Figure 12.42. The dip level of the two faults are all 80%, the grid fault angle is 75°, and the rotor speed is 1600 rpm. The DFIG is generating 0.2 pu active power before the faults. As for smaller-scale DFIG, the stator time constant is much smaller compared to an MW-rated DFIG, the duration of the two fault is reduced to about 80 ms. In Figure 12.42a, the first grid fault happens at t_0, and the crowbar is enabled. After about 100 ms, the transient rotor current decays and the crowbar is switched off at t_1, the RSC is switched on again with the vector control. The vector control method keeps working after the voltage recovery of the first voltage dip at t_2. The next grid fault happens at t_3; the operation of DFIG under the second fault is the same as in the case of the first grid fault. It can be found that as the stator natural flux decays slowly after voltage recovery, the rotor and stator transient currents are much larger in the second voltage dip. They reach about 26 A and 74 A under the second grid faults, even with the rotor-side crowbar, compared to about 18 A and 55 A under the first grid fault. The DFIG may fail to ride through the second fault as the transient rotor and stator currents are increased by more than 30%. In Figure 12.42b, the action of the DFIG is the same as that in Figure 12.42a under dip, from t_0 to t_2 and from t_3 to t_4. The crowbar is active again in Figure 12.42b after t_2 to accelerate the damping of stator natural flux, so that the transient stator and rotor currents under the second dip at t_3 are about the same with the first one (at t_0). However, it can be found that the fluctuations of electromagnetic torque after the voltage recovery (t_2) is about 1.0 pu, even larger than under voltage dips, which may reduce the lifetime of the mechanical systems.

The improved control method is then applied after the voltage recovery (t_2) in Figure 12.42c. In the test system, as the maximum output stator current is limited to

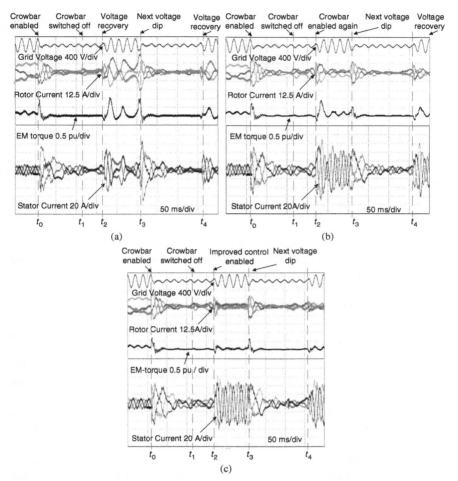

Figure 12.42   Test results of a DFIG under recurring grid faults with crowbar and (a) vector control; (b) crowbar active again; and (c) improved control after voltage recovery.

about 1.0 pu, the control parameters are designed to limit the stator current below 60 A. The action of DFIG is the same as that shown in Figure 12.42a under voltage dips, from t_0 to t_2 and from t_3 to t_4. After the voltage recovers at t_2, the improved control is applied. It can be concluded as the damping of stator natural flux is accelerated after voltage recovery at t_2, the transient stator and rotor currents under the second dips at t_3 are about the same as in the case of the first one (at t_0). Besides, the fluctuations of the electromagnetic torque after the voltage recovery during t_2 and t_3 are suppressed at the same time. It is reduced to no more than 0.2 pu with improved control method, while it is about 1.0 pu in Figure 12.42b. With the improved control method, the DFIG can ride through the recurring faults successfully, as long as it is able to ride through single grid faults. At the same time, the influence on the reliability of the mechanical system is smaller, compared to the control strategy used in Figure 12.42b.

The simulations and test results indicate that the DFIG can ride through the recurring faults as long as it can ride through the single grid fault with the same voltage dip level. The electromagnetic torque fluctuations can be suppressed after the voltage recovery as well, which will enhance the reliability of the mechanical system [19].

12.6 SUMMARY

In this chapter, important fault ride-through strategies are introduced. Starting with the introduction of improved PLL under grid faults, new control strategies for the DFIG to ride through grid fault are introduced [20]. They normally use software changes to apply the improved control, in order to realize the FRT of the DFIG. Three control strategies are selected to be discussed. Then, hardware solutions are analyzed. They use additional circuits to help the DFIG WPS to withstand the voltage dips [21]. The improved control strategies are easy to imply but may not be able to withstand serious grid faults [22], and the hardware solutions may have to be used together with the improved control strategies under serious voltage dips [23]. Corresponding simulation and test results are shown for the methods proposed. For the recurring fault ride-through, its challenges are analyzed and a corresponding FRT strategy is also evaluated [24–26]. The crowbar is used under voltage dips and an improved control strategy is applied during the voltage recovery after the voltage dips; a rotor natural current with rotor rotation frequency is generated in the opposite direction of the stator natural flux, so that the damping of the stator natural flux can be accelerated [27]. At the same time, the rotor-forced current with slip frequency is generated corresponding to the rotor natural current to ensure that the angle between the rotor current and the stator flux vector is always 180°; so the electromagnetic torque fluctuations can be suppressed [28]. With this improved control method, the transient rotor current and voltage under recurring faults will basically be the same as that under single grid fault; so the DFIG can ride through the recurring faults as long as it can ride through the single grid fault with the same voltage dip level. The electromagnetic torque fluctuations can be suppressed after the voltage recovery as well, and it will enhance the reliability of the mechanical system [29].

REFERENCES

[1] D. Xiang, L. Ran, P. J. Tavner, and S. Yang, "Control of a doubly-fed induction generator in a wind turbine during grid fault ride-through," *IEEE Trans. Energy Convers.*, vol. 21, no. 3, pp. 652–662, Sep. 2006.

[2] S. Xiao, G. Yang, H. Zhou, and H. Geng, "A LVRT control strategy based on flux linkage tracking for DFIG based WECS," *IEEE Trans. Ind. Electron.*, vol. 60, no. 7, pp. 2820–2832, Jul. 2013.

[3] J. Liang, W. Qiao, and R. G. Harley, "Feed-forward transient current control for low voltage ride-through enhancement of DFIG wind turbines," *IEEE Trans. Energy Convers.*, vol. 25, no. 3, pp. 836–843, Sep. 2010.

[4] L. Yang, Z. Xu, J. Ostergaard, Z. Y. Dong, and K. P. Wong, "Advanced control strategy of DFIG wind turbines for power system fault ride through," *IEEE Trans. Power Syst.*, vol. 27, no. 2, pp. 713–722, May 2012.

[5] R. Zhu, Z. Chen, X. Wu, and F. Deng, "Virtual damping flux-based LVRT control for DFIG-based wind turbine," *IEEE Trans. Energy Convers.*, vol. 30, no. 2, pp. 714–725, Jun. 2015.

[6] L. Zhou, J. Liu, and S. Zhou, "Improved demagnetization control of a doubly-fed induction generator under balanced grid fault," *IEEE Trans. Power Electron.*, vol. 30, no. 12, pp. 6695–6705, Dec. 2015.

[7] V. F. Mendes, C. V. de Sousa, S. R. Silva, B. C. Rabelo Jr., and W. Hofmann, "Modeling and ride-through control of doubly fed induction generators during symmetrical voltage sags," *IEEE Trans. Energy Convers.*, vol. 26, no. 4, pp. 1161–1171, Dec. 2011.

[8] J. Morren and S. W. H. de Haan, "Ride through of wind turbines with doubly-fed induction generator during a voltage dip," *IEEE Trans. Energy Convers.*, vol. 20, no. 2, pp. 435–441, Jun. 2005.

[9] G. Pannell, B. Zahawi, D. J. Atkinson, and P. Missailidis, "Evaluation of the performance of a DC-link brake chopper as a DFIG low-voltage fault-ride-through device," *IEEE Trans. Energy Convers.*, vol. 28, no. 3, pp. 535–542, Sep. 2013.

[10] A. O. Ibrahim, T. H. Nguyen, D. C. Lee, and S. C. Kim, "A fault ride-through technique of DFIG wind turbine systems using dynamic voltage restorers," *IEEE Trans. Energy Convers.*, vol. 26, no. 3, pp. 871–882, Sep. 2011.

[11] O. Abdel-Baqi and A. Nasiri, "A dynamic LVRT solution for doubly fed induction generators," *IEEE Trans. Power Electron.*, vol. 25, no. 1, pp. 193–196, Jan. 2010.

[12] A. Causebrook, D. J. Atkinson, and A. G. Jack, "Fault ride through of large wind farms using series dynamic braking resistors", *IEEE Trans. Power Syst.*, vol. 22, no. 3, pp. 966–975, 2007.

[13] X. Yan, G. Venkataramanan, P. S. Flannery, Y. Wang, Q. Dong, and B. Zhang, "Voltage-sag tolerance of DFIG wind turbine with a series grid side passive-impedance network," *IEEE Trans. Energy Convers.*, vol. 25, no. 4, pp. 1048–1056, Dec. 2010.

[14] W. Chen, D. Xu, N. Zhu, M. Chen, and F. Blaabjerg, "Control of doubly fed induction generator to ride through recurring grid faults," *IEEE Trans. Power Electron.*, vol. 31, no. 7, pp. 4831–4846, 2016.

[15] W. Chen, F. Blaabjerg, N. Zhu, M. Chen, and D. Xu, "Comparison of control strategies for DFIG under symmetrical grid voltage dips," in Proc. 39th Annual Conf. IEEE, IECON, Nov. 2013, pp. 1542–1547, 10–13.

[16] J. Lopez, P. Sanchis, X. Roboam, and L. Marroyo, "Dynamic behavior of the doubly fed induction generator during three-phase voltage dips," *IEEE Trans. Energy Convers.*, vol. 22, no. 3, pp. 709–717, Sep. 2007.

[17] J. Morren and S. W. H. de Haan, "Short-circuit current of wind turbines with doubly fed induction generator," *IEEE Trans. Energy Convers.*, vol. 22, no. 1, pp. 174–180, Mar. 2007.

[18] X. Kong, Z. Zhang, X. Yin, and M. Wen, "Study of fault current characteristics of the DFIG considering dynamic response of the RSC," *IEEE Trans. Energy Convers.*, vol. 29, no. 2, pp. 278–287, Jun. 2014.

[19] J. Lopez, E. Gubia, P. Sanchis, X. Roboam, and L. Marroyo, "Wind turbines based on doubly fed induction generator under asymmetrical voltage dips," *IEEE Trans. Energy Convers.*, vol. 23, no. 1, pp. 321–330, Mar. 2008.

[20] H. Geng, C. Liu, and G. Yang, "LVRT capability of DFIG-based WECS under asymmetrical grid fault condition," *IEEE Trans. Ind. Electron.*, vol. 60, no. 6, pp. 2495–2509, Jun. 2013.

[21] S. Hu, X. Lin, Y. Kang, and X. Zou, "An improved low-voltage ride-through control strategy of doubly fed induction generator during grid faults," *IEEE Trans. Power Electron.*, vol. 26, no. 12, pp. 3653–3665, Dec. 2011.

[22] F. K. A. Lima, A. Luna, P. Rodriguez, E. H. Watanabe, and F. Blaabjerg, "Rotor voltage dynamics in the doubly fed induction generator during grid faults," *IEEE Trans. Power Electron.*, vol. 25, no. 1, pp. 118–130, Jan. 2010.

[23] D. Xie, Z. Xu, L. Yang, J. Østergaard, Y. Xue, and K. P. Wong, "A comprehensive LVRT control strategy for DFIG wind turbines with enhanced reactive power support," *IEEE Trans. Power Syst.*, vol. 28, no. 3, pp. 3302–3310, Aug. 2013.

[24] J. Hu, Y. He, L. Xu, and B. W. Williams, "Improved control of DFIG systems during network unbalance using PI–R current regulators," *IEEE Trans. Ind. Electron.*, vol. 56, no. 2, pp. 439–451, Feb. 2009.

[25] J. Vidal, G. Abad, J. Arza, and S. Aurtenechea, "Single-phase DC crowbar topologies for low voltage ride through fulfillment of high-power doubly fed induction generator-based wind turbines," *IEEE Trans. Energy Convers.*, vol. 28, no. 3, pp. 768–781, Sep. 2013.

[26] G. Pannell, D. J. Atkinson, and B. Zahawi, "Minimum-threshold crowbar for a fault-ride-through grid-code-compliant DFIG wind turbine," *IEEE Trans. Energy Convers.*, vol. 25, no. 3, pp. 750–759, Sep. 2010.

[27] M. Rahimi and M. Parniani, "Coordinated control approaches for low-voltage ride-through enhancement in wind turbines with doubly fed induction generators," *IEEE Trans. Energy Convers.*, vol. 25, no. 3, pp. 873–883, Sep. 2010.

[28] C. Wessels, F. Gebhardt, and F. W. Fuchs, "Fault ride-through of a DFIG wind turbine using a dynamic voltage restorer during symmetrical and asymmetrical grid faults," *IEEE Trans. Power Electron.*, vol. 26, no. 3, pp. 807–815, Mar. 2011.

[29] J. Yang, J. E. Fletcher, and J. O'Reilly, "A series-dynamic-resistor-based converter protection scheme for doubly-fed induction generator during various fault conditions," *IEEE Trans. Energy Convers.*, vol. 25, no. 2, pp. 422–432, Jun. 2010.

CHAPTER 13

THERMAL CONTROL OF POWER CONVERTER IN NORMAL AND ABNORMAL OPERATIONS

The thermal stress of a power electronic component is closely related to some important wear-out failure mechanisms of the converter system. In this chapter, with the descriptions of the loss model and the thermal model of the power device, the thermal loading of the power converter can be investigated and analyzed in cases of the normal condition and the abnormal grid condition. It is concluded that the most stressed power device between back-to-back power converters is different in the normal grid condition. Besides, the occurrence of the abnormal grid leads to the varied thermal cycling between the IGBT and the freewheeling diode due to the existence of the reactive current injection. Furthermore, an optimized thermal control method of the grid-side converter can be achieved during a wind gust, while the thermal performance of the rotor-side converter is almost unaffected in a wind turbine system.

13.1   LOSS MODEL OF POWER CONVERTER

As the loss dissipation of the power semiconductor device is an important indicator of its thermal performance, the generic loss model of power converter used in the DC/AC application is first addressed. Then, on the basis of different interfaces of the doubly fed induction generator (DFIG) back-to-back power converters, loss models of the grid-side converter and the rotor-side converter are investigated and evaluated, respectively.

13.1.1   Loss Model of a Power Semiconductor Device

Normally, an IGBT and a freewheeling diode are both configured in a power switch for the bidirectional current flow. Loss consumption of each power semiconductor device consists of the conduction loss (the on-state loss) and the switching loss (the dynamic loss) [1,2].

Advanced Control of Doubly Fed Induction Generator for Wind Power Systems, First Edition.
Dehong Xu, Frede Blaabjerg, Wenjie Chen, and Nan Zhu.
© 2018 by The Institute of Electrical and Electronics Engineers, Inc. Published 2018 by John Wiley & Sons, Inc.

In the case of the two-level configuration, the same power loss is shared between the upper leg and the lower leg due to its symmetrical structure. If the upper IGBT and lower freewheeling diode are taken into account, the conduction loss of each power switch P_{con} is an average loss within a fundamental frequency f_a [3].

$$
P_{con} = f_a \underbrace{\sum_{n=n_1}^{n_1+N/2} v_{CE}(|i_a(n)|) |i_a(n)| T_1(n)}_{T_{con}} + f_a \underbrace{\sum_{n=n_1}^{n_1+N/2} v_F(|i_a(n)|) |i_a(n)| (T_s - T_1(n))}_{D_{con}}
$$

(13.1)

where the first term T_{con} is the conduction loss of the IGBT, and the second term D_{con} is the conduction loss of the freewheeling diode. i_a is the sinusoidal current through the power device, T_1 is the ON time of the upper leg within a switching period T_s. v_{CE}, v_F are voltage drops of the IGBT and the diode, which are normally given by the manufacturer. N is the carrier ratio—total switching times within a fundamental frequency, n_1 is the starting point of the positive current, and the subscript n is the nth switching pattern. It is worth noting that the conduction loss in (13.1) is aimed for the each IGBT and diode, which only takes up a half the entire leg.

As the v_{CE} and v_F curves are closely related to the junction temperature (25 °C, 125 °C, or 150 °C), the characteristic of voltage drops at 150 °C are chosen for the worst case, and they can be expressed in terms of polynomial expression. Besides, space vector modulation is widely used in three-phase three-wire systems due to its higher utilization of the DC-link voltage. In order to guarantee the minimum harmonic, a symmetrical sequence arrangement of the non-zero vector and the zero-vector is normally used, so the conduction time of the upper leg and lower leg can be calculated by the phase angle and the amplitude of the converter output voltage [4]. Furthermore, the phase angle between the output voltage and the output current of the converter determines the starting point of the current polarity, affecting the loss distribution between the upper leg and the lower leg.

On the other hand, the switching loss in each power switch P_{sw} can be calculated as

$$
P_{sw} = \underbrace{\frac{1}{2} \frac{V_{dc}}{V_{dc}^{ref}} f_a \sum_{n=1}^{N} (E_{on}(|i_a(n)|) + E_{off}(|i_a(n)|))}_{T_{sw}} + \underbrace{\frac{1}{2} \frac{V_{dc}}{V_{dc}^{ref}} f_a \sum_{n=1}^{N} E_{rr}(|i_a(n)|)}_{D_{sw}}
$$

(13.2)

Similar to (13.1), the first term is the switching loss for the IGBT T_{sw}, and the second term D_{sw} is the switching loss for the freewheeling diode. E_{on}, E_{off} are turn-on and turn-off energies dissipated by the IGBT, respectively, while the E_{rr} is the reverse-recovery energy loss consumed by the diode, which are usually tested by the manufacturer at a certain DC-link voltage V_{dc}^{ref}. It is assumed that the switching energy is proportional to the actual DC-link voltage V_{dc}. With the help of curve fitting of the dynamical energy loss by using polynomial expression, the switching loss of

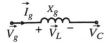

Figure 13.1   Single inductor used as the grid filter in the grid-side converter.

the power semiconductor can be calculated by the accumulation of the switching patterns within one fundamental frequency. It is noted that the switching loss in (13.2) is calculated for the each IGBT and diode.

13.1.2   Loss Model of Grid-Side Converter

As shown in Figure 13.1, a single inductor is used as a grid filter of the grid-side converter. If the current flowing into the converter is defined as the reference direction, the current and voltage relationship between the power grid and the converter output become

$$\vec{I}_g = I_{g\_Re} + jI_{g\_Im} \tag{13.3}$$

$$\vec{V}_C = V_{g\_Re} + X_g I_{g\_Im} + j(-X_g I_{g\_Re}) \tag{13.4}$$

where X_g denotes filter reactance, V_g denotes the grid voltage, I_g and V_c denote the current and voltage of the grid-side converter, respectively. Subscripts Re and Im represent the real and imaginary part of the phasor.

Due to the fact that only the slip power flows through the grid-side converter in a DFIG system, the bidirectional active power occurs because of the negative slip value during the super-synchronous operation and the positive slip value during the sub-synchronous operation. Assuming that the grid voltage is considered as the reference phasor, the phasor diagram of the voltage and current in the grid-side converter is shown in Figure 13.2. Due to the voltage drop across the inductor filter, the phase angle between the converter output voltage and current can be deduced.

During normal operation, the objective of the grid-side converter is to keep a constant DC-link voltage, which indicates that the active power is transferred from the DFIG rotor-side to the power grid. However, in cases of grid faults or weak grid

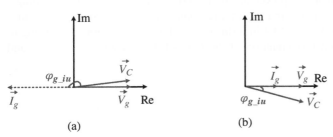

Figure 13.2   Phasor diagram of the grid-side converter voltage and current:
(a) Super-synchronous operation and (b) sub-synchronous operation.

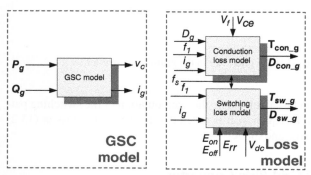

Figure 13.3   Block diagram to calculate the power loss for the grid-side converter (GSC: grid-side converter).

conditions, the grid-side converter is also responsible in providing certain amount of the reactive current. The components of the active power and reactive power are determined by the real part and the imaginary part of the converter current I_{g_Re} and I_{g_Im}. As a consequence, with the information of the active power P_g and the reactive power Q flowing over the grid-side converter, the voltage v_c and the current i_g can be calculated by using the grid-side converter model.

With the help of the aforementioned loss model of the power semiconductor, the process to calculate power loss of the grid-side converter can be summarized in Figure 13.3. For the conduction loss, due to the fixed switching frequency f_s and the current frequency f_1, as well as the converter current i_g, and the duty cycle in each switching pattern D_g related to the converter voltage phasor, the conduction loss of the IGBT T_{con_g} and the diode D_{con_g} can be calculated according to (13.1). For the switching loss of the grid-side converter, with the information of the switching frequency, the converter current and its frequency, the switching loss of the IGBT T_{sw_g} and the diode D_{sw_g} can be calculated according to (13.2).

13.1.3   Loss Model of Rotor-Side Converter

The rotor-side converter is connected to the rotor of the DFIG, and the stator-side active power and the reactive power are regulated by adjusting the rotor voltage. Neglecting the stator resistance and the rotor resistance, the steady-state DFIG equivalent circuit is shown in Figure 13.4 in terms of the phasor expression. According to the voltage equation and flux equation of the DFIG [5], the rotor-side voltage and

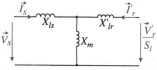

Figure 13.4   Single-phase DFIG equivalent circuit in phasor diagram.

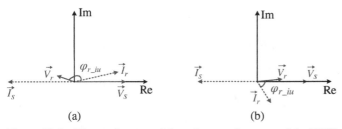

(a)   (b)

Figure 13.5   Phasor diagram of the voltage and current of the DFIG stator and rotor:
(a) Super-synchronous mode and (b) sub-synchronous mode.

current can be expressed by using the stator-side voltage and current:

$$\vec{I}'_r = -\frac{X_s}{X_m}I_{s\_Re} + j(sign(S_l))\left(-\frac{V_{s\_Re}}{X_m} - \frac{X_s}{X_m}I_{s\_Im}\right) \tag{13.5}$$

$$\vec{V}'_r = S_l\left(\frac{X_r}{X_m}V_{s\_Re} + \frac{\sigma X_r X_s}{X_m}I_{s\_Im}\right) - j(sign(S_l))\left(S_l\frac{\sigma X_r X_s}{X_m}I_{s\_Re}\right) \tag{13.6}$$

where X_s, X_m, and X_r denote stator reactance, the magnetizing reactance, and the rotor reactance, respectively. σ, the leakage coefficient, defined as $\left(X_s X_r - X_m^2\right)/X_s X_r S_l$ is the slip value of the induction generator. Moreover, the sign function $sign\left(S_l\right)$ means if S_l is positive, its value becomes 1. Alternatively, if S_l is negative, its value becomes -1.

According to (13.5) and (13.6), the phasor diagram of the stator and rotor in the DFIG is shown in Figure 13.5. In the case of the super-synchronous operation mode, the rotor voltage appears almost in opposite direction to the stator voltage due to the negative slip value. Moreover, the rotor current is almost lagging the rotor voltage by 180°, which indicates that the DFIG is providing the active power through the rotor-side, and the rotor-side converter is supplying the excitation energy to the induction generator. In the case of the sub-synchronous mode, the rotor current is lagging the rotor voltage by less than 90°, implying that the rotor-side converter provides both the active power and the reactive power to the induction generator.

If the stator voltage is set as the reference, the stator active power and reactive power are determined by the stator current real component I_{s_Re} and imaginary component I_{s_Im}. Afterwards, the phasor expressions of the rotor voltage v_r and the rotor current i_r can be deduced by using (13.5) and (13.6). On the basis of the DFIG model—deriving the analytical equations of rotor voltage and rotor current from the stator active power and reactive power, together with the aforementioned loss model of the power semiconductor, the flowchart to calculate the power loss of the rotor-side converter can be described in Figure 13.6. For the conduction loss, the conduction loss of the IGBT T_{con_r} and the diode D_{con_r} can be calculated with the information of the rotor current i_r, its frequency f_e, switching frequency f_s, and duty cycle D_g. It is worth noting that compared to the grid-side converter, the frequency of the rotor current varies with the changing slip values. For the switching loss of the rotor-side converter, the switching loss of the IGBT T_{sw_r} and the diode D_{sw_r} can be calculated

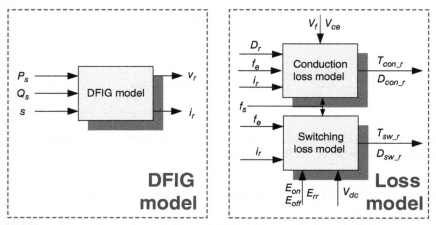

Figure 13.6   Block diagram to calculate the power loss for the rotor-side converter of the DFIG system.

with the information of the switching frequency of the rotor-side converter, the rotor current, and its varying frequency.

13.2   THERMAL MODEL OF POWER CONVERTER

As the thermal performance of the power device is closely related to the reliability and the cost of the power converter system, an appropriate thermal model needs to be developed in order to translate the power loss to the thermal stress of the power device. Besides, the analytical equation of the junction temperature is deduced by taking the thermal impedance of the power module and the operation frequency range of the power converter into account.

13.2.1   Thermal Impedance in Power Module

The power density of the power electronic converter is steadily being increased, pushing the improvement of an enhanced power range and a reduced impact on volume as well as the cost [6, 7]. With respect to the IGBT power module, many new techniques and novel materials are devoted to guarantee lower loss dissipation or higher operational junction temperature [8, 9].

The layout of a typical power semiconductor module is depicted in Figure 13.7. A number of power semiconductor chips—IGBTs and diodes are soldered onto the ceramic-based substrates like direct bond copper (DBC), which behaves as an electrical insulation. Then, the DBC can either be soldered onto a baseplate, or the bottom copper layer is directly mounted to the heat sink with a thermal interface material (TIM) in between. The electrical connections between the chips and the conductor tracks on the DBC are normally realized by a thick bonding wire, which is normally made of pure aluminum [9].

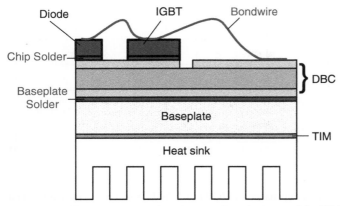

Figure 13.7   Basic structure of a power semiconductor module (DBC: Direct Bond Copper; TIM: Thermal Interface Material).

The thermal variables represented by electrical analogies are widely used [9–12], where a current source expresses the power dissipation either in the IGBT or the diode, the voltage source stands for the constant temperature level, and RC elements imitate the thermal impedance of the power device. Two kinds of RC networks are commonly adopted: the physical-meaning-based Cauer structure and the test-based Foster structure; the latter is actually more preferred by the industry [13, 14]. The thermal model of power devices, including the IGBT and the freewheeling diode are shown in Figure 13.8, in which the thermal impedance consists of the power module itself, the TIM, and the heat sink. For simplicity, the thermal coupling between the IGBT chip and diode chip is not considered.

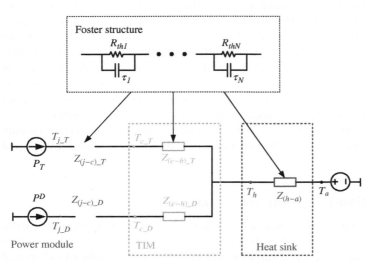

Figure 13.8   Thermal model of the power devices in a module with a diode (D) and an IGBT (T).

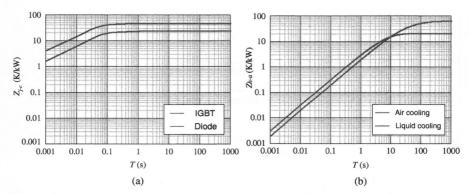

Figure 13.9 Dynamic thermal impedance: (a) IGBT and diode from junction to case, (b) air and liquid cooling from heat sink to ambient. *Note*: Air flow = 610 m^3/h, T_a = 25 °C, 500 m above sea level; liquid flow = 15 L/min, T_{fluid} = 40 °C, water/glycol ratio = 50%/50%.

The thermal impedances for the power module and the TIM are normally provided by the manufacturer datasheets. Figure 13.9 shows the dynamic thermal impedance of the IGBT and the diode in the selected 1 kA/1.7 kV power module. It is noted that the steady-state thermal resistance of the diode is higher than the IGBT, due to the smaller chip area of the diode in the standard power module. Moreover, it can be seen that the thermal time constant changes from hundreds of microseconds to hundreds of milliseconds.

However, the thermal impedance of the cooling method is uncertain, as the power modules can be used in different applications according to customer requirements. With respect to a full-scale power converter, the typical amount of heat that has to be transported away from the power modules may be from 50 to 100 kW for a 2 MW wind turbine [15]. Because cooling solutions take up considerable space available in the wind turbine nacelle, the forced air cooling and the liquid cooling cover 95% of all power module application [16]. Correspondingly, the dynamic thermal impedance of the air and liquid cooling from the heat sink to the ambient can be deduced from Semikron datasheets as shown in Figure 13.9b [17, 18]. It is evident that the steady-state thermal resistance of the air cooling system is three times higher compared to the liquid cooling, and it can also be seen that the maximum time constant of the thermal impedance is much higher compared to the power module itself—hundreds of seconds for the air cooling and dozens of seconds for the liquid cooling.

13.2.2 Junction-Temperature Calculation

Thermal profile of the power semiconductor for the wind power application usually contains the long-term thermal cycling and the short-term thermal cycling, which are imposed by the wind speed variation and the AC current within a fundamental frequency, respectively. At first glance, the lifetime and the reliability issues caused

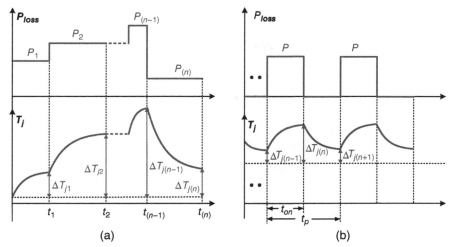

Figure 13.10   Power-loss profile and junction-temperature profile against time:
(a) Step-pulse of power dissipation and (b) periodical pulse of power dissipation.

by the wind turbulence may be more crucial due to the higher junction-temperature fluctuation of the power semiconductors. However, it is still essential to look at the influence of short-term thermal cycling due to their larger magnitude of numbers, since the wind speed normally varies in several seconds, while the fundamental period of the converter current changes only from dozens of milliseconds to hundreds of milliseconds.

With respect to the long-term thermal cycling, the corresponding step pulse of power loss is shown in Figure 13.10a. According to the Foster structure thermal model, the junction-temperature fluctuation $dT_{j(n)}$ during nth power-loss pulse can be calculated as [1]

$$dT_{j(n)} = dT_{j(n-1)} \sum_{i=1}^{k} e^{-\frac{t_{(n)}-t_{(n-1)}}{\tau_i}} + P_{(n)} \sum_{i=1}^{k} R_i(1 - e^{-\frac{t_{(n)}-t_{(n-1)}}{\tau_i}}) \qquad (13.7)$$

where the first item denotes the zero-input response of the previous junction-temperature fluctuation $dT_{j(n-1)}$ at time instant $t_{(n-1)}$, and the second term denotes the zero-state response of the power loss $P_{(n)}$ until the time instant $t_{(n)}$. R_i and τ_i indicate the ith thermal resistance and time constant in kth-order Foster structure. According to (13.7), the junction-temperature swing can easily be calculated from the power-loss profile.

With respect to the short-term thermal cycling, Figure 13.10b shows the thermal profile for the periodical-pulse power dissipation. According to (13.7), the junction-temperature fluctuation at the time instants $t_{(n)}$ and $t_{(n+1)}$ can be calculated based on previous states $t_{(n-1)}$ and $t_{(n)}$, respectively. Since $dT_{j(n+1)}$ has the same value as

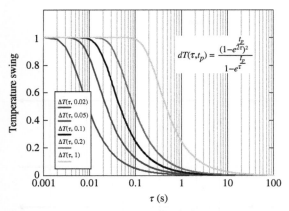

Figure 13.11   Temperature swing dependence on time constant of thermal impedance with various fundamental frequencies.

$dT_{j(n-1)}$ at the steady-state operation, the junction-temperature fluctuation dT_j can be expressed as

$$dT_j = P \sum_{i=1}^{k} R_i \frac{\left(1 - e^{-\frac{t_{on}}{\tau_i}} \right)^2}{1 - e^{-\frac{t_p}{\tau_i}}} \tag{13.8}$$

where P denotes the peak value of periodical power pulse, t_{on} denotes the on-state time, t_p denotes the fundamental period of the converter output current, and t_p normally has twice the value of t_{on}.

According to (13.8), Figure 13.11 shows the relationship between the temperature swing and the time constant of the thermal impedance with typical operational frequencies of the DFIG power converters. Even for a minimum frequency of 1 Hz, the temperature fluctuation caused by the cooling method (thermal time constant from dozens of seconds to hundreds of seconds) can almost be neglected compared to the effect from the thermal impedance of the power module (hundreds of milliseconds). Consequently, it can be concluded that, for the short-term thermal cycling, the junction-temperature swing is only close to the thermal resistance and thermal capacitance of the power module rather than the cooling method.

Accordingly, the mean junction temperature T_{jm} can be calculated as

$$T_{jm} = \sum_{i=1}^{k} R_i \frac{P}{2} + T_a \tag{13.9}$$

where T_a denotes the ambient temperature, and it is set to 50 °C as an indication of the worst case.

Based on (13.8) and (13.9), the thermal behavior of the power device at the steady state can be analytically calculated, avoiding the time-consuming simulation

TABLE 13.1   2 MW turbine and DFIG data

Rated wind speed v_{w_rate} (m/s)	12
Rated turbine speed (rpm)	19
Number of pole pairs p	2
Gear ratio	94.7
Rated shaft speed n_s (rpm)	1800
Rated fundamental frequency f_e (Hz)	10
Stator leakage inductance L_{ls} (mH)	0.038
Magnetizing inductance L_m (mH)	2.91
Rotor leakage inductance $L_{lr'}$ (mH)	0.064
Stator/rotor turns ratio k	0.369

due to the large difference between the switching period of the power device and its thermal time constant.

13.3   THERMAL LOADING DURING NORMAL OPERATION

As a case study, the calculation and simulation of the power loss and the junction temperature is performed in a 2 MW DFIG system. In the case of a normal operation, a set of typical wind speeds is taken into account. Afterward, the loss profile and the thermal profile of the back-to-back power converter can be obtained.

13.3.1   DFIG System in Case Study

It is known that a DFIG used in the wind power generation is normally realized by a partial-scale power converter. For simplicity, this configuration is named as the DFIG system, and a 2 MW wind turbine is selected for the case study.

On the basis of the DFIG generator data listed in Table 13.1 [19,20], in order to implement the low-voltage power module with 1.7 kV blocking capability, the DC-link voltage V_{dc} is kept at 1050 V as listed in Table 13.2. Moreover, the switching frequency f_s is usually very low for the multi-MW power converter, and is selected at 2 kHz.

Since the majority of the active power flows into the grid through the stator of the generator, the rotor-side converter handles the rest of the slip power as well as the excitation power. On the other hand, the grid-side converter is designed to keep the fixed DC-link in order to decouple back-to-back power converters from each other, and some amount of the reactive power can be provided in the case of the grid fault or in accordance with the demand from the transmission system operator. As listed in Table 13.2, it can be seen that the loading between power converters at the rated power are quite unequal, and the current of the rotor-side converter is much higher than the grid-side converter due to its lower output voltage and same active power to transfer. Besides, the rotor-side converter is also responsible for the ride-through

TABLE 13.2   Back-to-back power converter's data

Rated active power P_C (kW)	400
DC-link voltage V_{dc} (V_{dc})	1050
Switching frequency f_s (kHz)	2
Grid-side converter	
Rated output voltage (V_{rms})	704
Rated current (A_{rms})	328
Filter inductance (mH)	0.50
Rotor-side converter	
Rated output voltage (V_{rms})	374
Rated current (A_{rms})	618

of the fault situations [21]. Similar current stress of power converters leads to two paralleled 1 kA/1.7 kV power modules employed in each leg the rotor-side converter, and only one power module used in each leg of the grid-side converter as shown in Figure 13.12.

13.3.2   Loss Breakdown at Various Loading Conditions

With the DFIG parameters listed in Tables 13.1 and 13.2, the power loss of each power device can be calculated according to (13.1) and (13.2). In addition, the power loss is estimated at rated wind speed at normal grid condition and unity power factor.

Figure 13.13 indicates the loss distribution of each power semiconductor in terms of the grid-side converter and the rotor-side converter, respectively. The name of the power semiconductor can be found in Figure 13.12. As the turn-on and the turn-off loss of IGBT are higher than the recovery loss of the freewheeling diode at the same current, the switching loss in the diode is always lower. However, the conduction loss is mainly related to the power direction. At the rated wind speed, the rotor-side converter operates as a rectifier, therefore, more conduction losses can be observed in the freewheeling diode. On the contrary, the grid-side converter works as an inverter, where most conduction losses are dissipated in the IGBT. Generally speaking, the power loss dissipated in the rotor-side converter is more equal compared to back-to-back power converters.

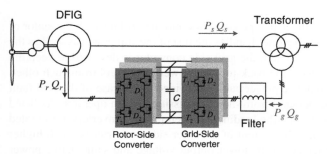

Figure 13.12   Two-level back-to-back power converter of the DFIG system.

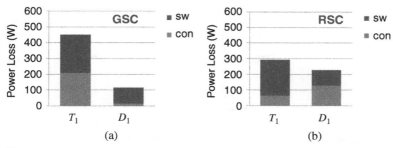

Figure 13.13   Loss breakdown of each power device at rated wind speed of 12 m/s: (a) Grid-side converter and (b) rotor-side converter. *Note*: "sw" and "con" are switching losses and conduction losses, respectively.

In order to investigate the loss behavior of the power device at different operation modes, several typical wind speeds are chosen with slip values from −0.3 to 0.2 and summarized in Table 13.3. It is noted that the wind speed of 8.4 m/s is regarded as the synchronous operating point.

The power loss of the grid-side converter at different wind speeds is shown in Figure 13.14a. It can be seen that the lowest power loss appears in the synchronous operating point due to the fact that no active power flows, and only the switching ripple current affects, while the highest occurs at rated wind power. Furthermore, the IGBT has more loss in the super-synchronous mode and the diode has more loss in the sub-synchronous mode due to the bidirectional active power.

The power loss of the rotor-side converter at different wind speeds is shown in Figure 13.14b. In the super-synchronous mode, the power loss increases with larger wind speed, and the conduction loss in the diode is dominating. In the sub-synchronous mode, the conduction loss in the IGBT is however dominating. Besides, it is also worth mentioning that a small frequency hysteresis of 1 Hz is introduced around the synchronous operation point in order to avoid an unequal load among the three phases caused by the DC current.

TABLE 13.3   Parameters of the DFIG and power converters at different wind speeds

Wind speed (m/s)	Generated power (MW)	Slip	Fundamental frequency (Hz)
5.9	0.26	0.3	15
6.8	0.39	0.2	10
7.6	0.55	0.1	5
8.4	0.74	0.02	1
9.2	0.98	−0.1	5
10.1	1.29	−0.2	10
12	2	−0.2	10
25	2	−0.2	10

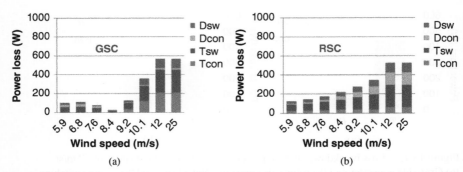

Figure 13.14   Power loss of each device at various wind speeds: (a) Grid-side converter and (b) rotor-side converter.

13.3.3   Thermal Profile at Various Loading Conditions

With the aid of the previous power-loss calculation and the established thermal model, the junction temperature of the power device of the DFIG power converters can be simulated using PLECS software [22]. The simulation result of the junction temperature in each power semiconductor is shown in Figure 13.15, where power converters operate at the rated power and in steady state. For the junction temperature of the grid-side converter, it can be seen that the IGBT becomes the hottest power semiconductor device. Moreover, the mean junction-temperature variation between the IGBT and the diode differs with 62.2 °C and 56.3 °C, respectively.

With respect to the junction temperature of the rotor-side converter, it is noted that the mean temperature of the diode is slightly higher than the IGBT's, and the thermal performance shows a more unequal distribution, where the difference of junction temperature between the IGBT and the diode is 12.4 °C compared to 5.1 °C. In

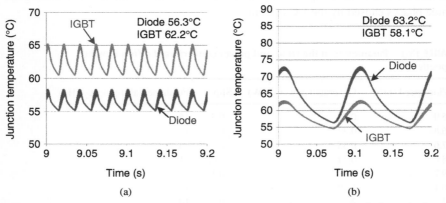

Figure 13.15   Junction temperature of the power semiconductor at rated wind speed of 12 m/s: (a) Grid-side converter and (b) rotor-side converter.

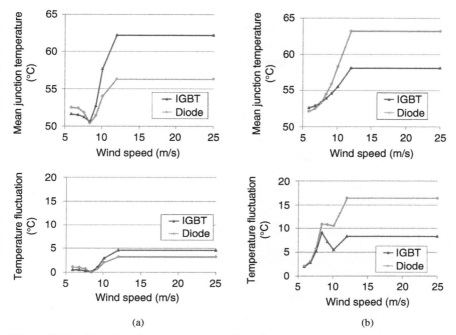

Figure 13.16   Mean junction temperature and junction-temperature fluctuation in the power device chip versus wind speed: (a) Grid-side converter and (b) rotor-side converter.

addition, it can also be observed that the frequency of the thermal cycling is the same with the converter current frequency.

A further comparison of the mean junction temperature and the junction temperature fluctuation with different wind speeds are shown in Figure 13.16. For the grid-side converter, the hottest device changes from the diode in the sub-synchronous mode to the IGBT in the super-synchronous mode. Moreover, the mean junction temperature as well as the junction temperature fluctuation becomes least crucial around the synchronous operating point due to the lowest power dissipation.

For the rotor-side converter, the hottest device changes from the IGBT in the sub-synchronous mode to the diode in the super-synchronous mode. Furthermore, although the mean junction temperature consecutively increases with higher wind speeds until the rated value, the temperature fluctuation of the power semiconductors becomes significantly important around the synchronous operating point because of its rather low fundamental frequency.

13.4   THERMAL LOADING IN ABNORMAL OPERATION

A lot of work has been devoted to control strategies of the wind power converter to satisfy the grid codes [21, 23, 24]. However, the loss and thermal performance under

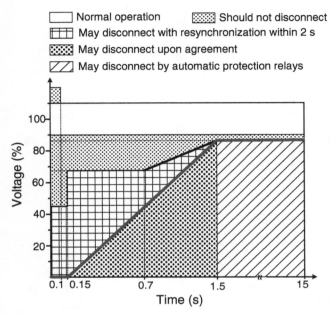

Figure 13.17   Low voltage ride-through requirement stated in German grid code.

this condition is another important and interesting topic that needs further investigation [25]. The scope of this section is to investigate and simulate the power loss and the thermal cycling of the DFIG system undergoing various balanced grid voltage dips. First, the typical configuration and relevant grid codes are introduced and addressed. Then, the operation behavior under abnormal grid conditions is described and investigated. Afterward, the loss distribution and thermal analysis are presented in terms of various voltage dips.

13.4.1   Grid Codes Requirements

As the wind power penetration into the power system increases in many countries, many transmission system operators are challenged by the impacts of maintaining reliability and stability of the power system. New grid codes stipulate that wind farms should contribute to the voltage control in the case of abnormal operations of the network (e.g., voltage dips due to network faults). A presentation of the most critical requirement imposed by E.ON Netz is realized in [26,27]. The behaviors under grid disturbance basically consist of two parts: the low voltage ride-through (LVRT) and the reactive current injection (RCI).

The LVRT requirement is given in Figure 13.17, where voltage drops within the area above the thickest line should not technically be disconnected. Even when the grid voltage drops to zero, the wind power plant must stay linked for 150 ms. During the period of the grid fault, the active current can be reduced in order to fulfill

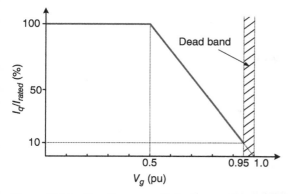

Figure 13.18 Reactive current injection requirement as a function of grid voltage for offshore wind turbines stated in German grid code.

the reactive power requirement. As described in Figure 13.18, various amounts of reactive current have to be injected along with different voltage dips. In the case that the voltage dip is above 0.5 pu, the wind power system is able to inject 1.0 pu overexcited reactive current to support and rebuild the grid voltage. Meanwhile, for the offshore wind farms, the grid voltage above 0.95 pu is regarded as the dead band boundary of RCI.

13.4.2   Operation Behavior under Voltage Dips

As shown in Figure 13.19, although both the grid-side converter and rotor-side converter are possibly able to support the RCI during grid voltage dips, due to the stator and rotor winding ratio as well as the derating design of power converters, it is better to compensate the reactive power from the rotor-side converter. As a consequence, the discussion will only focus on this part of the generation system.

As the stator of the DFIG is directly linked to the grid, the sudden grid voltage drop introduces a natural flux in the stator winding, which may produce large transient

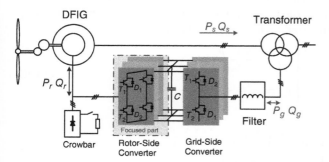

Figure 13.19   Crowbar-based DFIG system during the abnormal grid condition.

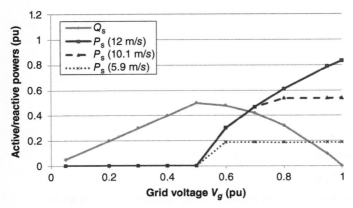

Figure 13.20 Active and reactive powers under various balanced grid voltage dips.

currents in the DFIG rotor side and may destroy the rotor-side converter. Although this dynamical behavior may be critical for the LVRT period, it is assumed that the operation behavior of the DFIG is investigated and evaluated under the condition that the stator flux is consistent with various voltage dips, which theoretically omits the transient period.

Figure 13.20 indicates the active power P_s and the reactive power Q_s delivered by the stator-side under various balanced grid voltage dips. According to the RCI description in Figure 13.18, if the dip level of the grid voltage is higher than 0.5 pu, the reactive power linearly increases along with the grid voltage. On the other hand, if the dip level of the grid voltage is less than 0.5 pu, it decreases rapidly mainly due to the less RCI demand. Moreover, the active power stays zero if the voltage dip is below 0.5 pu. The active power for the DFIG will normally follow the maximum power point tracking for the wind turbine. Conditions of 12 m/s (2 MW), 10.1 m/s (1.29 MW), and 5.9 m/s (0.26 MW) are indicated. Because of the relative lower active power at lower wind speed, the DFIG has an additional capability of the reactive current output. Consequently, it is noted that lower wind speeds affect the active power reference less.

On the basis of the steady-state equivalent DFIG model, if the stator active power and reactive power in d-axis and q-axis are introduced, the rotor current i_r' and the rotor voltage v_r' (referred to the stator-side) can be expressed as

$$\begin{cases} i_{rd}' = -\frac{2}{3}\frac{X_s}{X_m}\frac{P_s}{V_{sd}} \\ i_{rq}' = -\frac{V_{sd}}{X_m} + \frac{2}{3}\frac{X_s}{X_m}\frac{Q_s}{V_{sd}} \end{cases} \tag{13.10}$$

$$\begin{cases} v_{rd}' = S_l\left(\frac{X_r}{X_m}V_{sd} - \frac{2}{3}\frac{\sigma X_r X_s}{X_m}\frac{Q_s}{V_{sd}}\right) \\ v_{rq}' = -\frac{2}{3}S_l\frac{\sigma X_r X_s}{X_m}\frac{P_s}{V_{sd}} \end{cases} \tag{13.11}$$

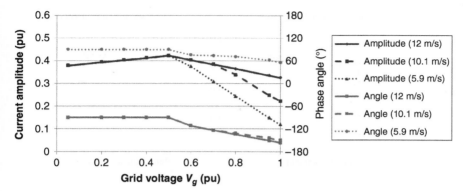

Figure 13.21   Amplitude and phase angle of the loading current during balanced grid voltage dips.

where X_s, X_r, and X_m denote the stator, rotor, and magnetizing reactance, respectively, at 50 Hz, S_l denotes the rotor slip value, σ denotes the leakage factor of the induction generator, and V_{sd} denotes the peak stator phase voltage of the induction generator.

The amplitude and phase angle of the rotor-side current during balanced LVRT are shown in Figure 13.21, where wind speeds of 12 m/s, 10.1 m/s, and 5.9 m/s independently evaluated. It can be seen that the maximum current amplitude appears at 0.5 pu grid voltage, and the amplitude of the rotor-side current decreases dramatically if the grid voltage dip is less than 0.5 pu for all three wind speeds. The phase angle between the rotor current and the rotor voltage is almost reverse between the sub-synchronous mode and the super-synchronous mode due to the opposite active power flow.

Simulation validation of loading characteristics can be realized based on PLECS blockset in Simulink. In the case of the wind speed of 10.1 m/s, an extreme voltage dip 0.05 pu is taken into account, and the voltage and the current of the induction generator's stator-side and rotor-side of the normal operation, and the LVRT condition at the steady state is shown in Figures 13.22a and 13.22b. According to grid code's requirement, the stator-side provides the reduced active power output but supplies the overexcited reactive power to support the grid voltage recovery during the LVRT, which is shown in the upper part of Figure 13.22. As a result, the amplitude of the stator current increases from 0.5 to 1.0 pu, and phase angle changes from 0 ° (releasing the active power) to 90 ° (injecting the reactive power). In the lower part of Figure 13.22, the corresponding rotor-side voltage and current changes as well as their phase angle. Moreover, it is noted that the amplitude of the rotor current becomes higher during the LVRT operation.

13.4.3   Loss Distribution and Thermal Behavior During LVRT

The loss comparison of each power semiconductor in the power converter under the normal condition and the LVRT operation can be seen in Figure 13.23, where the

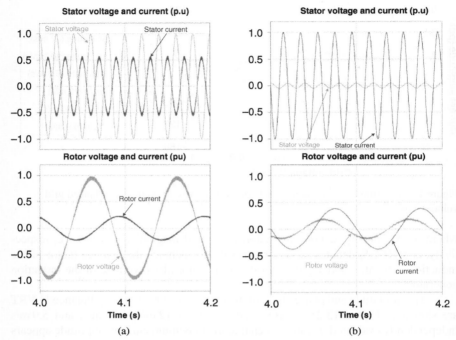

Figure 13.22 Simulated results of the DFIG system at wind speed of 10.1 m/s: (a) Normal operation and (b) LVRT = 0.05 pu.

power loss of three different wind speeds and one LVRT condition are considered. During normal operations, it is noted that the power loss consistently increases with higher wind speed. During the LVRT operation, it is noted that the power loss between the IGBT and the diode becomes more unequal, and it is the LVRT operation that has the highest power dissipation.

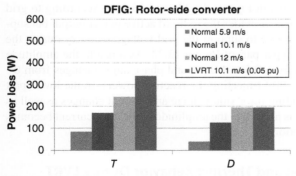

Figure 13.23 Loss comparison of each power semiconductor in the rotor-side converter under normal and LVRT situations. *Note*: *T* stands for the transistor and *D* stands for the diode.

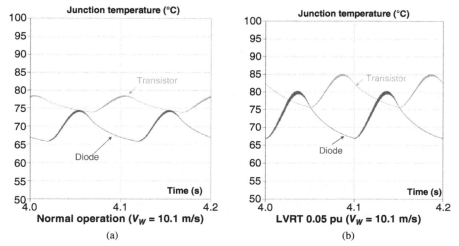

Figure 13.24   Junction temperature of the rotor-side converter in normal operation and LVRT situation at wind speed of 10.1 m/s: (a) Normal operation and (b) LVRT condition (0.05 pu).

As shown in Figure 13.24, at the wind speed of 10.1 m/s, the junction temperature in the rotor-side converter is compared between the normal operation and LVRT condition. It can be seen that the LVRT condition induces a higher junction temperature for both the IGBT and the diode, and it is consistent with the loss distribution in Figure 13.23.

Based on the well-known Coffin–Manson lifetime models, the mean junction temperature and the junction-temperature fluctuation of the power semiconductor are the most important two indicators. Hence, it is interesting to investigate the thermal excursion of the power device under various grid voltage dips as shown in Figure 13.25, where the wind speeds of 12 m/s, 10.1 m/s, and 5.9 m/s are studied, respectively.

The simulated mean junction temperature and the junction-temperature fluctuation of each switching device in the rotor-side converter in relation to the grid voltage are shown in Figures 13.25a and 13.25b, respectively. Both the mean junction temperature and the junction-temperature fluctuation vary slightly if the symmetrical grid dip is above 0.5 pu, while they change dramatically if the grid voltage dip is below 0.5 pu. It is noted that the most stressed power device appears in the case of the wind speed of 12 m/s. The highest mean junction temperature and the junction-temperature fluctuation of the IGBT appear around 0.5 pu grid voltage. However, from the diode point of view, the thermal stress becomes highest around 0.6 pu grid voltage. In the case of the wind speed of 5.9 m/s, the most stressed power device appears at 0.5 pu. Moreover, if the grid voltage is below 0.5 pu, although the mean junction temperature shows similar performance as other wind speeds, the junction-temperature fluctuation becomes less stressed due to the higher fundamental frequency of the loading current.

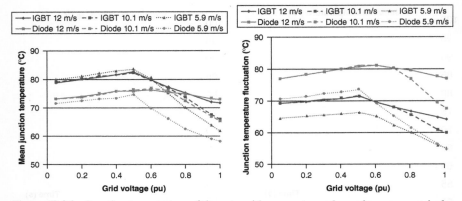

Figure 13.25 Junction temperature of the rotor-side converter under various symmetrical voltage dips: (a) Mean junction-temperature value versus grid voltage and (b) junction-temperature fluctuation versus grid voltage.

13.5 SMART THERMAL CONTROL BY REACTIVE POWER CIRCULATION

The scope of the section is, first, to calculate the allowable reactive power circulation between back-to-back power converters of the DFIG, where different operation modes are taken into account. Then, a control method is proposed to improve the reliable operation of the power module by reactive power circulation in the condition of wind gusts.

13.5.1 Effects of Reactive Power on Current Characteristic

As both the rotor-side converter and the grid-side converter have the ability to control the reactive power, it is possible to circulate the reactive power within the DFIG system. As shown in Figure 13.26, the reactive power delivered to the power grid will

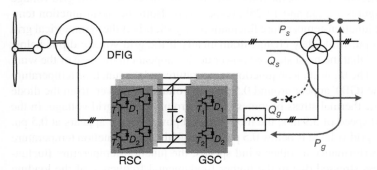

Figure 13.26 Compensation scheme of the reactive power in the DFIG wind turbine system.

not be changed in case the reactive power of back-to-back power converters is controlled in the opposite direction. However, some realistic limitations exist on using this control scheme.

According to (13.12), the analytical formula of the converter output voltage V_c can be expressed as

$$V_C = \sqrt{(V_g + i_{gq}X_g)^2 + (i_{gd}X_g)^2} \leq \frac{V_{dc}}{\sqrt{3}} \tag{13.12}$$

where V_g, V_{dc} denotes the rated peak phase grid voltage and DC-link voltage, respectively, X_g denotes filter reactance at 50 Hz, and i_{gd} and i_{gq} denote the grid-side converter peak current in the d-axis and q-axis, respectively.

It is evident that regardless of the operation modes, the amplitude of the converter voltage is increased in the case of the overexcited reactive power injection. Therefore, the maximum overexcited current should not saturate the modulation index as listed in (13.12).

The second restriction lies in the capacity of the power device:

$$\sqrt{i_{gd}^2 + i_{gq}^2} \leq I_m \tag{13.13}$$

where I_m denotes the peak current of the power module, which cannot be exceeded by the increasing reactive current.

The third limitation is that the capacity of the induction generator Q_s must be taken into account [28].

$$\frac{3}{2}V_g i_{gq} \leq Q_s \tag{13.14}$$

Similar restrictions apply to the rotor-side converter, that is, the linear modulation range, the rating limitation of the power device, and the induction generator capacity. With the aid of the winding ratio k between the stator and the rotor, the rotor voltage and the rotor current are able to transfer back to rotor-side variables. Three limitations can be expressed as

$$\frac{\sqrt{v'_{rd}{}^2 + v'_{rq}{}^2}}{k} \leq \frac{V_{dc}}{\sqrt{3}} \tag{13.15}$$

$$k\sqrt{i'_{rd}{}^2 + i'_{rq}{}^2} \leq I_m \tag{13.16}$$

$$-\frac{3}{2}V_{sm}i_{sq} \leq Q_s \tag{13.17}$$

If two typical wind speeds of 5.9 m/s and 10.1 m/s are selected for the subsynchronous and the super-synchronous operation, the range of the reactive power in the rotor-side converter can be calculated, as summarized in Table 13.4. It can be seen that the range of the reactive power in the rotor-side converter is limited by the induction generator capacity, while the range of the reactive power in the grid-side converter is restricted by the DC link and the induction generator capacity.

TABLE 13.4 Range of the reactive power for back-to-back power converter

	Sub-synchronous mode		Super-synchronous mode	
	Grid-side converter	Rotor-side converter	Grid-side converter	Rotor-side converter
Typical wind speed (m/s)	5.9		10.1	
Rated power (pu)	0.13		0.65	
Active power current (pu)	0.06	0.19	0.11	0.54
Range of reactive power (pu)	$(-0.23, 0.06)$	$(-0.29, 0.23)$	$(-0.23, 0.06)$	$(-0.29, 0.23)$

Considering aforementioned limitations, the possible range of the reactive power in the grid-side converter and the rotor-side converter can be depicted in Figure 13.27. Since the current amplitude and the phase angle of the power converter are two important indicators of the power device loading, both of them are evaluated and investigated.

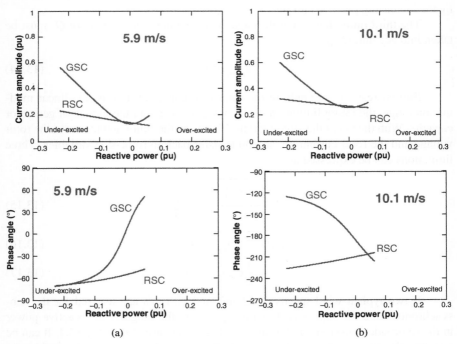

(a)　　　　　　　　　　　　　　　　(b)

Figure 13.27　Effects of reactive power circulation on back-to-back power converters in a DFIG system: (a) Sub-synchronous mode at wind speed of 5.9 m/s and (b) super-synchronous mode at wind speed of 10.1 m/s.

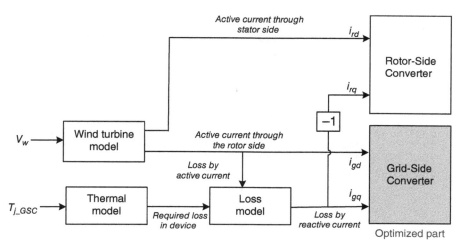

Figure 13.28   Thermal-oriented control diagram of back-to-back power converters during a wind gust.

From the characteristic of current amplitude, it is evident that minimum current appears almost under no reactive power for the grid-side converter, while for the rotor-side converter, the current decreases with higher overexcited reactive power. Unfortunately, this reactive power is much lower than the underexcited reactive power, which prevents the current of the rotor-side converter to reach its minimum value. From the characteristic of phase angle, regarding the grid-side converter, it is noted that the phase angle changes from unity power factor to leading or the lagging power factor significantly in response to the circulated amount of reactive power. Regarding the RSC, it is noted that with increasing capacitive reactive power, the power factor angle tends to be in phase under the sub-synchronous mode or inverse phase under the super-synchronous mode. However, the phase shift looks insignificant in both cases.

13.5.2   Thermal Performance Improvement by Reactive Power Control

During wind gusts, the abrupt change of the wind speed can be reflected by the adverse thermal cycling in the power converter, as investigated in [29]. Therefore, it is possible to control the junction-temperature fluctuation using a proper thermal-oriented reactive power control during wind gusts, as shown in Figure 13.28.

The typical one-year return period wind gust is defined in IEC, Mexican-hat-like curve [30]. The thermal cycling of the back-to-back power converters without and with thermal-oriented reactive power control is shown in Figures 13.29a and 13.29b, respectively.

As shown in Figure 13.29a, the active power reference becomes zero at the synchronous operation point. Moreover, it can be seen that the minimum junction temperature appears around the synchronous operation point and the maximum junction

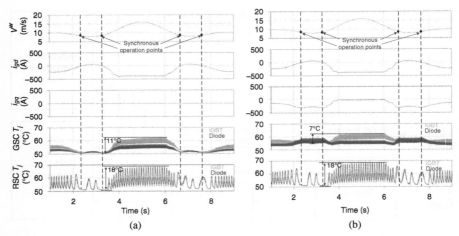

Figure 13.29   Thermal cycling of back-to-back power converters during wind gusts in a DFIG system: (a) Without thermal-oriented reactive power control and (b) with thermal-oriented reactive power control.

temperature appears above the rated wind speed. The thermal stress becomes least serious around the synchronous wind speed due to that no active power flow through the grid-side converter, while it becomes most serious at the extremely low frequency of the rotor-side converter current.

As shown in Figure 13.29b, it is noted that by injecting proper thermal-oriented reactive power, the maximum junction-temperature fluctuation in the grid-side converter is decreased from 11 °C to 7 °C, for the reason that the additional thermal-oriented reactive power is introduced under small active power in order to actively heat up the device, which enhances the lifetime of the power converters. Meanwhile, the maximum junction-temperature fluctuation in the rotor-side converter remains the same, 18 °C, due to rather higher active power reference in the entire wind speed. Generally speaking, when introducing additional reactive power to the wind turbine system, the thermal behavior of the diode fluctuates more than that of the IGBT in the grid-side converter, while the thermal behavior of the IGBT and the diode are slightly changed in the rotor-side converter.

13.6 SUMMARY

This chapter starts with the loss model of power semiconductor devices, which is the basis of the thermal stress evaluation of power electronic converters. An analytical calculation of the power loss is proposed to evaluate the effects of the power module selection, the modulation scheme, and the power factor of the converter specification. Then, according to the physical layout of the power module, the thermal model is established in terms of the power module itself and its cooling solutions. Afterward, aiming at back-to-back power converters used in the DFIG system, the thermal stress of the power semiconductor can be investigated and assessed in normal and abnormal grid conditions with considerations of the wind profile and the grid code requirements.

It is concluded that, due to the same direction of the active power flowing through back-to-back power converters, the most-stressed power device between them is different in the normal grid condition. Besides, the occurrence of the LVRT leads to the varied thermal cycling between the IGBT and the freewheeling diode due to the existence of the reactive current injection. Furthermore, it can be seen that due to the fact that both the grid-side converter and the rotor-side converter have the ability to provide the reactive power, an optimized thermal control method of the grid-side converter can be achieved during a wind gust, while the thermal performance of the rotor-side converter is almost unaffected.

REFERENCES

[1] "Semiconductors," ABB Applying IGBTs. [Online]. Available: http://new.abb.com/products/semiconductors

[2] T. B. Soeiro and J. W. Kolar, "Analysis of high-efficiency three-phase two- and three-level unidirectional hybrid rectifiers," *IEEE Trans. Ind. Electron.*, vol. 60, no.9, pp. 3589–3601, Sep. 2013.

[3] D. Zhou, F. Blaabjerg, M. Lau, and M. Tonnes, "Thermal analysis of multi-MW two-level wind power converter," in Proc. IECON, 2012, pp. 5862–5868.

[4] K. Zhou and D. Wang, "Relationship between space-vector modulation and three-phase carrier-based PWM: A comprehensive analysis," *IEEE Trans. Ind. Electron.*, vol.49, no. 1, pp. 186–196, Feb. 2002.

[5] G. Abad, J. Lopez, M. Rodriguez, L. Marroyo, and G. Iwanski, *Doubly Fed Induction Machine: Modeling and Control for Wind Energy Generation Applications*. Wiley-IEEE Press, 2011.

[6] J. Kolar, J. Biela, S. Waffler, T. Friedli, and U. Badstuebner, "Performance trends and limitations of power electronic systems," in Proc. Integr. Power Electron. Syst. Rec., 2010, pp. 17–36.

[7] Y. Song and B. Wang, "Survey on reliability of power electronic systems," *IEEE Trans. Power Electron.*, vol. 28, no. 1, pp. 591–604, Jan. 2013.

[8] G. Majumdar and T. Minato, "Recent and future IGBT evolution," in Proc. PCC, 2007, pp. 355–359.

[9] A. Wintrich, U. Nicolai, and T. Reimann, *Semikron Application Manual*. 2011.

[10] K. Ma and F. Blaabjerg, "Multilevel converters for 10 MW wind turbines," in Proc. EPE, 2011, pp. 1–10.

[11] H. Wang, K. Ma, and F. Blaabjerg, "Design for reliability of power electronic systems," in Proc. IECON, 2012, pp. 33–44.

[12] R. Schnell, M. Bayer, and S. Geissmann, "Thermal design and temperature ratings of IGBT modules," ABB Application Note, 5SYA 2093-00, 2011.

[13] T. Schutze, "Thermal equivalent circuit models," Infineon Application Note, AN2008-03, 2008.

[14] Q. Gao, C. Liu, B. Xie, and X. Cai, "Evaluation of the mainstream wind turbine concepts considering their reliabilities," *IET Renew. Power Gener.*, vol. 6, no. 5, pp. 348–357, Sep. 2012.

[15] K. Olesen, F. Osterwald, M. Tonnes, R. Drabek, and R. Eisele, "Designing for reliability, liquid cooled power stack for the wind industry," in Proc. IEMDC, 2011, pp. 896–901.

[16] D. Zhou, F. Blaabjerg, M. Lau, and M. Tonnes, "Thermal profile analysis of doubly-fed induction generator based wind power converter with air and liquid cooling methods," in Proc. EPE, 2013, pp.1–10.

[17] Semikron datasheet, SKiiP 1814 GB17E4-3DUL.

[18] Semikron datasheet, SKiiP 1814 GB17E4-3DUW.

[19] H. Li, Z. Chen, and H. Polinder, "Optimization of multibrid permanent-magnet wind generator systems," *IEEE Trans. Energy Convers.*, vol. 24, no. 1, pp. 82–92, Mar. 2009.

[20] R. Pena, J. C. Clare, and G. M. Asher, "Doubly fed induction generator using back-to-back PWM converters and its application to variable-speed wind-energy generation,", in Proc. Electr. Power Appl., May 1996, pp. 231–241.

[21] D. Xiang, L. Ran, P. J. Tavner, and S. Yang, "Control of a doubly fed induction generator in a wind turbine during grid fault ride-through," *IEEE Trans. Energy Convers.*, vol. 21, no. 3, pp. 652–662, Sep. 2006.

[22] User manual of PLECS blockset, Version 3.2.7, March 2011. [Online]. Available: http://www.plexim .com/files/plecsmanual.pdf

[23] J. Lopez, E. Gubia, E. Olea, J. Ruiz, and L. Marroyo, "Ride through of wind turbines with doubly fed induction generator under symmetrical voltage dips," *IEEE Trans. Ind. Electron.*, vol. 56, no. 10, pp. 4246–4254, Oct. 2009.

[24] F. Lima, A. Luna, P. Rodriguez, E. Watanabe, and F. Blaabjerg, "Rotor voltage dynamics in the doubly fed induction generator during grid faults," *IEEE Trans Power Electron.*, vol. 25, no. 1, pp. 118–130, Jan. 2010.

[25] K. Ma, F. Blaabjerg, and M. Liserre, "Thermal analysis of multilevel grid-side converters for 10-MW wind turbines under low-voltage ride through," *IEEE Trans. Ind. Appl.*, vol. 49, no. 2, pp. 909–921, Mar. 2013.

[26] E.ON-Netz, *Requirements for Offshore Grid Connections*. Apr. 2008.

[27] M. Tsili and S. Papathanassiou, "A review of grid code technical requirements for wind farms," *IET Renew. Power Gen.*, vol. 3, no. 3, pp. 308–332, Sep. 2009.

[28] C. Liu, F. Blaabjerg, W. Chen, and D. Xu, "Stator current harmonic control with resonant controller for doubly fed induction generator," *IEEE Trans. Power Electron.*, vol. 27, no. 7, pp. 3207–3220, Jul. 2012.

[29] D. Zhou, F. Blaabjerg, M. Lau, and M. Tonnes, "Thermal behavior optimization in multi-MW wind power converter by reactive power circulation," *IEEE Trans. Ind. Appl.*, vol. 50, no. 1, pp. 433–440, Jan. 2014.

[30] IEC 61400-1, "Wind turbines—Part 1: Design requirements," 3rd ed., International Electro-technical Commission, 2005.

DFIG TEST BENCH

CHAPTER 14

DFIG TEST BENCH

In this chapter, the designing and implementation of a reduced-scale DFIG test system in the laboratory is introduced. The focus is on introducing how to build up such a system in a laboratory, including the demands for the test system, the structure, the hardware design for each part, as well as the control design, but mostly about how to realize different control strategies. Additionally, information on how to test this system and keep it running is also introduced. In this chapter, the intention is to provide a guidance and reference for researchers who want to build a test system in the lab and verify their research.

14.1 INTRODUCTION

In the previous chapters, the normal and advanced control for DFIG under unbalanced grid voltage, harmonic distorted grid voltages, as well as under grid faults are introduced. To verify these advanced control strategies, the experiment or test verifications are necessary before they can be used on large-scale DFIG WPS. From an economic point of view, the advanced control strategies are often tested in a reduced scale DFIG test system in the laboratory. As a result, building up this kind of reduced-scale DFIG test system is also very important for the research on the advanced control for DFIG WPS under different grid conditions.

In this chapter, the design and implementation of a reduced-scale 30 kW DFIG test bench in the laboratory are introduced. The focus will be on the explanation about how to build up such a system in the laboratory, including the demands for the test system, the structure, the hardware design for each part and also the control design will be introduced. Additionally, information on how to test this system and keep it running is also introduced. The chapter provides a guidance and reference for researchers who want to build a test system in the lab, and to verify their ideas.

14.2 SCHEME OF THE DFIG TEST BENCH

The basic scheme of a DFIG WPS has been introduced in chapter 2 and it is shown here again in Figure 14.1. The corresponding scheme of the DFIG test bench in the

Advanced Control of Doubly Fed Induction Generator for Wind Power Systems, First Edition.
Dehong Xu, Frede Blaabjerg, Wenjie Chen, and Nan Zhu.
© 2018 by The Institute of Electrical and Electronics Engineers, Inc. Published 2018 by John Wiley & Sons, Inc.

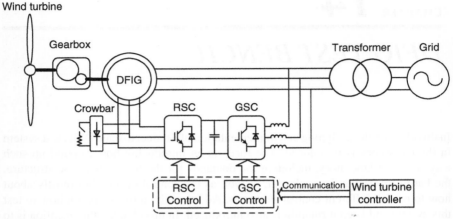

Figure 14.1 Basic scheme of a real DFIG WPS.

laboratory is shown in Figure 14.2. Comparing these two figures, it can be seen that the different part between the real DFIG WPS and the reduced DFIG test bench are:

- The wind turbine and the gearbox in a real DFIG WPS are replaced by the caged motor (CM) controlled by a driving inverter and the mechanical behavior of the wind turbine and gearbox need to be emulated by the CM and its driving inverter.

- The non-ideal grid in a real power system is emulated by a grid emulator. As in the lab, a grid emulator is necessary to generate these grid transient processes,

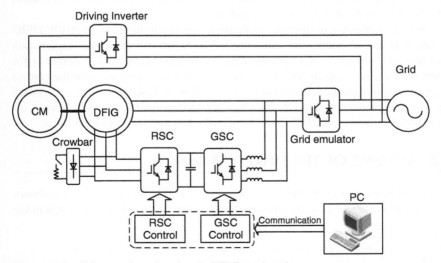

Figure 14.2 Scheme of a reduced-scale DFIG test bench.

in order to emulate the grid voltage unbalance, harmonic distortion, and the grid faults.

- The DFIG and the back-to-back converters in the test bench are reduced in power scale, which means the power rating of the DFIG, RSC, and GSC in the test bench will be much smaller compared to the real DFIG WPS. This may influence the control performance. In a small-scale power system, the DFIG stator and rotor resistances may be much larger, and for the RSC and GSC, the switching frequency will be larger as well. So in real application, the best should be done to design the test bench in the same way as it is in the real large power scale DFIG WPS.

- The up-level control signal in a real DFIG WPS is normally coming from the wind turbine controller. In the DFIG TEST bench, it can be emulated by a computer communicating with the RSC and the GSC controller.

In this chapter, a 30 kW DFIG test bench is used as an example. Building up each part of the DFIG test bench will be introduced. Finally, the starting and the protection of the total test system will be demonstrated in detail as well.

14.3 THE CAGED MOTOR AND ITS DRIVING INVERTER

The wind turbines in a real DFIG WPS is emulated by the caged motor driving by its driving inverter in a DFIG test bench. The shaft of the DFIG and the caged motor is connected, so the mechanical power generated by the caged motor can be sent to the DFIG, and then the DFIG is able to generate power to the grid. Normally, the caged motor and its driving inverter need to operate in three different operation modes:

- **Speed-Control Mode:** In this mode, the caged motor is controlled by the driving inverters to work with a fixed rotor speed. The DFIG is working under power control or torque control at a fixed rotor speed.

- **Torque-Control Mode:** In this mode, the caged motor is controlled with a fixed output torque. The rotor speed is controlled by the DFIG, and the DFIG is working under speed-control mode.

- **Inertia Emulation Mode:** In this mode, the steady state of caged motor is controlled in either speed-control mode or torque-control mode. During the speed or torque changes, the output torque and the rotor speed of the caged motor are controlled to emulate the transient process of the real wind turbines with very large inertia, in this case, for example, the rotor speed variation during grid faults can be emulated.

The driving inverter in this 30 kW DFIG test bench is a commercialized 30 kW inverter, as the above mentioned for the driving inverter can easily be achieved by commercialized inverters, or can be achieved by the commercialized inverters after some simple reprograming.

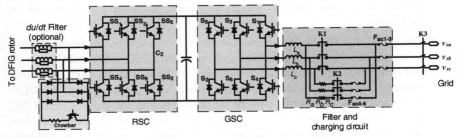

Figure 14.3   Scheme of the reduced scale DFIG converters.

14.4   DFIG TEST SYSTEM

The DFIG test system, with the DFIG, the RSC, the GSC, and the crowbar are the major part of the test bench, as the advanced control strategies will be carried out on them, and it is shown in Figure 14.3. It has been mentioned that the structure of the DFIG and converters in the test system are basically the same as in the case of a real DFIG WPS. However, as the power scale is reduced, the performance of the system may be different on the test bench, for example, the switching frequency in smaller-scale system will be larger, and the resistance will be larger as well. So, when the test system is designed, these aspects must be considered, in order to make the test bench more consistent with the real DFIG WPS [1].

14.4.1   DFIG

This book will not discuss the design of the generator. But, it should be noticed that in a real DFIG WPS, the number of turns of rotor winding is normally larger than that in the stator winding. This may be different from the commonly used induction machines. Also, the DFIG is demanded to operate in super-synchronous speed, up to 1.2 pu. These aspects must be considered when ordering the DFIG. The parameters of the 30 kW DFIG used in the test bench introduced in this chapter are given in Table 14.1 [2].

TABLE 14.1   Parameters of the DFIG in the 30 kW test bench compared with real 1.5 MW DFIG WPS

	1.5 MW	30 kW
Rated power	1.5 MW	30 kW
Rated voltage (L–L) (V_{rms})	690	380
Stator resistance R_s (mΩ)	2.139	92
Stator inductance L_s (mH)	4.05	26.7
Rotor resistance R_r (mΩ)	2.139	56
Rotor inductance L_r (mH)	4.09	26.2
Mutual inductance L_m (mH)	4.00	25.1
Turns ratio n_{sr}	0.369	0.32

14.4.2 Hardware Design of the GSC

14.4.2.1 Switching Devices

With the steady-state model of the DFIG introduced in Chapter 4, the rated active power on the rotor-side P_{r_rated} can be found as

$$P_{r\_rated} = S_l P_{s\_rated} \tag{14.1}$$

This is also the active power that the GSC has to deliver to the grid. As the GSC is connected to the grid directly, the output voltage of the GSC is the grid voltage, which is $V_{s(L-L)} = 380 \cdot V_{rms}$ in this case. So the rated current I_{g_rated} of the GSC can be calculated as

$$I_{g\_rated} = \sqrt{3}\frac{P_{r\_rated}}{V_{s(L-L)}} = 1.732 \times \frac{30000 \times 0.2}{380 \times 1.414} \, \text{A} \approx 19\,\text{A} \tag{14.2}$$

The current rating of the switching devices (IGBT in this case) is normally selected to be more than two times the rated current, so the current rating of the IGBT used in the test bench should be larger than 40A. For the voltage rating, the minimum DC-bus voltage for the GSC $V^G_{dc\_min}$ should be

$$V^G_{dc\_min} = V_{L-L} = 1.414 \times 380\,\text{V} \approx 537\,\text{V} \tag{14.3}$$

The DC-bus voltage should also consider the demands of the RSC. It can be found from (4.20) and (4.10) that the rated rotor voltage can be found:

$$V_{r\_rated(L-L)} = \frac{1}{n_{sr}}SV_{L-L} = \frac{1}{0.305} \times 0.2 \times 1.414 \times 380\,\text{V} \approx 352\,\text{V} \tag{14.4}$$

where n_{sr} is the turn ratio between the stator and the rotor. So, the minimum DC-bus voltage for the RSC $V^R_{dc\_min}$ should be

$$V^R_{dc\_min} = V_{r\_rated(L-L)} \approx 353\,\text{V} \tag{14.5}$$

Therefore, as the two-level topology is used for the GSC, IGBT chips with a voltage rating of 1200 V and current rating of 40 A is used for the GSC.

14.4.2.2 Switching Frequency

The switching frequency of the converters with a power rating of 10 kW with IGBT is normally between 10 and 50 kHz. However, as in this test bench, the behavior of the MW-rated wind power system is emulated, the switching frequency should be selected to be the same as that of the MW-rated wind power system, which is normally a few kHz. In this 30 kW test system, it is selected as 2 kHz.

14.4.2.3 Grid Filter Inductance

In a real application, the filter inductance of the three-phase grid-connected inverter is normally designed such that the voltage drop in the inductor is about 10–20% of the grid voltage. In this case, the grid filter inductance L_g will be

$$L_g = \frac{0.2V_{s(L-L)}}{2\sqrt{3}\pi f_g I_{g\_rated}} \approx 9.9\,\text{mH} \tag{14.6}$$

where f_g is the grid frequency, which is 50 Hz in this case.

14.4.2.4 DC-Bus Voltage and Capacitance

It has been calculated that the minimal DC-bus voltage for the GSC, $V^G_{dc\_\min} = 537\text{V}$ and that for the RSC, $V^R_{dc\_\min} = 352\text{V}$, so the targeted DC-bus voltage is chosen to be about 1.2 times the DC-bus voltage for the GSC $V^G_{dc\_\min}$, which is

$$V_{dc} = 650\,\text{V} \tag{14.7}$$

The design of the DC-bus capacitance will be carried out with the consideration of power balancing of the GSC and the RSC in a switching period. Normally, (14.7) is used to calculate the DC-bus capacitance C_d for the GSC and the RSC.

$$C_d \geq \frac{T_r \Delta P_{\max}}{2V_{dc}\Delta V_{\max}} \tag{14.8}$$

where T_r is the delay of the control system, normally five times the switching period, which is 0.5 ms in this case, so $T_r = 0.5 \times 5\text{ms} = 2.5\text{ms}$. $\Delta P_{\max}$ is the maximum power flow, which is the rotor rated power $P_{r\_rated} = S_l P_{s\_rated} \approx 6\text{kW}$. $\Delta V_{\max}$ is the maximum voltage fluctuations in the DC-bus, normally designed to be 10% of the DC-bus voltage V_{dc}. So, the DC-bus capacitance C_d can be calculated as

$$C_d \geq \frac{2.5\text{ms} \times 6\text{kW}}{2 \times 650\text{V} \times 0.1 \times 650\text{V}} = 177\text{uF} \tag{14.9}$$

So, according to the above calculation, the DC-bus capacitance should be larger than 177 μF. In this test bench, the DC-bus capacitance is selected to be 4700 μF, with a voltage rating of 900 V. In this case, the two converters are strongly decoupled.

14.4.2.5 Charging Circuits

The charging circuits, as shown in Figure 14.3, are designed to achieve soft starting of the GSC when it is connected to the grid. Before the GSC is started and connected to the grid, the DC-bus voltage is zero. If the main contactor K1 is directly closed without soft charge, the inrush current into the large DC capacitance may destroy the diode paralleled with the IGBTs.

The charging circuit is working in the following sequence:

Step 1: The soft-start contactor K1 is closed first, the soft-start resistors R_a, R_b, and R_c are connected to the charging circuit so that the charging current can be limited by these resistors. For example, the largest charging current I_{start} in phase A can be expressed as

$$I_{start} \approx \frac{1}{R_a} V_{L-L} \qquad (14.10)$$

The current rating of the IGBT and the diodes are 40 A, so I_{start} must be smaller than this. So, the smallest resistance can be calculated as

$$R_a \approx \frac{1}{I_{rated}} V_{L-L} \approx 13.5\Omega \qquad (14.11)$$

The soft-start resistance is chosen to be 15 Ω. It should be noticed the power or current rating of the resistance also need to be considered.

Step 2: After the DC-bus voltage has risen up to a certain value (500 V in this case), the main contactor K2 is closed and the filter inductance is connected into the circuit.

Step 3: The soft-start contactor K1 is opened a few milliseconds and the soft-start resistor is by-passed. So, the control system can be enabled and control the GSC to work in normal operation. In this way, the soft starting of the GSC is finished.

14.4.3 Control Design of the GSC

14.4.3.1 Control Scheme

The control scheme of a GSC is similar to the typical three-phase PFC boost rectifier, which has been introduced in Chapter 4 and it is shown here again in Figure 14.4.

The three-phase grid voltage $\vec{v}_{gabc}$, current $\vec{i}_{gabc}$, and the DC-bus voltage V_{dc} are sampled for the control scheme. The grid angler θ_s and angler speed ω_s are produced

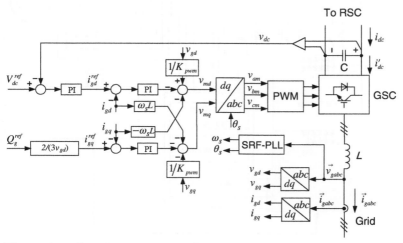

Figure 14.4   Control scheme of the GSC.

by the PLL. The control scheme consists of the current inner loop and the DC-bus voltage/reactive power outer loop. The DC-bus voltage reference V_{dc}^{ref} is provided by the wind turbine controller and compared with the sampled DC-bus voltage V_{dc}. The error is put into the outer loop PI controller, and the output of the PI controller is the rotor current reference in d-axis i_{gd}^{ref}. The reactive power loop is an open loop based on (5.10) and the rotor current reference in d-axis i_{gq}^{ref} is calculated. The inner current loop is designed based on (5.8), after decoupling, and the rotor currents in the dq reference frame i_{gd} and i_{gq} can be regulated by the output voltage of the GSC v_d and v_q. The PI controllers in the reference frames are used and the output of the PI controller after decoupling is the voltage reference of the GSC v_d^{ref} and v_q^{ref}. They are transferred back into the abc reference frame and the drive signals are generated after the SPWM modulation, or SVM modulation in some cases.

14.4.3.2 Control Circuits

To achieve the control scheme introduced above, the control circuits need to be designed and built. The basic scheme of the control circuits of the GSC is shown in Figure 14.5. The control circuits consist of the following parts:

- **Sampling Circuit:** The sampling circuit measures the output current and the DC voltage of the GSC, as well as the grid voltage, and transfers them into the analogy signals which are compatible with the DSP controller. In the GSC, the three-phase grid voltage, three-phase GSC current, and the DC-bus voltage need to be sampled and transferred, as shown in Figure 14.4. As a result, the

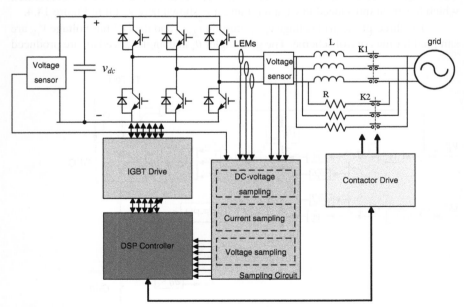

Figure 14.5   Scheme of the control circuit of the GSC.

sampling circuits can also be divided into three parts: the voltage sampling, the current sampling, and the DC-bus voltage sampling circuits, as shown in Figure 14.5. The output of the sampling circuits is provided as the feedback in the control loop and connected to the DSP controller.

- **DSP Controller:** The DSP controller is the central processor of the whole system. The related calculations in the control scheme shown in Figure 14.4 are realized in the DSP controller. The duty cycles of every switching devices are also generated in the DSP controller and sent to the IGBT drive. Furthermore, the DSP controller will send control signals to the contactor drive to control the contactor driving start up.

- **IGBT Drivers:** The IGBT drivers receive the duty cycle signals from the DSP controller and drive the IGBT. It can be regarded as part of the control circuits, but somewhere, it is also regarded to be a part of the IGBT device. With the driving of the IGBT drive, the GSC is able to generate the output voltage required by the DSP controller, so the whole control loop can be closed-loop connected.

- **Contactor Drivers:** The contactor driver circuit controls the contactor according to the DSP controller's demands. This part cannot be found in the control scheme shown in Figure 14.4, but it is related to the soft start of the GSC, as well as the protection of the system.

The soft-start sequence of the GSC can also be regarded as a part of the control design, and it has been introduced in the last section. A detailed schematic of each part is not given in this book, as a lot of recommended circuits can be found in the references and also in the data sheet of the current and the voltage sensors.

14.4.4 Testing of the GSC

After the main circuits and the control circuits have been built, and the control strategy has been programmed, the next step is to test and debug the GSC and make it work normally. This testing method or sequence can be different according to the user's different experiences. A testing sequence is provided here according to the author's experiences, just as a reference for the readers.

Step 1: Test all the control and protection circuits and make sure they are working normally.

Step 2: Open contactors K1 and K2, use a DC source to raise the DC-bus voltage, and use the DSP controller to give open-loop sinusoidal signal to every phase, and check the output voltage of the GSC, until the DC-bus voltage reaches the normal DC-bus voltage, in order to make sure if all the hardware and IGBT drives are working normally.

Step 3: Test the soft-start circuits and the related control, until the soft-start circuits are able to charge the DC bus to more than 500 V without enabling the IGBT.

Step 4: Enable the controller, and adjust the controller parameters according to the test results. A resistor load can be added to the DC bus in order to test the heavy load performance.

14.4.5 Hardware Design of the RSC

For the switching devices (the IGBT), the targeted DC-bus voltage of the GSC has been designed to be $V_{dc} = 650$V, and the rated rotor current of the DFIG I_{r_rated} can be calculated from the steady-state model of the DFIG, as expressed in (14.12), and this is also the rated current for the RSC.

$$I_{r\_rated} \approx n_{sr}I_{s\_rated} = n_{sr}\frac{P_{rated}}{\sqrt{3}V_{L-L}} \approx 19.6\text{A} \qquad (14.12)$$

The current rating of the IGBT is also chosen to be two times rated current, so the IGBT with the voltage rating of 1200 V and current rating of 40 A is selected for the RSC.

As the RSC is directly connected to the DFIG rotor, a filter inductance is not necessary for the RSC. However, in some cases, a *dv/dt* filter is installed at the AC side of the RSC, but in the test bench, it can be omitted.

14.4.6 Control Design of the RSC

14.4.6.1 Control Scheme

The widely used vector control scheme of the RSC and DFIG has been introduced in Chapter 5. The RSC will control the DFIG to operate in starting mode, power-control mode, or speed-control mode according to different working conditions. Their control scheme is shown in Figures 14.6–14.8 again here. The inner current loop is similar in

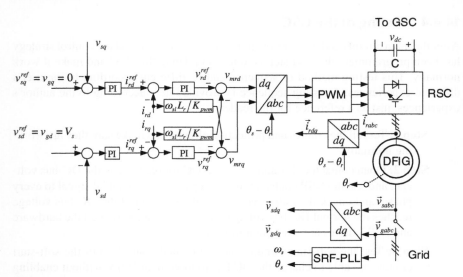

Figure 14.6   Control scheme of the RSC in starting mode.

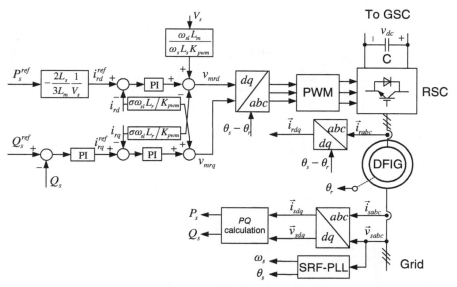

Figure 14.7   Control scheme of RSC and DFIG in power-control mode.

these three different control modes. For the starting mode, the outer loop is the stator voltage loop; for the power-control mode, the outer loop is the power loop; while in the speed-control mode, the outer loop is the speed-control loop. A detailed analysis of the control scheme and the design of the controller parameters can be found in Chapter 5, so it will not be repeated in this chapter.

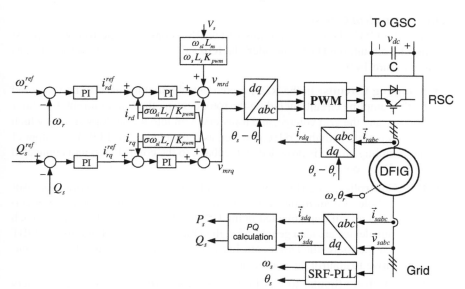

Figure 14.8   Control scheme of the DFIG in speed-control mode.

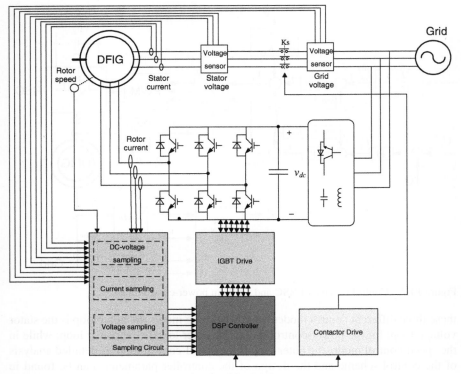

Figure 14.9   Scheme of the control circuit of the RSC.

14.4.6.2   *Control Circuits*

Similar to the GSC, the control circuits of the RSC consist of the following parts: the sampling circuit, the DSP controller, the IGBT driver, and the contactor drive, as shown in Figure 14.9.

- **Sampling Circuits:** In the RSC, more control variables need to be measured by the sampling circuit, including the rotor current, the stator current, the stator voltage, and the grid voltage, and transfers them into the analogy signals which are compatible with the DSP controller, as shown in Figure 14.9. The output of the sampling circuit is provided as feedback to the control loop and connected to the DSP controller.

- **DSP Controller:** The DSP controller is also working as the central processor of the whole system in the RSC. The related calculations in the three operation modes are all implemented in the DSP controller. The duty cycles of each switching device are also generated in the DSP controller and sent to the IGBT drive. Furthermore, the DSP controller will send the control signals to the contactor drive to control the contactor.

- **IGBT Driver:** Similar to that in the GSC, the IGBT drive receives the duty cycle signals from the DSP controller and drives the IGBT. So, the RSC is able to generate the output rotor voltage required by the control strategy and the whole control loop can be closed-loop connected.

- **Contactor Driver:** The contactor drive circuit drives the contactor according to the DSP controller's demands. In the RSC, it is the stator main breaker Ks, as shown in Figure 14.9. It is important in the mode-switching process between the starting mode and the power-control mode.

14.4.7 Testing of the RSC

As there are three operation modes for the RSC and the DFIG; the testing of the RSC will be more complicated compared to the case for the GSC. A testing sequence according to the experience is provided here as a reference for the readers.

> **Step 1:** Test all the control circuits as well as the protection circuits and make sure they are working normally. Open the stator breaker Ks, use a DC source to raise the DC-bus voltage, and use the DSP controller to give open-loop sinusoidal signals to every phase, and check the output voltage of the RSC, until the DC-bus voltage reaches the normal DC-bus voltage. So it can be confirmed that all the hardware and IGBT drivers are working normally.

> **Step 2:** The starting mode needs to be tested first. Set the driving inverter to speed-control mode, keep the stator breaker Ks open, and test the starting mode. Check the three-phase grid voltage and the stator voltage until the stator voltage in the starting mode has been controlled to track the grid voltage.

> **Step 3:** Then it is the power-control mode. Close the stator breaker Ks after stator voltage in the starting mode has been controlled to track the grid voltage. Operate the RSC and DFIG in power-control mode, keep increasing the output power and adjusting the control parameters, until the output power reaches the rated power.

> **Step 4:** The speed-control mode is normally tested last. Change the driving inverter to torque-control mode, and switch the RSC and DFIG to speed-control mode. Change the rotor speed references and adjust the controller parameters to get an appropriate response.

14.5 ROTOR-SIDE CROWBAR

In a real DFIG system, the rotor-side crowbar with a rectifier is commonly used as it is relatively cost-efficient compared to the structure of three bidirectional switches. The rotor-side crowbar with a rectifier is also used and its schematic is shown in Figure 14.10. The switch is achieved by an IGBT S_c, when S_c is turned on and the crowbar is enabled and the crowbar resistance R_{c_dc} is connected to the rotor circuit. The RC circuit consisting of the resistor R_{cn} and the absorbing capacitance C_{cn} is used to reduce the voltage peak when S_c is switched off. The resistance R_b is normally

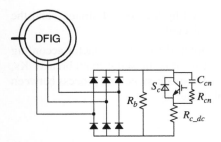

Figure 14.10   Schematic of the rotor-side crowbar in the DFIG.

selected to be higher than 1 kΩ; it is used to prevent overvoltage when the S_c is turned off and the crowbar is disabled.

The largest voltage and current in the rotor-side crowbar during the voltage dips can be calculated from the dynamic model of DFIG under grid faults, which has been introduced in Chapter 11. The maximum rotor current with rotor-side crowbar I_{rm} can be found as

$$I_{rm} \approx \frac{\omega_r}{\omega_s} \frac{pV_s}{\sqrt{R_c^2 + \omega_r \sigma L_r^2}} \tag{14.13}$$

Normally, I_{rm} is designed to be about two times the rated rotor current, the maximum voltage dip level p is 80%, and the maximum rotor speed ω_r is 1.2 pu. The required crowbar resistance R_c in this test bench can be derived from (14.13), and referred to the rotor side, as shown in (14.14).

$$R_c \approx \frac{1}{n_{sr}^2} \sqrt{\left(\frac{\omega_r}{\omega_s} \frac{pV_s}{I_{rm}}\right)^2 - (\omega_r \sigma L_r)^2} \approx 25\Omega \tag{14.14}$$

This crowbar resistance R_c here is the resistance in a linear circuit; its relationship with the real crowbar resistance in the crowbar shown in Figure 14.10 has been studied in Chapter 12, as expressed in

$$R_{c\_dc} = \sqrt{3}R_c \approx 43\Omega \tag{14.15}$$

The next step is to select the IGBT S_c. The maximum rotor current I_{rm} is designed to be 2 pu in this 30 kW test bench by considering the stator–rotor ratio:

$$I_{rm} \approx n_{sr} 2I_{rated} = 39\,\text{A} \tag{14.16}$$

As the crowbar is a nonlinear circuit, the maximum current flow in the IGBT S_c will be larger than I_{rm}, so more margin needs to be given while choosing the IGBT. When the crowbar is disabled, the voltage in the IGBT S_c is the DC-bus voltage of the RSC and GSC, which is about 650 V. So, the IGBT with a voltage rating of 1200 V and a current rating of 60 A is selected.

The resistance R_b is used to prevent overvoltage when S_c is turned off and the crowbar is disabled. It is selected to be 100 Ω in this test bench. The RC circuits to protect the IGBT S_c are selected according to the recommendation provided by the IGBT datasheets.

14.6 GRID EMULATOR

14.6.1 Demands of the Grid Emulator

As the non-ideal grid is somehow non-predictable and cannot be controlled, it is necessary to emulate the non-ideal grid with grid emulator in the lab. To cooperate with the DFIG test bench and finish the test of the DFIG under the non-ideal grid, the grid emulator should fulfill the following demands:

- The grid emulator needs to provide normal three-phase voltage to the DFIG under normal working conditions and at the same time, the grid emulator should be able to send the power produced by the DFIG back to the grid. This is the basis for the DFIG to work normally.

- The grid emulator is required to emulate non-ideal grid voltages, including grid harmonic distortions, grid voltage unbalances, and symmetrical, asymmetrical grid voltage dips introduced by the grid faults. The fault parameters, including the voltage dip levels, the fault durations, and the faulted phase should be adjusted freely.

- The grid emulator should also emulate the grid behavior, for example, the non-instant voltage recovery after the grid faults, introduced by the action from the protection breaker, and the recurring grid faults.

So, according to these demands, it can be found that the grid emulator should be a three-phase voltage source which is able to provide the output voltage similar to the grid voltage. At the same time, the grid emulator should be able to control the output voltage freely and with a fast dynamic response speed, in order to emulate the three-phase non-ideal grid voltage [3, 4].

14.6.2 Hardware Design

14.6.2.1 Scheme of the Grid Emulator

According to the above-listed demands, a back-to-back converter is used to build up the grid emulator, as shown in Figure 14.11. The grid emulator consists of two converters which are back-to-back connected. It provides a three-phase output voltage to emulate the grid voltage of the DFIG test system. Also, the non-ideal grid conditions, including the grid voltage harmonic distortions, grid voltage unbalance, and the grid voltage dips and recovery introduced by the grid faults can be emulated by controlling the output voltage of the grid. The fault-emulated converter shares a common DC bus with the grid converter. The DC-bus voltage is provided by the grid converter. The grid converter is a fully controlled rectifier; it has the ability to work bidirectionally.

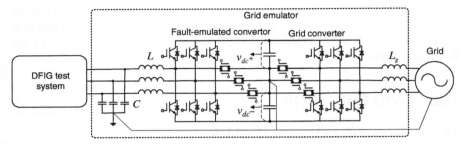

Figure 14.11 Schematic of the grid emulator.

Therefore, the active power generated by the DFIG test system can be sent back to the grid by the grid converter [5].

In this test system, the T-type three-level voltage source topology is used in the fault-emulated converter and the grid converters. The traditional two-level voltage source topology could also be a choice [6].

14.6.2.2 Switching Devices

To select the switching devices, the rated voltage and current of the grid emulator should be determined first. The grid converter is connected to the grid, and the output voltage of the fault-emulated converter emulates the grid voltage, so the rated AC-side voltage V_{rated_AC} of both converters is at the grid voltage, as shown in (14.16).

$$V_{rated\_AC} = V_s = 220V_{rms} = 311\,\text{V} \tag{14.17}$$

With SPWM modulation, the smallest required DC-bus voltage V_{dc_min} can be calculated as

$$V_{dc\_min} = 2V_s = 622\,\text{V} \tag{14.18}$$

With 20% margin, the DC-bus voltage V_{dc} is selected to be

$$V_{dc\_min} = 780\,\text{V} \tag{14.19}$$

Regarding the rated power, as all the power generated by the DFIG test system will flow through the grid emulator, the rated power of the grid emulator should not be less than the rated power of the DFIG test system. Furthermore, as in grid faults, the stator current of the DFIG may reach two times the rated current, so the grid emulator should withstand a short-time overcurrent. The rated power, DC-bus voltage, and the selected switching devices are listed in Table 14.2. The normal rated power of the grid emulator is designed to be 30 kW and it has the ability to withstand short-time transient currents up to 120 A. So, the voltage rating of the IGBT is chosen to be 1200 V for the switches in the main bridge and 600 V for the switch in the middle bridge, and the current rating is selected to be 200 A.

TABLE 14.2   The selection of switching devices in the grid emulator

Rated AC-side output voltage	311 V
Rated DC-bus voltage	780 V
Rated normal power	30 kW
Maximum short-time current	120 A
Selected voltage rating of the IGBT	600 V/1200 V
Selected current rating of the IGBT	200 A

14.6.2.3   Switching Frequency

The switching frequency of the grid emulator is chosen according to its rated power, as for the converters with the rated power of tens of killowatts, the switching frequency is normally selected to be more than 10 kHz. It should also be considered that in three-level converters, the switching frequency can be smaller compared to the two-level converter. In this test bench, the switching frequency of the grid emulator is selected as 16 kHz, for both the fault-emulated converter and the grid converter.

14.6.2.4   Filter Inductance and Capacitance

The output filter of the grid converter is similar to that of the GSC, the L filter is used. For the fault-emulated converter, LC filter is used to provide a high quality AC output voltage. The filter inductance can be designed by the empirical formula, as shown in (14.6). It can also be designed according to the current ripple limit on the output current. For the three-level converter, the largest current ripple ΔI_{max} can be expressed as

$$\Delta I_{max} = \frac{1}{2} \frac{V_{dc}}{2Lf_{sw}} \tag{14.20}$$

where V_{dc} is the DC-bus voltage, f_{sw} is the switching frequency, and L is the inductance. Normally, the current ripple ΔI_{max} is designed to be 20% of the rated current, so the inductance can be calculated as

$$L = \frac{1}{2} \frac{V_{dc}}{2\Delta I_{max}f_{sw}} = \frac{1}{4} \frac{760V}{0.2 \times 64A \times 10000Hz} \approx 1.4mH \tag{14.21}$$

For the filter capacitance, the LC filter is normally designed so that the corner frequency of the LC filter is about 10% of the switching frequency, which means

$$f_n = \frac{1}{2\pi\sqrt{LC}} = 0.1f_{sw} = 1000Hz \tag{14.22}$$

Thereby, the capacitance C can be calculated:

$$C \approx 180uF \tag{14.23}$$

14.6.2.5 DC-Bus Capacitance

The design method of the DC-bus capacitance is the same as that in the case of the GSC; according to (14.8), the smallest DC-bus capacitance should be

$$C_d \geq \frac{0.1\,\text{ms} \times 30\text{kW}}{2 \times 750\text{V} \times 0.1 \times 750\text{V}} = 260\text{uF} \tag{14.24}$$

This is a very small value, so in real applications, the DC-bus capacitance is selected to be 9400 µF, so it can be ensured that the DC-bus voltage will not become too high/low at abnormal condition. And the voltage rating is 900 V.

14.6.3 Control Design

14.6.3.1 Control Scheme of the Grid Converter

The grid converter operates as a fully controlled PFC rectifier, so its control scheme is similar to that described in the GSC of the DFIG test system, as shown in Figure 14.12. As for the three-level converter, the middle-point voltage needs to be controlled, and the upper-bus voltage v_{dc+} and the lower bus voltage v_{dc-} are both connected in the outer DC-bus voltage loop. The sum of v_{dc+} and v_{dc-} are controlled to track the voltage reference V_{dc}^{ref}, and the output of the sum loop is the d-axis current reference i_{gd}^*. While the difference of v_{dc+} and v_{dc-} are controlled to track zero, and the output of the difference loop is the o-axis current reference i_{go}^{ref}. As no reactive current output is needed for the grid converter, the q-axis current reference i_{gq}^{ref} is zero. The inner current loop is basically the same as that in the GSC, except that an o-axis current loop is added to the control scheme for the grid converter, as it is a

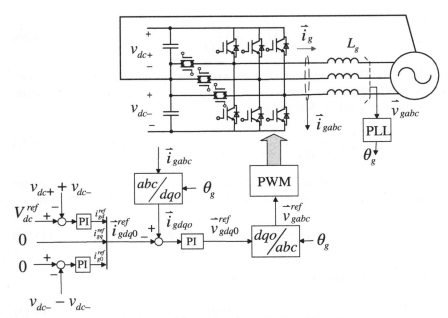

Figure 14.12 Control scheme of the grid converter in the grid emulator.

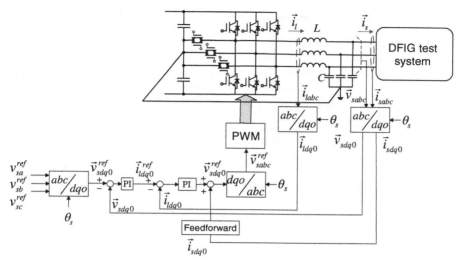

Figure 14.13   Control scheme of the fault emulator converter.

three-phase four-line system and thereby the zero-sequence current in the neutral line needs to be controlled as well. The grid converter also needs a soft-start control before it starts to operate, which is the same as that for the GSC, so it will not be repeated here.

14.6.3.2 *Control Scheme of the Fault-Emulated Converter*

The fault-emulated converter needs to control the three-phase output voltage to emulate the non-ideal grid voltages and the three-phase output voltage needs to be controlled independently and with a good dynamic response. The control scheme of the fault-emulated converter is shown in Figure 14.13. The vector control is also used to achieve a good control performance. The outer loop is the output voltage loop; the three-phase voltage references v_{sa}^{ref}, v_{sb}^{ref}, and v_{sc}^{ref} are transferred to the *dqo* reference to be the voltage reference vector $\vec{v}_{sdqo}^{ref}$ and the output voltage is controlled to track $\vec{v}_{sdqo}^{ref}$ in the outer voltage loop. The output of the outer voltage loop is the current reference of the inner current loop. The inductor current $\vec{i}_l$ is controlled in the inner current loop, the output of the inner loop $\vec{v}_{sdqo}^{ref}$ is transferred back into the *abc* reference frame, and $\vec{v}_{sabc}^{ref}$ is the voltage reference for SPWM modulation. A feedforward control of the load current or the output current of the DFIG test system $\vec{i}_s$ is added to the control scheme, in order to reduce the output impedance of the fault-emulated converter, so the output voltage will not be seriously influenced by the transient current introduced by the DFIG under grid faults.

In normal situations, the voltage references v_{sa}^{ref}, v_{sb}^{ref}, and v_{sc}^{ref} are the ideal three-phase AC voltages. To emulate the non-ideal grid voltage, the voltage references v_{sa}^{ref}, v_{sb}^{ref}, and v_{sc}^{ref} need to be adjusted to the harmonic voltage, unbalanced voltage, or set a voltage dips on the reference.

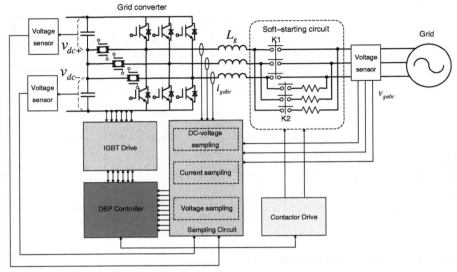

Figure 14.14   Control circuits of the grid converter in the grid emulator.

14.6.3.3 *Control Circuits*

The control circuits of the grid converter and the fault emulating converter are shown in Figures 14.14 and 14.15, respectively. The structure of the control circuits is similar to the GSC and the RSC for the DFIG system. The voltages and currents are measured

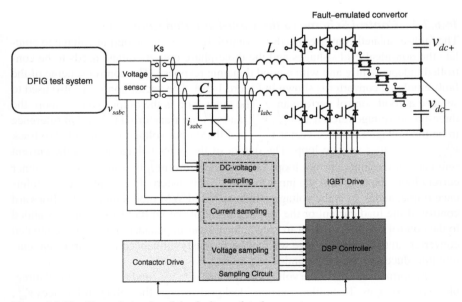

Figure 14.15   Control circuits of the fault-emulated converter.

by the voltage or current sensors, including the three-phase grid voltage v_{gabc}, grid current i_{gabc}, DC-bus voltages V_{dc+} and V_{dc-} for the grid converter. For the fault emulating converter, the three-phase output voltage v_{sabc}, inductor current i_{labc}, and the DFIG current, or load current i_{sabc} are measured too. The corresponding signals are sent to the sampling circuits. The sampling circuit transfers the voltage and current into the signal which is compatible with the DSP controller, and is sent to the DSP controller as the feedback of the control loop. The calculation of the control strategy is realized in the DSP control. The feedback signals from the sampling circuits are used to calculate the voltage reference, and the duty circuits are also generated in the DSP controllers by SPWM modulation. The duty cycle signals are sent to the IGBT drive circuits, and they are used to drive the converter. The DSP controllers also send control signals to the contactor drive circuit to control the contactor and to realize the soft-start or enable the protection.

14.6.4   Testing of the Grid Emulator

A test sequence for the grid emulator is also given here as a reference.

Step 1: The grid converter needs to be tested at first, as it provides the DC-bus voltage for the fault-emulated converter. The testing method of the grid fault converter is almost the same as that of the GSC of the DFIG system. It should be noticed that for the three-level converter, the neutral point voltage of the DC bus should be checked carefully.

Step 2: After the grid converter has operated normally, the grid emulating converter needs to be tested with no load firstl, then with a resistor load.

Step 3: When the grid converter works normally with resistor load, it should be connected to the DFIG. Some stability problem may occur when the stator of the DFIG is connected to the grid emulator; we should adjust the controller parameters or possibly adjust the load current feedforward parameters in the fault-emulated converter.

14.6.5   Test Waveforms of Grid Emulator

Some test results from the grid emulator are shown here as examples. The test waveforms of the grid emulator to emulate non-ideal grid is shown in Figure 14.16, which includes the symmetrical voltage dips and voltage swells, asymmetrical voltage dips, the frequency variations, and the harmonically distorted grid voltage. It is shown that the grid emulator is able to emulate many kinds of non-ideal grid voltage conditions for the DFIG test system, and the dynamic response is within 5 ms.

The performances of the DFIG under the grid faults emulated by the grid emulator are shown in Figures 14.17 and 14.18. In Figure 14.17, two symmetrical voltage faults are emulated by the grid emulator connected to the DFIG test system. The three-phase grid voltages and the output voltage of the grid emulator drops to 20% of the normal values at t_0, and back to normal at t_1. The second fault is emulated at t_2,

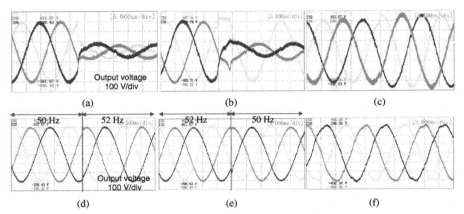

Figure 14.16 Non-ideal grid voltages emulated by the grid emulator: (a) Symmetrical voltage dips; (b) asymmetrical voltage dips; (c) voltage swells; (d, e) frequency variations; and (f) harmonic distorted grid voltage.

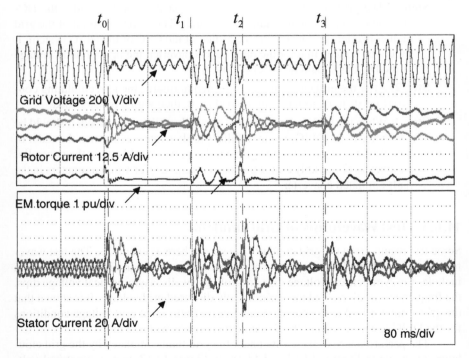

Figure 14.17 Performance of the DFIG test system during the symmetrical grid faults emulated by the grid emulator.

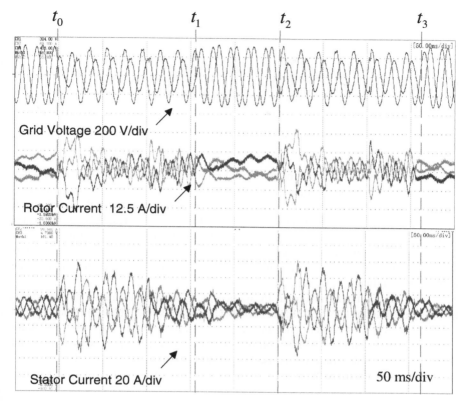

Figure 14.18 Performance of the DFIG test system under the asymmetrical grid faults emulated by the grid emulator.

the voltage drops again and recovers at t_3. The grid emulator is able to emulate the voltage drops introduced by the grid faults, and the response time is within 10 ms. It can be found that large transient stator currents are introduced by the voltage dips on the DFIG, which is also the output current of the grid emulator, and the grid emulator operates normally under this load current disturbance. In Figure 14.18, two asymmetrical voltage faults are emulated by the grid emulator: one-phase grid voltage is controlled to drop to 50% of the normal value at t_0, and back to normal at t_1, the second fault is emulated at t_2 and the voltage drops again and recovers at t_3. It can be found the grid emulator can emulate the asymmetrical voltage dips introduced by the asymmetrical grid faults as well.

14.7 COMMUNICATIONS AND UP-LEVEL CONTROL

In the real WTS, the up-level control demands of the DFIG converters are given by the wind turbine controller according to the wind speed and the grid demands. They

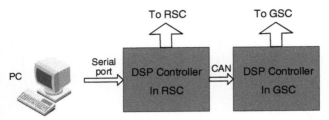

Figure 14.19   Overall communication scheme of the DFIG test bench.

can be power/speed reference of the DFIG converters, the start or stop command, or the mode-switching command. In the DFIG test bench, these control commands are directly given by a PC software and sent to the DSP controller of the wind power converters by CAN communication, as shown in Figure 14.19. The PC communicates to the DSP controller of the RSC through the serial port, and the RSC's DSP controller communicates with the GSC's DSP controller with CAN. The control commands come from the PC and they are sent to the GSC which needs to be handled by the RSC first, and then sent to the GSC.

The communication from the PC to the RSC is achieved by a serial port communication; the baud rate is 9600 bps. A control software is made based on the Modbus protocol and the main function of the control software in the PC is shown in Figure 14.20. The main function of this control software includes display, debugging, and auxiliary. The display function is used to monitor system status, including the system state, the real-time voltage, current and power output, as well as the fault status.

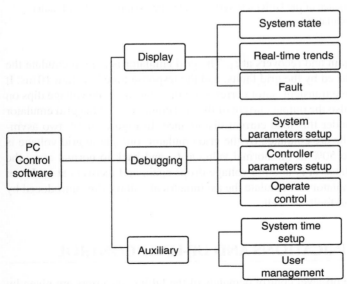

Figure 14.20   Main functions of the control software in PC in Figure 14.19.

The debugging function is used for debugging of the DFIG test system, for example, to start or stop the system, to adjust the system and controller parameters online. The auxiliary function is to manage the auxiliary functions of the system, like user management or system time set up.

The monitor and debugging interfaces of the PC software are shown in Figures 14.21a and 14.21b, respectively. It can be found in Figure 14.21a that the important system real-time electrical variables can be displayed on the interface, including the grid voltage, stator voltages, rotor currents, stator currents, and the DC voltage. The debugging interface is shown in Figure 14.21b. The control software is able to control the DFIG test system to start, to stop, to change the control mode, as well as to adjust the controller parameters in the rotor current loop, reactive power loop, and also the voltage loop.

14.8 START-UP AND PROTECTION OF THE SYSTEM

14.8.1 Start-Up of the System

The DFIG test bench consists of five power electronic converters and two motors or generators, so the starting of the system must be planned carefully. A starting sequence is introduced here, together with different test waveforms for better understanding. As shown in Figure 14.22, the grid emulator is started first to provide the three-phase grid voltage to the system. Then, the driving inverter and the driving motor is started to make the DFIG operate under normal rotating speed. At last, the DFIG system is started and connected to the grid emulator and the related test can then be carried out.

The grid emulator needs to be started first, to provide the three-phase grid voltage for the DFIG test system. The starting waveforms of the grid emulator are shown in Figure 14.23. The grid output currents, DC-bus voltage, and the output voltages of the grid emulator are shown. The soft-starter circuit charges the DC-bus capacitor first and when the DC-bus voltage is larger than 560 V, the soft-start finishes, and then the grid converter is enabled and finally controls the DC-bus voltage to be stabilized to about 780 V. The fault emulator measures the DC-bus voltage as well. When then the DC-bus voltage is stabilized, the fault emulator converter is then enabled and the output voltage is gradually increased to the rated value (about 311 V).

Then, the driving inverter and driving motor are started and the motor should be controlled in speed-control mode and the rotor speed should be adjusted to 1200–1800 rpm, which is also the rotor speed within the operation range of the DFIG.

The DFIG test system is started at last. The start-up of the DFIG system has been introduced in Chapter 5 and the start-up waveforms are given in Figure 14.24. The GSC is started first to provide the DC-bus voltage of the RSC. The soft-start circuit charges the DC bus first and then the GSC is started to control the DC-bus voltage to be about 650 V. Then, the RSC is started in starting mode, the stator voltage is controlled to track the grid voltage (or the emulated grid voltage generated by the grid emulator). The rotor current is increased gradually as the demagnetizing current

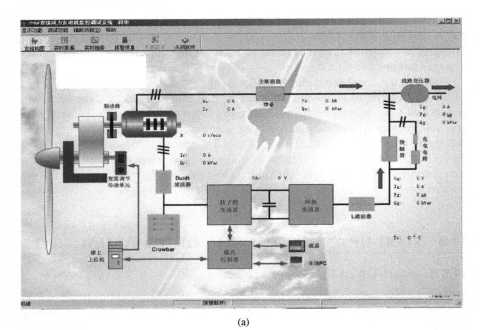

(a)

(b)

Figure 14.21 User interface of the PC software (a) system monitor interface and (b) the debugging interface.

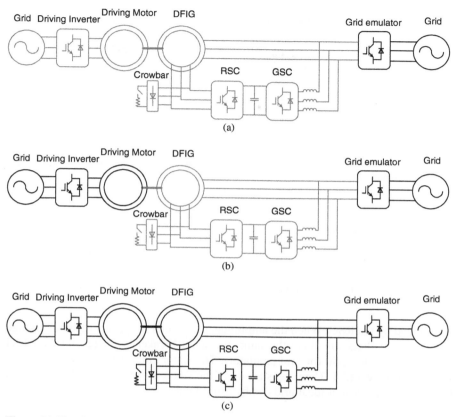

Figure 14.22 Start-up sequence of the DFIG test bench: (a) Start the grid emulator; (b) start the driving inverter and (c) start the DFIG test system.

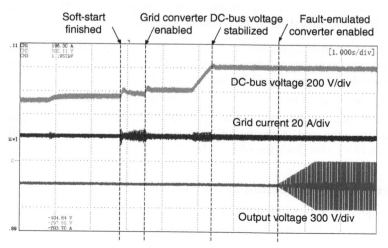

Figure 14.23 Start-up of the grid emulator.

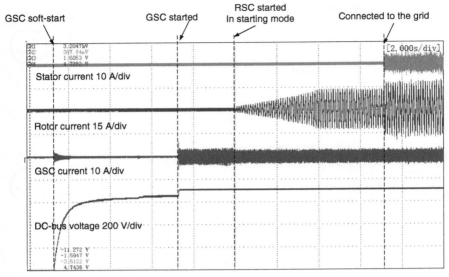

Figure 14.24   Start-up of the DFIG test system.

is provided by the RSC. When the amplitude, phase, and frequency of stator voltage and the grid voltage are about the same, the stator breaker is closed and the DFIG is connected to the grid (or grid emulator). The stator current is no longer zero, as shown in Figure 14.24. Then, the start-up of the whole system is completed and the test of DFIG under non-ideal grid can be carried out.

14.8.2   Shutdown of the System

The shutdown of the system can be regarded to be the reverse process of the start-up of the system. It will be carried out in the following sequence:

- The DFIG test system should be shutdown first. The active power of the DFIG should be reduced to zero and the stator breaker can be opened. After the stator voltage is reduced to zero, the GSC can be stopped.
- Reduce the rotor speed to zero and shut down the driving inverter.
- Reduce the output of the grid emulator to zero, and then disconnect the grid emulator from the grid. After the DC voltage of the grid emulator is small enough (<10 V), shut down the grid emulator.

14.8.3   Overcurrent Protection of the System

An overcurrent protection system is available in each power electronic converter and their structures are about the same. So, the protection in the RSC is taken as an example in this section and the next.

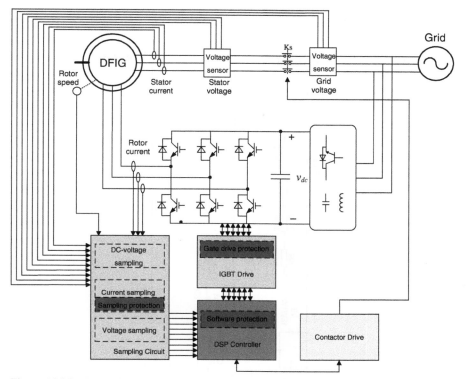

Figure 14.25   Protection scheme of the DFIG test system.

As shown in Figure 14.25, the overcurrent protection of the RSC is achieved in three levels: the gate drive protection, the sampling protection, and the software protection. The gate drive protection and sampling protection are achieved through protection circuits in IGBT drives and sampling circuits, while the software protection is achieved by the software program in the DSP controller.

14.8.3.1 Gate Drive Protection

The gate driving protection circuits are placed on the IGBT driver board. It is achieved by measuring the collector–emitter voltage of the IGBT when the IGBT is switched on. If the current in the IGBT is too large, the large voltage drop will be measured between the collector–emitter of the IGBT, and the gate drive is locked in this case. This protection is normally integrated into the IGBT driving chips. Take HCPL-316 J as an example; the overcurrent protection circuits in the IGBT drive based on HCPL-316 J is shown in Figure 14.26. When the DESAT pin of the HCPL-316 J is larger than that for 7 V, the gate drive signal will be forced to low voltage level (−9 V) and the IGBT will be switched off. So, the DESAT is connected to the collector of the IGBT, and two diodes with the forward voltage to be 1.7 V is used to adjust the protection

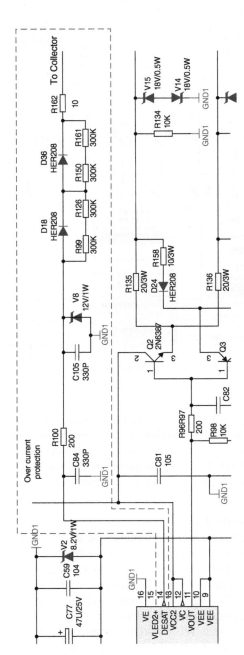

Figure 14.26 Gate drive protection circuit based on HCPL-316J.

threshold to be 3.6 V. In this case, when the collector–emitter voltage of the IGBT reaches 3.6 V, the IGBT will be forced to turn off in order to prevent the overcurrent.

The dynamic response speed of the gate drive protection is fast, as it directly measures the voltage and no signal transmission is needed. The protection can be achieved within 10 μs.

14.8.3.2 Sampling Protection

The sampling protection is placed on the sampling circuits of the RSC. It uses the measured three-phase current to achieve the protection. The sampled three-phase current signal is used to compare it with the protection threshold. If one of the three-phase currents is larger than the threshold, the IGBT driving signal is blocked and all the IGBTs are switched-off. The schematic of the sampling protection on the RSC is shown in Figure 14.27.

As shown in Figure 14.27, the phases A, B, and C current sampling signals are compared with the threshold in three comparators, respectively. The output of the three comparators are connected to an OR gate and then connected to an RS trigger. So, if one of the three-phase currents exceeds the threshold, the output of the RS trigger will be switched and it blocks the IGBTs. The IGBTs will remain blocked until the fault is reset by the reset signal from the DSP controller.

The response speed of the sampling protection is smaller compared to the gate-drive protection, as it needs to measure the current and the signals needed to go through a few gate circuits. The response speed of the sampling protection is around a few tens of microseconds, depending on the speed of the related gate drives.

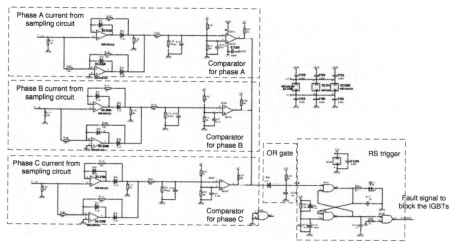

Figure 14.27   Sampling protection circuit used for overcurrent protection.

14.8.3.3 Software Protection

The software protection is achieved by the DSP controller, the current signal in DSP is compared with the threshold in software. If the current is too large, the DSP will block the IGBTs. The response speed of the software controller is about one interrupt cycle. So in this system, it is about 0.5 ms.

14.8.4 Overvoltage Protection of the System

The overvoltage protection in a DFIG system usually refers to the overvoltage protection of the DC bus. This function is achieved by software program—as the DC-bus capacitance is very large, the DC voltage will not rise very fast, the response speed of the software protection is fast enough.

14.9 SUMMARY

In this chapter, the designing and implementation of a reduced-scale 30 kW DFIG test bench in the laboratory is introduced that includes the demands for the test system, the structure, the hardware design, the control design, and also about how to realize the control strategies. Additionally, information on how to test this system and keep it running is also introduced. The communication and protection of the system are introduced as well. We have also discussed the grid emulator. Through this chapter, guidance and reference are provided to researchers who want to build a test system in the lab and to verify their research on the DFIG system [7, 8].

REFERENCES

[1] R. Pena, J. C. Clare, and G. M. Asher, "Doubly fed induction generator using back-to-back PWM converters and its application to variable speed wind-energy generation," *Proc. Inst. Elect. Eng. Elect. Power Appl.*, vol. 143, no. 3, pp. 231–241, 1996.

[2] J. Xu, "Research on converter control strategy of doubly fed induction generator system for wind power," Ph.D. thesis, Zhejiang University, Hangzhou, China, 2011.

[3] R. Zeng, H. Nian, and P. Zhou, "A three-phase programmable voltage sag generator for low voltage ride-through capability test of wind turbines," in *Proc. ECCE*, 2010, pp. 305–311.

[4] R. Lohde, and F. W. Fuchs, "Laboratory type PWM grid emulator for generating disturbed voltages for testing grid connected devices," in *Proc. EPE/PEMC*, 2009, pp. 1–9.

[5] I. Iyoda, Y. Abe, T. Ise, N. Shigei, and K. Hayakawa, "A new parameter of voltage sags and its effect on inverters of PV systems surveyed by a newly developed voltage sag generator," in *Proc. IECON*, 2010, pp. 2759–2764.

[6] B. Gong, D. Xu, and B. Wu, "Cost effective method for DFIG fault ride-through during symmetrical voltage dip," in *IECON 2010—36th Ann. Conf. IEEE Ind. Electron. Soc.*, Nov. 2010, pp. 3269–3274.

[7] C. Wessels, R. Lohde, and F. W. Fuchs, "Transformer based voltage sag generator to perform LVRT and HVRT tests in the laboratory," in *Proc. EPE/PEMC*, 2010, pp. T11-8–T11-13.

[8] L. Dongyu, Z. Honglin, X. Shuai, and Y. Geng, "A new voltage sag generator base on power electronic devices," in *Proc. IEEE Int. Symp. Power Electron. Distrib. Gener. Syst.*, 2010, pp. 584–588.

APPENDIX

A.1 FLUX EQUATIONS IN $\alpha\beta$ REFERENCE FRAME

Stator flux in ABC reference frame and rotor flux in *abc* reference frame are as follows:

$$\boldsymbol{\psi}_s = \boldsymbol{L}_{ss}\boldsymbol{i}_s + \boldsymbol{L}_{sr}\boldsymbol{i}_r^r \tag{A.1}$$

$$\boldsymbol{\psi}_r^r = \boldsymbol{L}_{rs}\boldsymbol{i}_s + \boldsymbol{L}_{rr}\boldsymbol{i}_r^r \tag{A.2}$$

where $\boldsymbol{\psi}_s = \begin{bmatrix} \psi_{sa} & \psi_{sb} & \psi_{sc} \end{bmatrix}^{\mathrm{T}}$, $\boldsymbol{\psi}_r^r = \begin{bmatrix} \psi_{ra}^r & \psi_{rb}^r & \psi_{rc}^r \end{bmatrix}^{\mathrm{T}}$, $\boldsymbol{i}_s = \begin{bmatrix} i_{sa} & i_{sb} & i_{sc} \end{bmatrix}^{\mathrm{T}}$ and $\boldsymbol{i}_r^r = \begin{bmatrix} i_{ra}^r & i_{rb}^r & i_{rc}^r \end{bmatrix}^{\mathrm{T}}$,

$$\boldsymbol{L}_{ss} = \begin{bmatrix} L_{ms} + L_{ls} & -\dfrac{1}{2}L_{ms} & -\dfrac{1}{2}L_{ms} \\ -\dfrac{1}{2}L_{ms} & L_{ms} + L_{ls} & -\dfrac{1}{2}L_{ms} \\ -\dfrac{1}{2}L_{ms} & -\dfrac{1}{2}L_{ms} & L_{ms} + L_{ls} \end{bmatrix} \tag{A.3}$$

$$\boldsymbol{L}_{rr} = \begin{bmatrix} L_{mr} + L_{lr} & -\dfrac{1}{2}L_{mr} & -\dfrac{1}{2}L_{mr} \\ -\dfrac{1}{2}L_{mr} & L_{mr} + L_{lr} & -\dfrac{1}{2}L_{mr} \\ -\dfrac{1}{2}L_{mr} & -\dfrac{1}{2}L_{mr} & L_{mr} + L_{lr} \end{bmatrix} \tag{A.4}$$

$$\boldsymbol{L}_{sr} = \boldsymbol{L}_{rs}^{\mathrm{T}} = L_{ms} \begin{bmatrix} \cos\theta_r & \cos\left(\theta_r + 120°\right) & \cos\left(\theta_r - 120°\right) \\ \cos\left(\theta_r - 120°\right) & \cos\theta_r & \cos\left(\theta_r + 120°\right) \\ \cos\left(\theta_r + 120°\right) & \cos\left(\theta_r - 120°\right) & \cos\theta_r \end{bmatrix} \tag{A.5}$$

where L_{ms} is the mutual inductance of stator phase winding with maximum mutual flux linkage, and L_{mr} is the mutual inductance of rotor phase winding with maximum mutual flux linkage. If we refer the rotor side variables to the stator side so that both

Advanced Control of Doubly Fed Induction Generator for Wind Power Systems, First Edition.
Dehong Xu, Frede Blaabjerg, Wenjie Chen, and Nan Zhu.
© 2018 by The Institute of Electrical and Electronics Engineers, Inc. Published 2018 by John Wiley & Sons, Inc.

stator windings and rotor windings have the same turns, then $L_{ms} = L_{mr}$. L_{ls} and L_{lr} are the leakage inductances of the stator and rotor windings, respectively.

By introducing the transformation from the ABC (abc) reference frame to the complex plane, the above equation is rewritten as

$$\frac{2}{3}\begin{bmatrix} 1 & \alpha & \alpha^2 \end{bmatrix} \boldsymbol{\psi}_s = \frac{2}{3}\begin{bmatrix} 1 & \alpha & \alpha^2 \end{bmatrix} L_{ss} i_s + \frac{2}{3}\begin{bmatrix} 1 & \alpha & \alpha^2 \end{bmatrix} L_{sr} i_r^r \qquad (A.6)$$

The first term of the right side of equation (A.6) can be deducted as follows:

$$
\frac{2}{3}\begin{bmatrix} 1 & \alpha & \alpha^2 \end{bmatrix} L_{ss} i_s = \frac{2}{3}\begin{bmatrix} 1 & \alpha & \alpha^2 \end{bmatrix}
\begin{bmatrix}
L_{ms} + L_{ls} & -\frac{1}{2}L_{ms} & -\frac{1}{2}L_{ms} \\
-\frac{1}{2}L_{ms} & L_{ms} + L_{ls} & -\frac{1}{2}L_{ms} \\
-\frac{1}{2}L_{ms} & -\frac{1}{2}L_{ms} & L_{ms} + L_{ls}
\end{bmatrix} i_s
$$

$$
= \frac{2}{3}\begin{bmatrix} \frac{3}{2}L_{ms} + L_{ls} & \left(\frac{3}{2}L_{ms} + L_{ls}\right)\alpha & \left(\frac{3}{2}L_{ms} + L_{ls}\right)\alpha^2 \end{bmatrix} i_s \qquad (A.7)
$$

$$
= \left(\frac{3}{2}L_{ms} + L_{ls}\right)\frac{2}{3}\begin{bmatrix} 1 & \alpha & \alpha^2 \end{bmatrix} i_s
$$

$$
= \left(\frac{3}{2}L_{ms} + L_{ls}\right)\vec{i}_s
$$

where $\vec{i}_s$ is the complex vector in $\alpha\beta$ frame transformed from vector i_s in ABC frame, i.e. $\vec{i}_s = \frac{2}{3}\begin{bmatrix} 1 & \alpha & \alpha^2 \end{bmatrix} i_s$.

The second term of the right side of equation (A.6) can be deducted as follows:

$$
\frac{2}{3}\begin{bmatrix} 1 & \alpha & \alpha^2 \end{bmatrix} L_{sr} i_r^r
$$

$$
= \frac{2}{3}\begin{bmatrix} 1 & \alpha & \alpha^2 \end{bmatrix} L_{ms}
\begin{bmatrix}
\cos\theta_r & \cos(\theta_r + 120°) & \cos(\theta_r - 120°) \\
\cos(\theta_r - 120°) & \cos\theta_r & \cos(\theta_r + 120°) \\
\cos(\theta_r + 120°) & \cos(\theta_r - 120°) & \cos\theta_r
\end{bmatrix} i_r^r
$$

$$
= \frac{2}{3}L_{ms}\begin{bmatrix} 1 & \alpha & \alpha^2 \end{bmatrix}
\begin{bmatrix}
\cos\theta_r & \cos(\theta_r + 120°) & \cos(\theta_r - 120°) \\
\cos(\theta_r - 120°) & \cos\theta_r & \cos(\theta_r + 120°) \\
\cos(\theta_r + 120°) & \cos(\theta_r - 120°) & \cos\theta_r
\end{bmatrix} i_r^r \qquad (A.8)
$$

$$
= \frac{2}{3}L_{ms}\begin{bmatrix} \frac{3}{2}e^{j\theta_r} & \frac{3}{2}e^{j\theta_r}\alpha & \frac{3}{2}e^{j\theta_r}\alpha^2 \end{bmatrix} i_r^r
$$

$$
= \frac{3}{2}L_{ms}e^{j\theta_r}\frac{2}{3}\begin{bmatrix} 1 & \alpha & \alpha^2 \end{bmatrix} i_r^r
$$

$$
= \frac{3}{2}L_{ms}e^{j\theta_r}\vec{i}_r^r
$$

where $\vec{i}_r^r$ is the complex vector in rotor $\alpha\beta$ frame transformed from vector i_r^r in abc frame on the rotor, i.e. $\vec{i}_r^r = \frac{2}{3}\begin{bmatrix} 1 & \alpha & \alpha^2 \end{bmatrix} i_r^r$.

Besides following three relations are used in the above derivation:

$$[1 \quad \alpha \quad \alpha^2] \begin{bmatrix} \cos(\theta_r) \\ \cos(\theta_r - 120°) \\ \cos(\theta_r + 120°) \end{bmatrix}$$

$$= [1 \quad \alpha \quad \alpha^2] \begin{bmatrix} 1/2\left(e^{j\theta_r} + e^{-j\theta_r}\right) \\ 1/2\left(e^{j(\theta_r - 120^0)} + e^{-j(\theta_r - 120^0)}\right) \\ 1/2\left(e^{j(\theta_r + 120^0)} + e^{-j(\theta_r + 120^0)}\right) \end{bmatrix} \tag{A.9}$$

$$= \frac{1}{2}\left[e^{j\theta_r}\left(1 + \alpha e^{-j120^0} + \alpha^2 e^{j120^0}\right) + e^{-j\theta_r}\left(1 + \alpha e^{j120^0} + \alpha^2 e^{-j120^0}\right)\right]$$

$$= \frac{3}{2}e^{j\theta_r}$$

$$[1 \quad \alpha \quad \alpha^2] \begin{bmatrix} \cos(\theta_r + 120^0) \\ \cos(\theta_r) \\ \cos(\theta_r - 120^0) \end{bmatrix} = \alpha\frac{3}{2}e^{j\theta_r} \tag{A.10}$$

$$[1 \quad \alpha \quad \alpha^2] \begin{bmatrix} \cos(\theta_r - 120^0) \\ \cos(\theta_r + 120^0) \\ \cos(\theta_r) \end{bmatrix} = \alpha^2\frac{3}{2}e^{j\theta_r} \tag{A.11}$$

By using equations (A.7) and (A.8), equation (A.6) can be expressed as

$$\vec{\psi}_s = \left(\frac{3}{2}L_{ms} + L_{ls}\right)\vec{i}_s + \frac{3}{2}L_{ms}e^{j\theta_r}\vec{i}_r^r \tag{A.12}$$

By defining $L_m = \frac{3}{2}L_{ms}$, $L_s = L_m + L_{ls}$, the above equation can be simplified as

$$\vec{\psi}_s = L_s\vec{i}_s + L_m e^{j\theta_r}\vec{i}_r^r \tag{A.13}$$

Similarly, the rotor flux in the complex plane is derived:

$$\vec{\psi}_r^r = L_m e^{-j\theta_r}\vec{i}_s + L_r\vec{i}_r^r \tag{A.14}$$

where $L_r = L_m + L_{lr}$.

A.2   TYPICAL PARAMETERS OF A DFIG

Key parameters of a DFIG with four different power scale and voltage level are given in Table A.1 as the examples. Here all rotor-side parameters have been referred to the stator side with the same turns for both the stator and the rotor windings.

TABLE A.1 Key parameters of the DFIG with different power scales

Rated power	1.5 MW	250 kW*	30 kW	5 kW*
Rated phase-to-phase voltage (rms)	690 V	400 V	380 V	380 V
Stator resistance	2.139 mΩ	20 mΩ	92 mΩ	0.72 Ω
Stator inductance	4.05 mH	4.4 mH	26.7 mH	91.6 mH
Rotor resistance	2.139 mΩ	20 mΩ	56 mΩ	0.75 Ω
Rotor inductance	4.09 mH	4.4 mH	26.2 mH	91.6 mH
Mutual inductance	4.00 mH	4.2 mH	25.1 mH	85.6 mH
Turns ratio	0.369	1	0.32	0.54

Generally speaking, DFIG machines with higher power ratings have smaller stator resistance and rotor resistance in comparison to the lower power ones. Higher-power-rated DFIG machines also have smaller mutual inductance between the stator and rotor windings.

REFERENCES

[1] B. K. Bose, *Power Electronics and Drives*. Elsevier, 2006.
[2] S. J. Chapman, *Electric Machinery Fundamentals*. McGraw-Hill, 2005.
[3] G. Abad, J. Lopez, M. A. Rodriguez. L. Marroyo, and G. Iwanski, *Doubly Fed Induction Machine: Modeling and Control for Wind Energy Generation*. Wiley-IEEE Press, 2011.
[4] P. Kunder, *Power System Stability and Control*. McGraw-Hill, 1994.
[5] Y. He, J. Hu and L. Xu, *Operation Control of Grid Connected Doubly Fed Induction Generator*. China, Electric Power Press, 2012.

INDEX

Advanced Control of Doubly Fed Induction Generator for Wind Power Systems, First Edition.
Dehong Xu, Frede Blaabjerg, Wenjie Chen, and Nan Zhu.
© 2018 by The Institute of Electrical and Electronics Engineers, Inc. Published 2018 by John Wiley & Sons, Inc.

453

IEEE Press Series on Power Engineering

Series Editor: **M. E. El-Hawary,** Dalhousie University, Halifax, Nova Scotia, Canada

The mission of IEEE Press Series on Power Engineering is to publish leading-edge books that cover the broad spectrum of current and forward-looking technologies in this fast-moving area. The series attracts highly acclaimed authors from industry/academia to provide accessible coverage of current and emerging topics in power engineering and allied fields. Our target audience includes the power engineering professional who is interested in enhancing their knowledge and perspective in their areas of interest.